Chemistry of Peptides and Proteins
Volume 2

Chemistry of Peptides and Proteins

Volume 2

Proceedings of the
Fourth USSR-FRG Symposium
Tübingen, Federal Republic of Germany
June 8 – 12, 1982

Editors
Wolfgang Voelter · Ernst Bayer
Yuri A. Ovchinnikov · Erich Wünsch

Walter de Gruyter · Berlin · New York 1984

Editors

Wolfgang Voelter, Professor, Dr. rer. nat.
Head
Department of Physical Biochemistry
Institute for Physiological Chemistry
Hoppe-Seyler-Straße 1
D-7400 Tübingen, Federal Republic of Germany

Ernst Bayer, Professor, Dr. rer. nat.
Director
Institute for Organic Chemistry
Auf der Morgenstelle 18
D-7400 Tübingen, Federal Republic of Germany

Yuri A. Ovchinnikov, Professor, D. Sc.
Director
Shemyakin Institute of Bioorganic Chemistry
USSR Academy of Sciences
Moscow, USSR

Erich Wünsch, Professor, Dr. rer. nat., Dr. med. h.c.
Director
Department of Peptide Chemistry
Max-Planck-Institute for Biochemistry
D-8033 Martinsried, Federal Republic of Germany

CIP-Kurztitelaufnahme der Deutschen Bibliothek

Chemistry of peptides and proteins : proceedings of the
... USSR-FRG symposium. – Berlin ; New York : de Gruyter
ISSN 0723-6271
Vol. 2. Proceedings of the fourth USSR-FRG symposium :
Tübingen, Federal Republic of Germany, June 8-12, 1982. – 1984
ISBN 3-11-009580-7

Library of Congress Cataloging in Publication Data

USSR-FRG Symposium (4th : 1982 : Tübingen, Germany)
Chemistry of peptide and proteins, volume 2.
Includes bibliographical references and indexes.
1. Peptides--Congresses. 2. Proteins--Congresses.
I. Voelter, W.
QD431.A1U88 1982 547.7'5 84-4294
ISBN 3-11-009580-7

 Printing: Gerike GmbH, Berlin. – Binding: Lüderitz & Bauer GmbH, Berlin. Printed in Germany.

PREFACE

The 4th USSR-FRG Symposium on Chemistry of Peptides and Proteins took place in June 1982 in Tübingen.

The growing interest in all aspects of biologically active peptides is very well reflected in the topics discussed during the symposium. The molecular and structural basis for the biological activity of polypeptides has attracted the increasing interest of peptide chemistry. Many contributions are devoted especially to the fields of neurochemistry, neurotoxicology, immunology and membrane-active peptides and proteins. The isolation and structure analysis of new, biologically active peptides has entered a new era with the increasing sensitivity of the methods now available. Structure elucidation with a minimal amount of substance is no longer a dream. This miniaturization of methods is also important for gene technology and it seems as if the problems and progress in gene technology will give new impulses also to peptide chemistry.

The synthesis of polypeptides still poses problems despite the progress in this field, and the improved arsenal of methods for purity control steadily reveals new insights. For the first time the semisynthesis of peptides may reach production scale, as the papers on the semisynthesis of insulin show.

This symposium gives an excellent overview of the status of peptide chemistry. It should be mentioned that lively and open discussions both in and outside the lecture halls contributed to the success of the symposium. At this meeting many young scientists from the USSR and the FRG had the opportunity of participation for the first time. This certainly is important for the promotion of better understanding between the countries.

The organizers of the symposium gratefully acknowledge the generous support and sponsorship given by the Deutsche Forschungsgemeinschaft and the Soviet Academy of Sciences and they hope that this important and fruitful series of symposia can be continued for the benefit of the scientific community working in this field.

For the Editors
E. Bayer

Tübingen, February 1984

CONTENTS

PEPTIDE SYNTHESIS AND ANALYTICAL PROBLEMS OF SYNTHETIC PEPTIDES

Enzyme-Catalyzed Semisyntheses with Porcine Insulin

R. Obermeier, G. Seipke 3

Preparation, Properties and Application of Photo-Reactive Insulins

D. Saunders, P. Thamm, G. Klotz, D. Brandenburg 11

Synthesis of A19-[3,5-Diiodo-Tyrosine] Porcine Insulin

H.-J. Wieneke, E. Büllesbach, H.-G. Gattner, H. Zahn, W. Danho 17

Semisynthesis of Linear Proinsulin Analogues with Shortened C-Peptides

E.E. Büllesbach 23

Synthesis of Unsymmetrical Cystine Peptides: Directed Disulfide Pairing with the Sulfenohydrazide Method

S. Romani, W. Göring, L. Moroder, E. Wünsch 29

Synthesis of Mid Regional Parathyrin Segments

M. Casaretto, H. Zahn, W. Danho, R.-D. Hesch 35

Application of Modern Methods in the Synthesis of Cyclic Analogues of Bradykinin

F. Mutulis, I. Mutule, G. Chipens, V. Grigoryeva, M. Breslav 41

Total Synthesis of Neurotoxin II from Naja Naja Oxiana Cobra Venom

V.T. Ivanov, V.I. Deigin, V.V. Ulyashin, I.I. Mikhaleva .. 47

The Synthesis of the N-Terminal Amino Acids of the Peptide-Nucleoside Antibiotic Nikkomycin and their Use in Peptide Synthesis

W.A. König, W. Hass, H. Faasch 55

Synthesis and Activity of Peptides Analogues to the Toxic Peptides Isolated from Amanita Virosa Mushrooms

J.-U. Kahl, T. Miura, T. Wieland 63

Synthesis of a New Photolabile Tritiated Antamanide-Analogue

M. Nassal, P. Buć, T. Wieland 71

Synthesis and Properties of ($NLe^{56,60}$) 34-65 Bacteriorhodopsin Fragment

A.T. Kozhich, B.V. Vaskovsky, I.I. Mikhaleva, V.T. Ivanov 79

Synthetic Peptide Derivatives of Bacterial Lipoprotein and their Interaction with Lymphocyte Plasma Membranes

W.G. Bessler, K. Wiesmüller, W. Scheuer, R.B. Johnson 87

Synthesis of Defined Peptide Derivatives by Aminolysis of 3-[N^{α}-Benzyloxycarbonyl Peptidyloxy]-2-Hydroxy-N-Methyl-Benzamides at Elevated Temperatures II. Synthesis of the Peptide Derivatives Z-Ala-X-Gly-$N(Et)_2$, X=Phe, Leu, Val, Ser (Bu^t), Glu (OBu^t)

H. Berndt, C. Turck 97

Recently Developed Amino Protecting Groups

W. Voelter, H. Kalbacher, C. Beni, W. Heinzel, J. Müller 103

Syntheses and Reaction of Peptide Cyclols

M. Fähnle, M. Rothe 115

Synthesis of Cyclic Tripeptides Containing CONH Groups

M. Rothe, M. Fähnle, W. Mästle 121

Conformational Preferences of Model Peptides and their Impact in Peptide Synthesis

F. Maser, K. Bode, B. Klein, M. Mutter 127

Analytical Studies on Side Products in the Synthesis of Secretin

W. Göhring, R. Scharf, E. Wünsch, C. Gespach 137

High Performance Liquid Chromatography of Protected Peptides

V.V. Ylyashin, V.I. Deigin 145

Field Desorption Mass Spectra and Fast Atom Bombardment Spectra of Partially Protected Peptides

H. Hagenmaier, G. Vossler, L. Grotjahn 155

STRUCTURAL FEATURES OF PEPTIDES AND PROTEINS

Isolation and Structural Analysis of a New Neuropeptide, the Head Activator

H. Bodenmüller 165

Biochemical, Biological and Immunological Characterization of Bovine Nerve Growth Factor

G.P. Harper, H. Thoenen 171

Embryonic Haemoglobins in Mammals Haemoglobin Components in Human and Pig Embryos; Primary Structures of the Human and Pig ζ-Chains and their Phylogenetic Relationship to the α-Chains

H. Aschauer, F. Bieber, S. Bektas, T. Kleinschmidt, G. Braunitzer, T. Sanguansermsri 179

The Complete Primary Structure of Elongation Factor G from E. coli

Y.B. Alakhov, L.P. Motuz, N.V. Dovgas, L.M. Vinokurov 187

The Primary Structure of E. coli RNA Polymerase. Nucleotide Sequence of the rpoC Gene and Amino Acid Sequence of the ß'-Subunit

V.V. Gubanov, G.S. Monastyrskaya, S.O. Guryev, I.S. Salomatina, T.M. Shuvaeva, A.P. Bogachuk, V.M. Lipkin, E.D. Sverdlov 195

Determination of Amino Acid Sequences of Large Hydrophobic Peptides of Rhodopsin

M.B. Kostina, A.S. Zolotarev 203

Three-Dimensional Structure of Actinoxanthin at High Resolution

A. Kuzin, S. Trakhanov, V. Pletnev 209

Primary and Secondary Structures of Peptides which Form Voltage-Dependent Pores in Lipid-Bilayers

G. Jung, G. Boheim, R. Bosch, H. Brückner, P. Hartter, E. Katz, W.A. König, H. Schmitt, K.-P. Voges, W. Winter .. 215

Purification and Characterization of a Second Retinal Binding Protein in Halobacterium Halobium

P. Hegemann, M. Steiner, D. Oesterhelt 227

Homo- and Heteronuclear Two-Dimensional NMR Spectroscopy of Cyclic Peptides
H. Kessler, R. Schuck 233

Miscellaneous and Biological Activity of Peptides and Proteins

Bioorganic Chemistry of Rhodopsins
Y.A. Ovchinnikov 241

The Role of Tyrosine 26 and Tyrosine 64 in Bacteriorhodopsin Function
H.D. Lemke, D. Oesterhelt 249

Structure-Function Studies on Gastrin Related Peptides
L. Moroder, G. Borin, A. Lobbia, J.-P. Bali, E. Wünsch 255

Analogues of Angiotensin with Vasodepressor Activity
G. Chipens, J. Ancans, D. Berga, I. Vosekalna, N. Mishlyakova, A. Krikis 261

Regulatory Influence of Peptide Hormones on Neural Cells in Culture
B. Hamprecht, F. Löffler, G. Reiser, D. van Calker 267

Enzyme Immunoassay of Human Urinary Kallikrein
M. Franke, S. Rohrschneider, R. Geiger 273

The Alamethicin Pore is Formed by a Voltage-Gated Flip-Flop of α-Helix Dipoles
G. Boheim, W. Hanke, G. Jung 281

Conformations of the Transmembrane Channel Formed by Gramicidin A
S.V. Sychev, V.T. Ivanov 291

Study of Veratridine Receptors Associated with Sodium Channels
N.M. Soldatov, T.K. Prosolova, N.I. Kiyatkin, E.V. Grishin 301

The Role of Neurotoxins in Studying Sodium Channels
E.V. Grishin 309

Structure and Mechanism of Action of Neurotoxins from the Venom of Latrodectus Spiders

S. Salikhov, T. Slavnova, M. Tashmukhamedov, M. Adylbekov, A. Korneyev, J. Abdurakhmanova, A. Sadykov 319

Topography of the Acetylcholine Receptor - Neurotoxin Binding Sites Probed by Spectroscopy and Photolabeling

V.I. Tsetlin, K.A. Pluzhnikov, V.T. Ivanov 327

Peptides as Iron Ionophores (Siderophores)

G. Winkelmann 333

Interaction of Proteinases and their Protein Inhibitors as Studied by the Spin-Label Technique

H.R. Wenzel, H. Tschesche, E. von Goldammer, J. Paul 343

An Inhibitor of Elastase from Anemonia sulcata

H. Tschesche, H. Kolkenbrock 349

Phosphorylated Membrane Proteins PP 105 and PP 135 as Markers of Metastasizing Cells?

F.A. Anderer, J. Pfeifle 357

The Acrosomal Membrane System and its Role in Mammalian Fertilization

E. Töpfer-Petersen, A. Hinrichsen-Kohane, C. Schmoeckel, W.-B. Schill 363

Penicillin-Like Action of a Bactericidal Protein (Colicin M) of Escherichia coli

V. Braun, R. Dreher 371

Structure, Function and Biosynthesis of the Sperm Proteinase Acrosin

W. Müller-Esterl, H. Fritz, R. Fock-Nüzel, F. Lottspeich, A. Henschen 377

Structure-Functional Studies of DNA-Dependent RNA-Polymerase

V.M. Lipkin, O.Y. Chertov, I.A. Makarova, G.S. Monastyrskaya, E.D. Sverdlov, Y.A. Ovchinnikov 387

Affinity Labeling of the Binding Site for Initiation Substrate in E. coli DNA-Dependent RNA Polymerase by Adenosine-5'-Tri-Metaphosphate

Y.V. Smirnov, V.M. Lipkin, Y.A. Ovchinnikov, M.A. Grachev, A.A. Mustaev 395

Localization of Mutations Resulting in Rifamicin-Resistancy of *E. coli* RNA Polymerase

S.O. Guryev, G.S. Monastyrskaya, V.V. Gubanov, N.F. Kalinina, V.M. Lipkin, E.D. Sverdlov, A.I. Gragerov, V.G. Nikiforov, I.F. Kiver, I.A. Bass, O.N. Danilevskaya, S.Z. Mindlin 403

Role of Cyclic AMP in the Regulation of Genome Expression of Eukaryots

E.S. Severin 413

Cloning and Expression of the Human Leucocyte Interferon Gene

E.D. Sverdlov, S.A. Tsarev, E.M. Khodkova, G.S. Monastyrskaya, V.A. Efimov, O.G. Chakhmakhcheva, V.D. Solovyev, V.P. Kusnetsov 421

LIST OF CONTRIBUTORS 427

AUTHOR INDEX 429

SUBJECT INDEX 463

Peptide Synthesis and Analytical Problems of Synthetic Peptides

ENZYME-CATALYZED SEMISYNTHESES WITH PORCINE INSULIN

Rainer Obermeier, Gerhard Seipke
Hoechst AG, D-6230 Frankfurt/M., West-Germany

Introduction

Semisyntheses became the method of choice for chemical variations of large peptides or proteins from natural sources (1). Since ten years various semisynthetic procedures (2-8) have been reported which allow transformation of porcine into human insulin. Both species of insulin differ in the only position B30, where alanine is the C-terminal end of porcine insulin - and threonine of human insulin-B-chain (fig. 1).

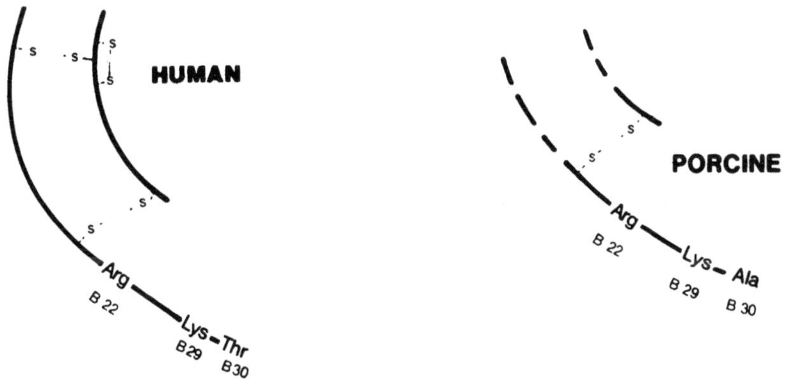

Figure 1. Schematic formula of human and porcine insulin B-chain C-terminus.

As most insulins contain only one Arg(B22) and one Lys(B29), the C-terminal region of the B-chain can be split off selectively by trypsin, which cleaves peptide bonds specifically at the carboxyl site of basic amino acids. During the last years a highly specific lysylendopeptidase (9) has been made commer-

Chemistry of Peptides and Proteins, Vol. 2

cially available, which even allows selective cuts only between -Lys(B29)-Ala(B30). The resulting des-Ala-B30-insulin (DAI) can be prepared alternatively by a very critical digest with carboxypeptidase A (10), however, degradation of the C-terminal Asn-A21 occurs to some extent. Thus, two readily available starting materials can be used for the semisynthetic exchange of the C-terminal sequence of porcine insulin B-chain namely desoctapeptide-B23-30-insulin (DOI) or DAI. Starting from DOI, chemical coupling methods (2, 3) with protected octapeptide corresponding to the human insulin sequence B23-30 resulted in poor yields and therefore extensive puri-

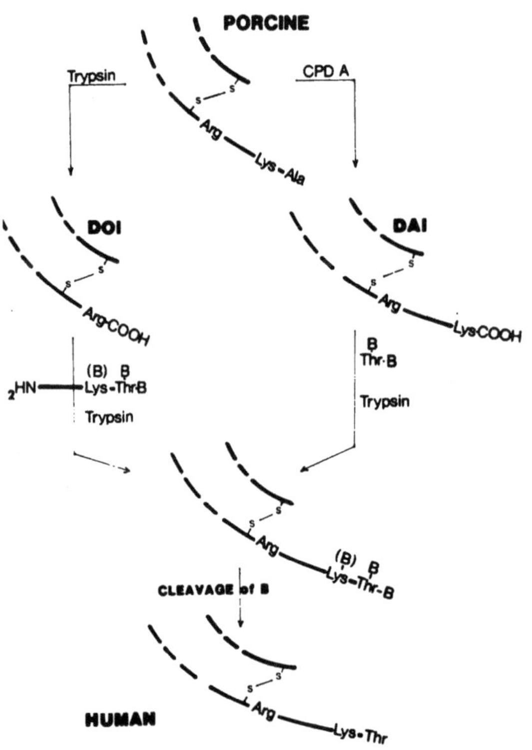

Figure 2. Multiple step enzymatic conversions of porcine into human insulin.

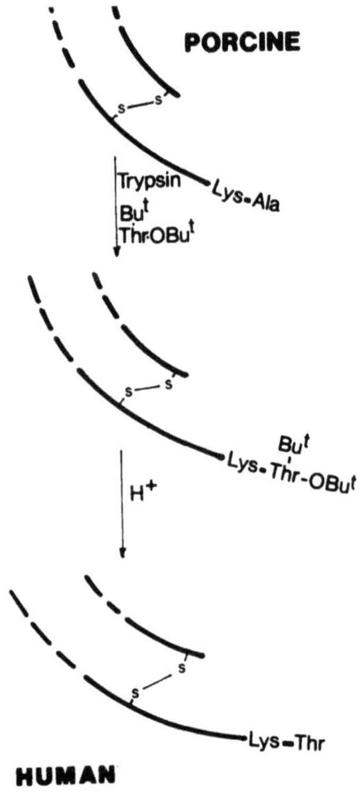

Figure 3. One step enzymatic conversions of porcine into human insulin

fication procedures. Substitution of the chemical coupling reagents by trypsin, however, raised yields up to 60-80 % (4). Enzymatic coupling of DAI with threonine esters (5) was an interesting simplification, based on the observation that under certain conditions the cleavage of the -Arg(B22)-Gly (B23) bond by trypsin is suppressed. When lysylendopeptidase became available, even that potential side reaction could be excluded (11) (fig. 2).

Very recently various groups (6, 7, 8) have independently found an essential improvement of the enzymatic semisynthesis with insulin. Under slightly variable conditions a one-step trypsin-catalyzed transamidation can be performed at the -Lys(29)-Ala(B30) peptide bond (fig. 3). In the following paper we would like to report some of our results on large-scale conversion of porcine to human insulin, insulin analogs and reaction rate and equilibria.

Results

The conversion of porcine to human insulin is exemplified by the following reaction. Porcine insulin (1 g, 0,17 mMol) is dissolved in 5 ml of 20 % acetic acid/water. Crystalline Thr-(But)OBut·Hac (6 g, 21 mMol) is added and the pH is adjusted to 4.5 by addition of acetic acid. 0,5 ml of an aqueous trypsin solution (80 mg) is stirred into the chilled reaction mixture. The reaction is kept overnight in the cold room. Conversion rates are monitored by HPLC. The reaction is finished, when about 90 % of the porcine insulin has been converted to di-tert-butylthreonine-B30-human insulin. The crude material is isolated by precipitation with methanol/ether, extensive washing with ether and drying in vacuo. HPLC of this material (fig. 4) exhibits 1-3 % DOI, 5-8 % unreacted starting material and 87 % human insulin ester. The difference to 100 % is covered mainly by desamidoinsulin respectively ester. Separation of the human insulin ester can be achieved by various chromatographic methods e.g. partition or ion exchange chromatography (3). In each case the less acidic and more hydrophobic insulin esteris clearly separated from unreacted porcine insulin and traces of DOI (fig. 4).

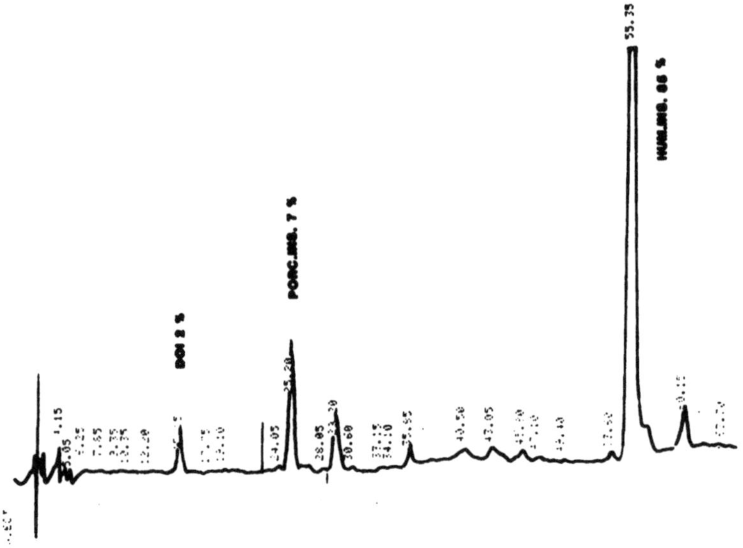

Figure 4. HPLC of crude reaction product (RP-8, Waters RCCS, 10 um, 8 x 100 mm, 0,05 M tetraethylammonium phosphate, 0,15 M $NaClO_4$, pH 3.0, gradient acetonitrile).

Cleavage of the protective groups is performed in the usual way by treatment with trifluoroacetic acid. The deblocked human insulin is isolated be precipitation with ether. The product, which already shows full biological activity, is then purified by means of ion exchange chromatography. The main fraction contains highly purified human insulin in an overall yield of 55-60 % based on the insulin content of the starting material. The insulin purity exceeds 98 %. HPLC reveals only trace amounts of desamidoinsulin. Further

impurities cannot be detected neither by HPLC nor by RIA. This material is presently produced in kilogram amounts for therapeutic application.
However, based on this transamidation type of reaction we also could easily prepare a number of B30-insulin analogs in order to investigate their immunogenic potency in pigs. These experiments are presently under way. Identical activity in blood sugar lowering tests and fat cell assay, compared to porcine and human insulin, have been determined and expected for DAI, Gly-, Ser-, Val-, Leu-, Phe- and Tyr-B30-insulin.
So far we have not been able to detect any unexpected biological properties in the series of B30-analogs. However, the synthesis of analogs showed some interesting characteristics, regarding the reactivity of -Lys-B29-X and amino acid derivate.
Conversion rates of porcine insulin and insulin esters with various amino acid esters are listed in table 1.

Insulin-Lys(B29)-X(B30)+Y $\xrightarrow[\text{pH 4.5}]{\text{Trypsin}}$ Insulin-Lys(B29)-Y(B30)+DOI			
X	Y	-Lys-Y (%)	DOI (%)
-	Thr(But)OBut	30	10
Ala	-	-	-
Thr(But)OBut	-		70
Ala	AlaOBut	25	10
Ala	Thr(But)OBut	85	3
Ala	ThrOMe	65	15
Ala	GlyOBut	25	5
Ala	Ser(But)OBut	30	15
Ala	ValOBut	85	4
Ala	LeuOBut	90	3
Ala	PheOBut	80	5
Ala	Tyr(But)OBut	75	10
Ala	D-Glu(OBut)$_2$	65	10
Ala	Leu-ProOBut	75	5
ThrOBut	Thr(But)OBut	40	4
ThrOMe	Thr(But)OBut	95	1
Thr	AlaOBut	5	1

Table 1. Conversion rates of various B30-insulin analogues.

Porcine insulin without amine component is not markedly attacked by trypsin at pH 4.5. Furthermore, besides hydrolysis enzymatic aminolysis of -Lys(B29)-X(B30) is dramatically influenced by the lipophilic character of the amine reactant as it is seen from the conversion rates rising from GlyOBut, AlaOBut, ThrOMe to Thr(But)OBut, LeuOBut and even Leu-ProOBut. Regulation of the transamidation during the enzymatically catalyzed transition state of the -Lys-X amide bond can also be seen if the reaction is carried out in the presence of equimolar amounts of two competing amino acid esters. The results measured by HPLC are shown in table 2. From our results it is quite obvious that the affinity of the -Lys-X bond to the active centre of the enzyme is regulated by the lipophilic character of the C-terminal end of the B-chain, too. This effect is suggested by the low reaction yields from DAI/Thr-(But)OBut and the extremely high conversion rate of human insulin-B30-methylester/Thr(But)OBut. Further support of this kind of influence is given by the difference of the hydrolytic activity of trypsin under our conditions. While porcine insulin remains essentially intact, lipophilic human insulin-B30-$(But)_2$ hydrolyzes to DOI.

In case of ThrOMe/ThrOBut it cannot be excluded that the reaction leads via insulin-B30-methylester to the final product. However, the positive influence of the more lipophilic character of LeuOBut and Thr(But)Obut in the competitive reaction with AlaOBut is clearly expressed by the high differences of the reaction products.

In regard to the surprising aminolytic specificity of trypsin at the -Lys-Ala site of insulin no other influence than pH has been detected under our experimental conditions. Though traces of -Arg(B22)-Thr(But)OBut can be isolated from large amounts of crude reaction product, none of the above mentioned amino acid esters or other nucleophils shows any preferred reaction site other than -Lys(B29)-. Thus, the specificity is determined by a tertiary structure of insulin, where the -Arg(B22)-Gly(B23)- bond is not disposable to the catalytic

$$\text{-Lys-Ala} + Y_1 + Y_2 \xrightarrow[\text{pH } 4.5]{\text{Trypsin}} \text{-Lys-}Y_1 + \text{-Lys-}Y_2 + (\text{DOI})$$

$$\frac{[Y]}{2} = [Y_1] = [Y_2]$$

Y_1	Y_2	-Lys-Y_1 (%)	-Lys-Y_2 (%)
Thr(But)OBut	AlaOBut	52	2
LeuOBut	AlaOBut	47	5
ThrOMe	Thr(But)OBut	6	50
LeuOBut	Thr(But)OBut	26	24
ValOBut	Thr(But)OBut	29	25
LeuOBut	PheOBut	30	30
Thr(But)OBut	PheOBut	25	25
LeuOBut	ValOBut	31	28

Table 2. Competition of amine components during tryptic transamidation of Ala-B30-insulin.

site of trypsin at low pH-values. However, cleavable conformation may be recovered, if energy can be generated and delivered to the molecule by additional lipophilic interactions between enzyme and substrate as it is the case with human insulin -B30-Thr-(But)OBut. This compound is rapidly hydrolyzed by trypsin at low pH in contrast to porcine insulin (table 1).

Conclusion

The one-step conversion of porcine insulin to human insulin is an excellent example of enzyme-catalyzed semisyntheses, which may be applied to various other peptides and proteins. However, we could demonstrate in case of insulin, that a high yield of enzymatic transamidation is no general principle and special conditions have to be investigated for each amino acid component and protein. Thus, we have been fortunate that our objective was conversion of porcine insulin into human insulin but not vice versa (table 1).

References

1. Offord, R.E. & Di Bello, C.: Semisynthetic Peptides and Proteins, Academic Press, London - New York - San Francisco (1978).
2. Ruttenberg, M.A., Science 177, 623 (1972).
3. Obermeier, R., Geiger, R., Hoppe-Seyler's Z. Physiol. Chem. 357, 759 (1976).
4. Inouye, K., et al., J.Am.Chem.Soc. 101, 751 (1979).
5. Morihara, K., Oka, T., Tsuzuki, H., Nature 280, 412 (1979).
6. Obermeier, R., Seipke, G., Grau, U., Semisynthetic Variations at the B-Chain C-Terminus of Porcine Insulin: Human Insulin Semisynthesis, in Kerp, L., Schlüter, K.J., Petersen, K.-G. (eds.): New Insulins, Proceedings of the Int. Symp., Freiburg i.Br., West Germany, 1981, in press.
7. Markussen, J., ibid.
8. Jonczyk, A., Gattner, H.-G., Hoppe-Seyler's Z. Physiol. Chem. 362, 1591 (1981).
9. Masaki, T., Nakamura, K., Isono, M., Soejima, M., Agric. Biol.Chem. 42 1443 (1978).
10. Schmitt, E., Gattner, H.-G., Hoppe-Seyler's Z.Physiol. Chem. 359, 799 (1978).
11. Kobayashi, M., Ohgaku, S., Iwasaki, M., Shigeta, G., Oka, T., Morihara, K., Diabetes 30, 519 (1981).

PREPARATION, PROPERTIES AND APPLICATION OF PHOTO-REACTIVE INSULINS

Derek Saunders, Peter Thamm, Gernot Klotz, Dietrich Brandenburg
Deutsches Wollforschungsinstitut
D-5100 Aachen, Federal Republic Germany

Introduction

The first step in the action of insulin is binding to specific receptors in the membrane of target cells (1). Our present understanding of the hormone and its receptor is almost exclusively based on studies with the separate partners or their *reversible* interaction. It was to be expected that new information could be obtained by transforming the reversible hormone/ receptor system into an *irreversible* complex. To this end, photo-affinity labelling (2) appeared to be particularly promising. We have applied this technique with the following aims:

1) Synthesis of a wide variety of homogeneous photo-sensitive insulin derivatives (P-insulins).
2) Direct localisation of binding sites on the insulin molecule.
3) Application of radio-iodinated P-insulins for the labelling of receptors for studies of receptor structure and turnover.
4) Covalent coupling of P-insulins to viable cells and observation of biological responses.
5) Determination of the intrinsic potency of ^{125}I-P-insulin-receptor complexes.

The purpose of this paper is to outline the strategy and tactics developed for the preparation of photo-insulins in our Institute and to summarize some of the results obtained in collaborative studies with other groups.

Chemistry of Peptides and Proteins, Vol. 2

Results

Four photo-sensitive labels were synthesized:

Apa-ONp (Apa = 4-azido-phenylacetyl)
Napa-ONp (Napa = 2-nitro-4-azido-phenylacetyl)
Nap-Gly-ONSu (Nap = 2-nitro-4-azidophenyl)
Msc-Phe(N_3)-ONp (Phe(N_3) = 4-azido-L-phenylalanine)

Preparation of Photo-insulins

All photo-reactive derivatives have been prepared by modification of the native hormone. Reaction of unprotected insulin with Apa-ONp and fractionation of the resulting mixture by ion exchange chromatography at acidic and then alkaline pH gave the three derivatives 1, 2 and 8 in low yields (2-5%). Intermediate partial protection of amino groups allowed the selective introduction of photo-labels and was used in all other cases. The Msc group proved to be of particular advantage. Preparation of the P-insulins listed in Table 1 followed the general strategies developed for the directed modification of the native insulin structure (3). Acylations of bis-Msc-insulins were performed with an excess (2-6 eq.) of active ester in dimethylsulfoxide/N-methylmorpholine. The homogeneous photo-insulins were isolated by a combination of gel filtration in acetic acid, ion exchange chromatography on SP-Sephadex at pH 3 and/or DEAE-Sephadex or Sephacel at pH around 7.5.

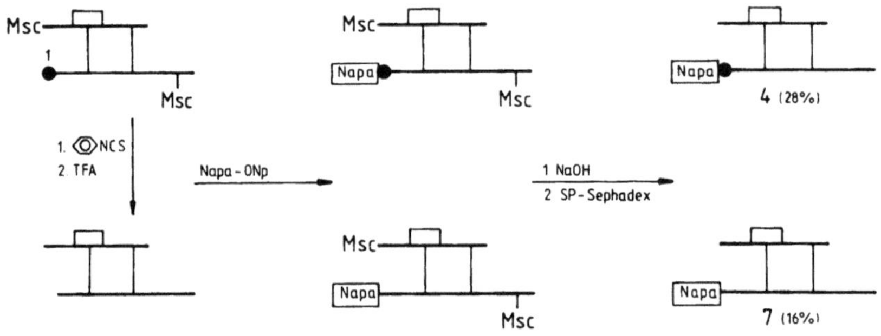

Figure 1. Preparation of B1/B2 photo-insulins.

Position B29 (side-chain of lysine). Compounds 1 and 2 were obtained via A1,B1$(Msc)_2$-insulin in yields of 27 and 7%, respectively. Introduction of the Napa moiety was accompanied with some side reactions and necessitated more extensive purification of compound 2 than usual.

N-Terminus of the B-chain. Besides direct acylation of A1,B29-$(Msc)_2$-insulin (yields of 3 and 4: 29%), two compounds were obtained in which the N-terminal Phe was removed. The preparation of 4 and 7 is outlined as examples of the procedures in Fig. 1. In contrast to the strong aggregation tendency of 4, the removal of Phe led to a derivative with good solubility properties. The compound 6 differs from insulin only by the replacement of H by N_3, and the maintenance of native structure is confirmed by its ability to give well-shaped rhombohedral 2-zinc crystals.

N-Terminus of the A-chain. Replacement of A1-glycine necessitates a rather elaborate strategy (3). The amino group was first protected by a Boc residue, then B1 and B29 were blocked with Msc groups, A1 was deblocked, and Gly^{A1} removed by Edman degradation. Subsequent acylation with the appropriate esters gave, after deprotection and purification, compounds 9 and 10 in yields of ca. 3%.

Besides these monofunctional P-insulins, two compounds bearing 2 photo groups have also been obtained. They have been designed to investigate structural features of insulin receptors by means of introducing bridges which are not cleaved by disruption of insulin disulfide groups. Compound 11 was obtained from A1-trifluoroacetyl insulin by acylation and deprotection. The dimer 12 was prepared by reacting A1-Tfa-insulin with 4-sulfophenylisothiocyanate at B1, linking the B29 amino groups with adipoyl-bis-p-nitrophenylester, removal of B1-phenylalanines by Edman degradation, introduction of the Apa groups, and base treatment to remove the Tfa groups.

The P-insulins were characterized by cellulose acetate electrophoresis at pH 2.2 and 8.6, electrophoresis after oxidative cleavage of disulfide bonds, TLC, end group determination with dansyl chloride, amino acid analysis, UV and IR spectra (characteristic band at 2120 cm^{-1}).

Biological activity was determined under *reversible* conditions in the dark by measuring insulin-stimulated lipogenesis in isolated rat fat cells. The P-insulins exhibit potencies which are comparable to other analogous insulin derivatives. Thus, photolabels at the N-terminus of the B-chain affect potency only slightly, at B29 to a somewhat greater extent, and at A1 substantially. However, a photo-label based on D-diaminobutyric acid can, as expected, be tolerated rather well.

Table 1. Photo-reactive insulins.

	Photo-insulin	Ref.	Potency	Affinity labelling
1	B29-Apa*	5,6	54	rat liver membr. a,b
2	B29-Napa*	5,7	35-60	rat liver membr. 10 adipocytes 9
3	B1-Apa*	5,6	85	rat liver membr. a
4	B1-Napa	5,7	81	adipocytes
5	B1-Nap[Gly^{B1}]	7	62	adipocytes d
6	[$Phe(N_3)^{B1}$]*	5,6	77	rat liver membr. b
7	B2-Napa-des-Phe^{B1}*	5,7	75-100	solub. pork liver membr. c rat liver membr. adipocytes d,e,11,13 hepatocytes 12
8	A1-Apa	5,6	17	
9	A1-Nap	7	16-20	adipocytes 9
10	[$D\text{-}Dab(Apa)^{A1}$]*	8	66	rat liver membr. b
11	B1,B29-$(Nap\text{-}Gly)_2$	7		
12	B2,B2'-(Apa_2)-B29-B29'-adipoyl-dimer	7		

* ^{125}I-labelled (peroxidase or chloramine T method) for structural studies.

References (a - e: see ref. 9): a) Wisher et al.,1978; b) Wisher et al.,1980; c) Kühn et al.,1980; d) Diaconescu et al.,1980; Brandenburg et al.,1980.

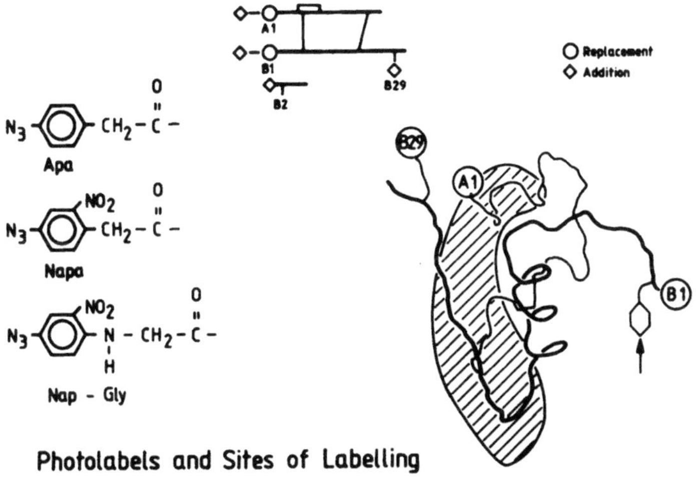

Figure 2. Schematic drawing of the spatial structure of insulin with the putative receptor binding region (based on Blundell et al., 14).

Except for the azidophenyl-insulins of Yip (15) these are the only homogeneous photo-insulins described and applied for receptor-labelling so far. Specific covalent binding was found with A1,B1,B2 and B29 photo-insulins. B1 and B2 were unexpected sites since they are remote from the binding area of Fig. 2. Analysis by SDS gel electrophoresis of ^{125}I-photo-insulin-labelled liver and adipocyte receptors revealed subunits of MW 125-130 K (corrected 90 K) and gave size estimates of the receptor (glyco)protein of ca.300 K, and under non-denaturing conditions 600 K. Internalization was demonstrated in hepatocytes (12) and adipocytes (13). Covalent binding of 7 and 2 to viable adipocytes gave a near-maximal long term stimulation of lipogenesis and glucose transport.

Photo-insulins such as the ones described are thus becoming valuable tools in research on the mechanism of insulin action. Molecular homogeneity, while not an absolute must in qualitative work, clearly is essential for more refined quantitative studies.

Acknowledgement

The authors wish to thank all colleagues participating in the joint work for a most stimulating and enjoyable collaboration, Mrs. C. Diaconescu for bioassays, Mr. K. Freude for technical and Mrs. W. Croon for secretarial assistance. We are grateful to SFB 113 Diabetesforschung, Düsseldorf, for financial support.

References

1. Czech, M.P.: Ann. Rev. Biochemistry 46, 359-384 (1977).

2. Bailey, J.L., Knowles, J.R.: Methods Enzymol. 46, 69 (1977).

3. Brandenburg, D., Weimann, H.-J., Trindler, P., Schüttler, A.: in Basic and Clinical Aspects of Immunity to Insulin (eds Keck, K., Erb, P.) W. de Gruyter, Berlin, New York, 375-393 (1981).

4. Brandenburg, D., Saunders, D.: in Amino-acids, Peptides and Proteins. Specialist Periodical Report (Sheppard, R.C., Senior Reporter), The Royal Soc. of Chemistry, London 12, 498-513 (1981).

5. Thamm, P., Dr. rer. nat.Thesis, Tech. Hochsch. Aachen 1980.

6. Thamm, P., Brandenburg, D.: in Peptides, Proc. 15th Europ. Peptide Symp. (eds. Siemion, Z., Kupryszewski, G.)Wrocław University Press, Poland, 639-642 (1979).

7. Thamm, P., Saunders, D., Brandenburg, D.: in Insulin, Chemistry, structure and function of insulin and related hormones (eds. Brandenburg, D., Wollmer, A.) W. de Gruyter, Berlin, New York, 309-316 (1980).

8. Saunders, D., Brandenburg, D.: Hoppe-Seyler's Z. Physiol. Chem. 362, 1237-1245 (1981).

9. Brandenburg, D., Diaconescu, C., Klotz, G., Saunders, D., Thamm, P., Uschkoreit, J.: in Current Views on Insulin Receptors (eds. Andreani, D., de Pirro, R., Lauro, R., Olefsky, J., Roth, J.) Academic Press, London, New York, 303-316 (1981).

10. Baron, M.D., Wisher, M.H., Thamm. P., Saunders, D., Brandenburg, D., Sönksen, P.H.: Biochemistry 20, 4156-4161 (1981).

11. Uschkoreit, J., Brandenburg, D.: in Current Views on Insulin Receptors (eds. Andreani, D., de Pirro, R., Lauro, R., Olefsky, J., Roth, J.) Academic Press, London, New York, 317-322 (1981).

12. Fehlmann, M., Carpentier, J.-L., le Cam, A., Thamm, P., Saunders, D., Brandenburg, D., Orci, L., Freychet, P.: J. Cell Biology 93, 82-87 (1982).

13. Berhanu, P., Olefsky, J.M., Tsai, P., Thamm, P., Saunders, D., Brandenburg, D.: Proc. Natl. Acad. Sci. USA 79, 4069-4073 (1982).

14. Blundell, T., Dodson, G., Hodgkin, D., Mercola, D.: Adv. Protein Chemistry 26, 279-402 (1972).

15. Yip, C.C., Moule, M.L., Yeung, C.W.T.: Biochemistry 21, 2940-2945 (1982),

SYNTHESIS OF A19-[3,5-DIIODO-TYROSINE] PORCINE INSULIN

H.-J. Wieneke, E. Büllesbach, H.-G. Gattner, H. Zahn
Deutsches Wollforschungsinstitut
D-51oo Aachen, Federal Republic of Germany

W. Danho
Chemical Research Department Hoffmann-La Roche Inc.,
Nutley, New Jersey, USA

Introduction

Iodination of the tyrosines of insulin has been principally investigated for the preparation of labelled tracers used in the study of insulin action and metabolism as well as for immunoassay of insulin present in blood or other body fluids. Direct iodination of insulin yields a heterogenous product with various distributions of iodine between the A and the B chains (1). Definitive data on the chemical and biological effects of the iodination on the insulin molecule cannot be drawn from such materials. We have used 3,5-diiodo-tyrosine as amino acid unit in the synthesis of homogenous iodinated peptide chains for the production of a defined iodinated porcine insulin analogue.

A19 – tyrosine is on the surface of the monomer and may be in= volved in the interaction between the hormone and its recep= tor (2). Pure A19-[3,5-diiodo-tyrosine] insulin is required to obtain information about the role of A19 – tyrosine in the receptor-binding region.

In this communication we wish to describe the synthesis of the protected [Tyr$(I_2)^{A19}$] A chain, its deprotection with trifluoro= acetic acid/thiophenol, conversion to the tetra-S-sulphonate form, and its combination with native di-S-sulphonate B chain to yield A19-[3,5-diiodo-tyrosine] porcine insulin.

Chemistry of Peptides and Proteins, Vol. 2

Scheme for the synthesis of [$Tyr(I_2)^{A19}$]porcine insulin

$HCl \cdot Tyr(I_2)$-OEt

↓ DCC/HOBt 68 %
1. N-Ethylmorpholine
2. Bpoc-Asn-OH
3. $CHCl_3$/c-hexane

Bpoc-Asn-$Tyr(I_2)$-OEt

↓ M.A. 65 %
1. CH_3COOH (8o %) pyridine/HBr
2. Bpoc-$Glu(OBu^t)$-OH
3. Ether

Bpoc-$Glu(OBu^t)$-Asn-$Tyr(I_2)$-OEt

↓ DCC/HOBt 76 %
1. $NaOH/H_2O$/dioxane/N_2
2. H-Cys(Trt)-Asn-OBu^t
3. LH-2o in MeOH

(A17-21) Bpoc-$Glu(OBu^t)$-Asn-$Tyr(I_2)$-Cys(Trt)-Asn-OBu^t

↓ DCC/HOBt 52 %
1. CH_3COOH (8o %) pyridine/HBr
2. Bpoc-Leu-$Tyr(Bu^t)$-Gln-Leu-OH
3. LH-2o in MeOH

(A13-21) Bpoc-Leu-$Tyr(Bu^t)$-Gln-Leu-$Glu(OBu^t)$-Asn-$Tyr(I_2)$-Cys(Trt)-Asn-OBu^t

DCC/HOBt

1. CH_3COOH (8o %) pyridine/HBr
2. Triethylamine
3. LH-2o in DMF
4. A1-12

Boc-Gly-Ile-Val-Glu(OBu^t)-Gln-Cys(Trt)-Cys(Trt)-Thr(Bu^t)-Ser(Bu^t)-Ile-Cys(Trt)-Ser(Bu^t)-Leu-Tyr(Bu^t)-Gln-Leu-Glu(OBu^t)-Asn-Tyr(I_2)-Cys(Trt)-Asn-OBu^t

1. CF_3COOH/Thiophenol
2. $Na_2SO_3/Na_2S_4O_6$
3. Sephadex G-25 in o.o5 M NH_4HCO_3
4. DEAE-Cellulose pH 5.6
5. Sephadex G-25 in o.o5 M NH_4HCO_3

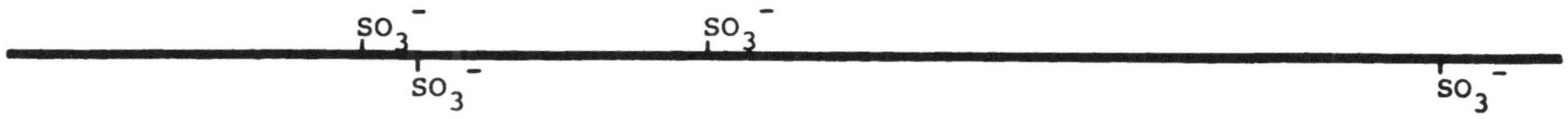

1. Di-S-sulphonated-B-chain
2. Mercaptoethanol/glycine buffer pH 1o.6
3. Sephadex G-25/oxidation buffer
4. Dialyze/lyophilization
5. CM-Cellulose / NaCl gradient
6. Sephadex G-25 in o.o5 M NH_4HCO_3
7) -"- G-5o sf in CH_3COOH (1o %)

[Tyr(I_2)^{A19}]porcine insulin

Results

The strategy of solution condensation of fragments using acid labile protecting groups was selected for the synthesis. C-terminal and γ-carboxyl functions were protected as tert-butyl esters, the hydroxyl functions of tyrosine, threonine and serine as tert-butyl ethers except for the 3,5-diiodo-tyrosine residue for which no protection was needed. For the temporary protection of α-amino groups, the benzyloxycarbonyl- and 2-(4-biphenylyl)-isopropyloxycarbonyl groups were used. Cysteine side chains were protected with trityl groups.

Synthesis of fragment A17-21

Bpoc-Asn-OH was coupled with $Tyr(I_2)$-OEt by the DCC/HOBt method. The crude product was recrystallized from $CHCl_3$/c-hexane to give the dipeptide Bpoc-Asn-$Tyr(I_2)$-OEt in 68% yield. This was deprotected with HBr/CH_3COOH and coupled with Bpoc-$Glu(OBu^t)$-OH by the mixed anhydride method. After purification by ether ex= traction a yield of 65% was obtained. Quantitative saponifi= cation in dioxane/water solution yielded Bpoc-$Glu(OBu^t)$-Asn-$Tyr(I_2)$-OH, which was then coupled with H-Cys(Trt)-Asn-OBu^t via the DCC/HOBt method to give the pentapeptide A17-21 in a yield of 76%. Purification was achieved on Sephadex LH-2o/MeOH.

Synthesis of fragment A13-21

The Bpoc-group of Bpoc-$Glu(OBu^t)$-Asn-$Tyr(I_2)$-Cys(Trt)-Asn-OBu^t was removed with 8o% acetic acid/pyridine hydrobromide. The re= sulting HBr·$Glu(OBu^t)$-Asn-$Tyr(I_2)$-Cys(Trt)-Asn-OBu^t was coupled to Bpoc-Leu-$Tyr(Bu^t)$-Gln-Leu by the DCC/HOBt method. After purification by gel filtration on Sephadex LH-2o in MeOH a yield of 52% of the nonapeptide derivate was obtained.

Synthesis of protected A19-[3,5-diiodo-tyrosine] A chain

The deprotection of Bpoc-function from Bpoc-Leu-Tyr(Bu^t)-Gln-Leu-Glu(OBu^t)-Asn-Tyr(I_2)-Cys(Trt)-Asn-OBu^t was carried out as described above. The free base was obtained by neutralization with triethylamine and gel filtration on Sephadex LH-2o in MeOH. This peptide derivate was coupled with fragment A1-12 by the DCC/HOBt method. Since the sequence A1-21 represents the final step of the synthetic procedure, no purification was performed on the crude coupling mixture.

Deprotection, conversion to the A19-[3,5-diiodo-tyrosine] S-sulphonated A chain and purification

The deprotection of the crude synthetic product was achieved in a single step with trifluoroacetic acid using thiophenol as trityl-cation scavenger (3). The conversion to the tetra-S-sulphonate form was accomplished by reaction with sodium sulphite and sodium tetrathionate. Purification was achieved by gel filtration on Sephadex G-25 in o.o5 M NH_4HCO_3 and then ion-exchange chromatography on DEAE-cellulose at pH 5.6 using a linear NaCl-gradient. The purity of the iodinated A chain was checked by:

1. Amino acid analysis after acid hydrolysis gave a composition in good agreement with the theoretically expected values.

2. Paper electrophoresis at pH 2.4 and 4.8. A single band was detected by Pauly reaction (4) which migrated to the same position as the natural S-sulphonated A chain.

Synthesis and purification of A19-[3,5-diiodo-tyrosine] porcine insulin

Tetra-S-sulphonated A19-[diiodo-tyrosine] A chain and native di-S-sulphonated B chain were reduced together with mercaptoethanol at pH 5.6 and oxidized at this pH for 48 h at 4° to yield the crude iodinated insulin (5). This was purified by ion exchange chromatography on CM-cellulose using a linear

NaCl gradient. Final step was the gel filtration on Sephadex G-5o sf in 1o% acetic acid. The purity of $[Tyr(I_2)^{A19}]$ porcine insulin was checked by:

1. Amino acid analysis. Acid hydrolysis gave a composition in good agreement with the theoretically expected values:
   Asp 3.oo (3); Thr 1.68 (2); Ser 2.78 (3); Glu 7.17 (7); Gly 4.09 (4); Ala 2.oo (2); Val 3.62 (4); Ile 1.59 (2); Leu 5.76 (6); Tyr 3.64 (4); Phe 2.93 (3); Lys 1.oo (1); His 1.98 (2); Arg 1.oo (1).

2. Cellulose acetate electrophoresis at pH 2.4 and 4.8. A single band was detected (Pauly reaction) which migrated to the same position as insulin.

Biological activity of A19-[3,5-diiodo-tyrosine] porcine insulin

The biological activity of the insulin analogue as determined in vitro with rat epididymal adipocytes by the method of Moody et al. (6) was 2.7%.

References

1. Massaglia, A., Rosa, U., Rialdi, G., Rossi, C. A.: Biochem. J. 115, 11-18 (1969).
2. Pullen, R. A., Lindsay, D. C., Wood, S.P., Tickle, I. J., Blundell, T., Wollmer, A., Krail, G., Brandenburg, D., Zahn, H., Gliemann, J., Gammeltoft, S.: Nature 259, 369-373 (1976).
3. Büllesbach, E.: Dissertation TH Aachen (1978).
4. Pauly, H.: Hoppe-Seyler's Z. Physiol. Chem. 42, 517 (19o4).
5. Rüegg, U., Gattner, H.-G.: Hoppe-Seyler's Z. Physiol. Chem. 356, 1527-1533 (1975).
6. Moody, A. J., Stan, M. A., Stan, M., Gliemann, J.: Horm. Metabol. Res. 6, 12-16 (1974).

SEMISYNTHESIS OF LINEAR PROINSULIN ANALOGUES WITH SHORTENED C-PEPTIDES

Erika E. Büllesbach

Deutsches Wollforschungsinstitut an der RWTH Aachen
D 5100 Aachen, Federal Republic of Germany

Introduction

The structure of proinsulin is characterized by the sequence B-chain-Arg-Arg-C-peptide-Lys-Arg-A-chain. The connecting peptide facilitates the formation of the disulphide links[1,2,3] although a high species variability was observed in length (27 - 35 amino acids) as well as in sequence.
X-ray studies of crystalline insulin have shown that the α-amino group Gly^{A1} and the α-carboxyl group Ala^{B30} were only separated by a distance of 8-10 Å[4]. These results suggested the introduction of a smaller connecting peptide. The stabilisation of the structure can also be achieved by a linkage between the two amino groups Gly^{A1} and Lys^{B29} via one of several dicarbonic acid derivatives (for a review see Ref. 5).
Now there are several biological and chemical features for the preparation of proinsulins with shortened C-peptides:

i) influence of the modified shortened C-peptide to the insulin structure and consequently the formation of the correct disulphide bridges,
ii) biological and immunological properties compared to native insulin and proinsulin,
iii) conversion of such proinsulin models to insulin,
iv) possibility of preparation by gene technology.

The present paper describes the preparation of open-chain proinsulin derivatives via semisynthesis.

Chemistry of Peptides and Proteins, Vol. 2

Semisynthesis

Five proinsulin analogues (Table 1) were prepared from suitably modified insulin chains[6,7]. The strategy of preparation was: the N-terminal elongation of native insulin A-chain disulphide with synthetic peptides, the modification of the B-chain disulphide to permit selective activation of the α-carboxyl group, condensation of both partially-protected large fragments, and deprotection and purification of the open-chain proinsulin analogue.

Table 1. Proinsulin models with shortened C-peptides.

B-chain-Arg-Arg-	Gly	-Lys-Arg-A-chain
B-chain-Arg-Arg-	Ala-Gly	-Lys-Arg-A-chain
B-chain-Arg-Arg-	Gly-Ala-Gly	-Lys-Arg-A-chain
B-chain-Arg-Arg-	Ala-Gly-Ala-Gly	-Lys-Arg-A-chain
B-chain-Arg-Arg-	Gly-Pro-Gly	-Lys-Arg-A-chain

Preparation of Thr-Arg-Arg-"C-peptide"-Lys(Msc)-Arg-A-chain

The starting material was Arg-A-chain disulphide monomer prepared by tryptic catalyzed reaction[6,8]. Further condensation with N-terminal protected amino acids or synthetic peptides were performed via mixed anhydride reactions in aqueous organic solvents[8,9,10] (Fig 1). All reactions were controlled by paper electrophoresis at pH 2.2 and 4.8 and end group determination[11]. After purification by ion exchange chromatography Thr-Arg-Arg-"C-peptide"-Lys(Msc)-Arg-A-chain was isolated in which the N-terminal amino acid Thr corresponded to the C-terminal amino acid of human B-chain.

A-chain

60.0% ↓ 1. Boc-ArgOH / Trypsin
2. TFA
3. SP-Sephadex pH 3

Arg-A-chain

73.1% ↓ 1. Boc-Gly-Lys(Msc)OH / M.A.
2. TFA
3. SP-Sephadex pH 3

Gly-Lys(Msc)-Arg-A-chain

47.5% ↓ 1. Boc-Thr(But)-Arg-Arg-AlaOH·2 TosOH
M.A.
2. TFA
3. SP-Sephadex pH 3

Thr-Arg-Arg-Ala-Gly-Lys(Msc)-Arg-A-chain

Fig. 1. N-terminal elongation of A-chain to prepare Thr-Arg-Arg-"C-peptide"-Lys(Msc)-Arg-A-chain (i.e. C-Peptide = Ala-Gly).

$B_{1-30}(SSO_3^-)_2$

96.5% ↓ Carboxypeptidase A

$B_{1-29}(SSO_3^-)_2$

81.0% ↓ 1. Reduction
2. iodine / 40% acetic acid
3. Sephadex G-50f / 10% CH_3COOH

$B_{1-29}(SS)$

74.7% ↓ 1. BF_3 / CH_3OH
2. Trypsin
3. Sephadex G-50f / 10% CH_3COOH

$B_{1-29}(SS)\text{-}O^{\gamma 13}O^{\gamma 21}(OMe)_2$

40.2% ↓ 1. MscONSu
2. SP-Sephadex pH 3
3. CM-Cellulose pH 4.8
4. Sephadex G-25f / 0.0001 M HCl

$N^{\alpha 1}N^{\varepsilon 29}Msc_2B_{1-29}(SS)\text{-}O^{\gamma 13}O^{\gamma 21}(OMe)_2 \cdot 3\ HCl$

Fig. 2. Preparation of $N^{\alpha 1}N^{\varepsilon 29}$-di-(methylsulphonylethyloxycarbonyl)-des-alanine^{B30}-B-chain-disulphide-$O^{\gamma 13}O^{\gamma 21}$-dimethyl ester trihydrochloride.

Modification of insulin B-chain

For subsequent N-terminal elongation of the A-chain derivative, des-B30-B-chain had to be activated at the C-terminus. For the selective activation the protection of the side chain carboxyl groups was necessary[7,9,10]. The preparation of such a protected native B-chain derivative was possible by a route involving several steps outlined in Fig 2. The resulting $N^{\alpha 1}N^{\varepsilon 29}$-di-(methylsulphonylethyloxycarbonyl)-des-B30-B-chain disulphide-$O^{\gamma 13}$ $O^{\gamma 21}$-dimethyl ester as trihydrochloride was isolated in an overall yield of 25%.

Preparation and isolation of open-chain proinsulins

To prepare the shortened proinsulin model 17.5 µmol B-chain derivative were dissolved in DMF. The activation was achieved by the mixed anhydride reaction by addition of 4 equivalents N-methylmorpholine as base and 2 equivalents isobutyl chloroformate at -10°C for 30 sec. The coupling was followed with 5 µmol Thr-Arg-Arg-"C-peptide"-Lys(Msc)-Arg-A-chain, dissolved in DMF/water 4:1 ("pH 7.2"). After 2 min the coupling reaction was stopped by addition of a solution of 1M $NH_2OH \cdot HCl$, dissolved in DMF / water 4:1 (adjusted to "pH 7.2" with N-methylmorpholine). The purification was achieved by gel filtration on Sephadex G-50f equilibrated in 10% acetic acid. The deprotection of the open-chain proinsulin model was performed in 0.1M NaOH for 30 min at 0°C under N_2 and a protein concentration of 10 mg/ml. The reaction was stopped by addition of glacial acetic acid and the deprotected derivative was purified by gel filtration on Sephadex G-50f equilibrated in 1M acetic acid. The open-chain proinsulin derivative was isolated as monomer disulphide or after oxidative sulphitolysis as hexa-S-sulphonate[10]. The final product was purified by ion exchange chromatography, the disulphide derivative on CM-cellulose at pH 7.3 or the hexa-S-sulphonate on CM-cellulose at pH 4[10] and eluted with a linear NaCl gradient.

Characterization

The shortened open-chain proinsulin derivatives were characterized i) by electrophoresis at pH 2.2, 4.8, and 8.6, ii), by tryptic digestion followed by electrophoresis at pH 2.2 (Pauly positive fragments were A_{1-21}, B_{1-22}. B_{23-29}) iii), by amino acid analysis after acid hydrolysis and iv) digestion with trypsin / carboxypeptidase B[12] and determination of 1 Lys and 3 Arg by amino acid analysis.

In another reaction the partially protected B-chain derivative was activated by the mixed anhydride method and coupled with Thr-Arg[7]. The deprotected and purified human-B-chain-peptidyl-arginine showed in amino acid analysis of the amino acid released by digestion with carboxypeptidase Y[13] that the C-terminus was extended with the sequence Thr-Arg and no racemization had taken place[7].

Scopes and limits of the mixed anhydride reaction in semisynthesis

Isobutyl chloroformate or its mixed anhydrides with carboxylic acids had shown three side reactions, first the irreversible formation of isobutyloxycarbonyl derivative of the α-amino groups, second the reversible formation of N^{im}-isobutyloxycarbonyl derivative (specially in activation of the insulin-B-chain derivative[7,9,10]) and third the reversible O-acyl derivatives.

Side reactions on the imidazole and hydroxyl side chains were reversible by treatment with a solution of hydroxylamine. Advantages of the mixed anhydride reaction was the high speed and and lack of racemization.

Acknowledgement

This work was financially supported by Deutsche Forschungsgemeinschaft.

References

1. Steiner, D. F., Clark, J. L.: Proc. Natl. Acad. Sci. USA 60, 622-629 (1968).

2. Markussen, J., Heding, L. G.: Int. J. Peptide Protein Res. 6, 245-252 (1974).

3. Frank, B. H., Pettee, J. M., Zimmerman, R. E., Burck, P. J. Proc. 7th American. Peptide Sympos. (Rich, D. E., Gross, E., eds.) Pierce Chemical Co. Rockford Ill. 1981 in press.

4. Blundell, T., Dodson, G., Hodgkin, D., Mercola, D.: Advances Prot. Chem. 26, 279-402 (1972).

5. Geiger, R.: Chemiker Zeitung 100, 111-129 (1976).

6. Naithani, V. K., Gattner, H. G.: Hoppe Seyler's Z. Physiol. Chem. 362, 685-695 (1981).

7. Büllesbach, E. E., Schmitt, E. W., Gattner, H. G.: Int. J. Peptide Protein Res. 20 in press (1982).

8. Naithani, V. K., Gattner, H. G., Büllesbach, E. E., Föhles, J., Zahn, H.: "Peptides" Proc. 6th American Peptide Sympos. (Gross, E., Meienhofer, J., eds.) pp 571-576 Pierce Chemical Co, Rockford Ill. 1979

9. Büllesbach, E. E., Schmitt, E. W., Gattner, H. G., Naithani, V. K., Föhles, J.: Proc. 3rd USSR-FRG Sympos. on Chemistry of Peptides and Proteins (1980) (Voelter, W., Wünsch, E., Ivanov, V. T., Ovchinnikov, Yu. A., eds) Walter de Gruyter in press.

10. Büllesbach, E. E.: Tetrahedron Lett. 23, 1877-1880 (1982).

11. Gray, W. R.: Methods Enzymol. 11, 139-151 (1967).

12. Kemmler, W., Peterson, J. D., Steiner, D. F.: J. Biol. Chem. 246, 6786-6791 (1971).

13. Lee, H. M., Diordan, J. F.: Biochem. Biophys. Res. Commun. 85, 1135-1142 (1978).

SYNTHESIS OF UNSYMMETRICAL CYSTINE PEPTIDES: DIRECTED DISULFIDE PAIRING WITH THE SULFENOHYDRAZIDE METHOD

Sigrun Romani, Walter Göhring, Luis Moroder and Erich Wünsch
Max-Planck-Institut für Biochemie, Abteilung Peptidchemie,
8033 Martinsried bei München, BRD.

For the selective synthesis of unsymmetrical cystine-peptides various methods have been proposed so far (1). Among these, the use of alkoxycarbonylsulfenyl cysteine derivatives proved to be the most promising approach: it allowed an unambiguous synthesis of human insulin (2). Nevertheless this method of Kamber (3) may not be of general use, particularly for tryptophan-containing peptides; sulfenyl halides are known to react with indole functions at high rates (4,5,6).
To bypass this critical drawback we have recently proposed the sulfenohydrazides as cysteine activated intermediates to be selectively converted via thiolysis into the unsymmetrical disulfides according to the scheme shown in Fig. 1.

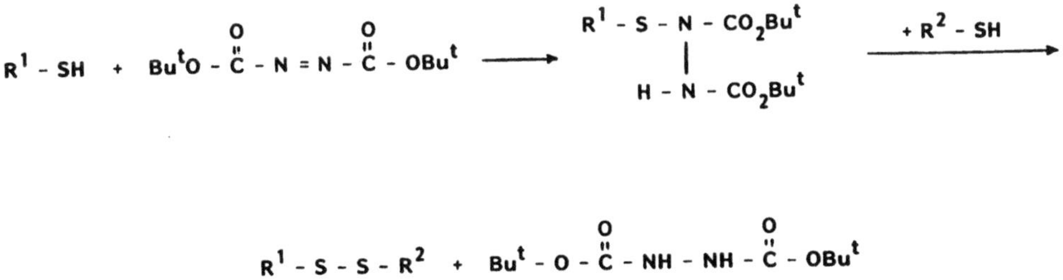

Fig. 1. Reaction scheme of the sulfenohydrazide method for the synthesis of unsymmetrical cystine peptides.

Chemistry of Peptides and Proteins, Vol. 2

Reaction components A	+ B	Solvent	Molar ratio A : B	Reaction time [h]	Ratio of reaction products in [%] [1]	[3]	[4]
[1*]	[2]	DMF	1 : 1.35	24	54	26	20
[2]	[1*]	DMF	1.35 : 1	24	5	0	95
[1]	[2*]	DMF	1 : 1	24	37	0	63
[1]	[2*]	TFE	1 : 1	24	74	10	16
[1]	[2*]	DMF	1.1 : 1	24	23	0	77
[1]	[2*]	DMF	1 : 1	24	37	0	63
[1]	[2*]	DMF	1 : 1.5	24	36	0	64
[1]	[2*]	DMF	1 : 3	24	0	0	100
[1]	[2*]	DMF	1 : 5	24	0	0	100

[1] BOC-Thr(Bu^t)-Ala-Cys-Gly-Gln-Lys(BOC)-Ser(Bu^t)-Pro-OBu^t

[1*] BOC-Thr(Bu^t)-Ala-Cys(NCO_2Bu^t-$NHCO_2Bu^t$)-Gly-Gln-Lys(BOC)-Ser(Bu^t)-Pro-OBu^t

[2] BOC-Ala-Gly-Cys-OH

[2*] BOC-Ala-Gly-Cys(NCO_2Bu^t-$NHCO_2Bu^t$)-OH

[3] [BOC-Thr(Bu^t)-Ala-Cys-Gly-Gln-Lys(BOC)-Ser(Bu^t)-Pro-OBu^t]$_2$

[4] BOC-Ala-Gly-Cys-OH
BOC-Thr(Bu^t)-Ala - Cys-Gly-Gln-Lys(BOC)-Ser(Bu^t)-Pro-OBu^t

[5] [BOC-Ala-Gly-Cys-OH]$_2$

Table 1. Correlation between ratio of reaction products and experimental conditions in the synthesis of unsymmetrical cystine peptides via the sulfenohydrazide procedure. For experimental details see Ref. 7.

The unsymmetrical cystine peptides are obtained with variable success depending upon the reaction conditions as summarized in Table 1. This systematic investigation has clearly shown that optimal yields in unsymmetrical disulfides are obtained by operating throughout the reaction with a larger excess of the sulfenohydrazide components. Under these conditions thiol-induced disproportionation to symmetrical disulfides is largely suppressed as well documented by the preparative example shown in Fig. 2. After the cysteine component was totally consumed as determined by hplc of the reaction mixture and by Ellman's test, the unsymmetrical cystine peptide was isolated by simple precipitation procedures (the activated tripeptide derivative is ether-soluble) as homogeneous compound in 84% yield [tlc: butanol/acetic acid/water/pyridine

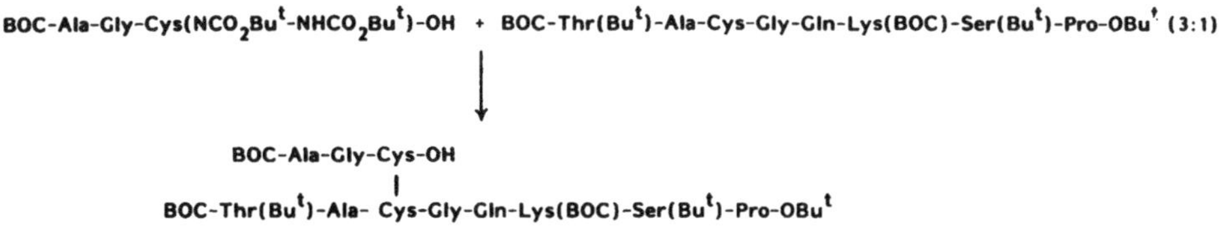

Fig. 2. Scheme of synthesis of an unsymmetrical cystine peptide.

60:6:24:20; hplc: µ-Bondapak C18 (0.4 x 30 cm), eluent: 0.1M ammonium acetate (pH 6.8)/acetonitrile, 52:48 (v/v), isocratic elution at a flow rate of 1.5 ml/min, absorbance at 210 nm; amino acid analysis (6N HCl, 24h, 110°C): Thr 1.04 (1), Ser 1.02 (1), Glu 1.04 (1), Pro 1.05 (1), Gly 1.99 (2), Ala 1.99 (2), Cys_2 0.92 (1), Lys 0.92 (1); peptide content (M_r 1533.96): 96%).

By lowering the molar ratio of the reaction components thiol-induced disproportionation of the unsymmetrical disulfide to the symmetrical compounds was found to become a competing side reaction. Hplc on µ-Bondapak C18 (for experimental details see above) was used to monitor the formation of unsymmetrical cystine peptide and its disappearance with concomitant increase in content of symmetrical cystine peptide (its concentration after 24h was still negligible).

This observation prompted us to examine the use of thiol-trapping reagents to block in such cases the thiol-interchange at the maximum content of unsymmetrical disulfide and additionally to simplify the work-up procedures. Besides soluble SH-reagents, e.g. N-alkyl-maleimides, nitrostyrene, 4-chloro-7-sulfobenzofurazan or 4-chloro-7-nitrobenzofurazan, we paid particular attention to thiol-reagents on polymeric matrix. For this purpose 3-maleimidopropionic acid was coupled via its N-hydroxysuccinimido ester (8) to aminopolyoxyethylene glycol (M_r = 6000), a generous gift of Prof. Mutter (Mainz), to yield 3-maleimidopropionic acid amido-PEG (0.02 µmol ß-Ala/mg PEG) as well as to L-leucyl-oxymethylpolystyrene (0.51 µmol

L-Leu/mg resin) to produce 3-maleimido-propionyl-L-leucyl-oxymethylpolystyrene (ß-Ala/L-Leu 1:1 as determined by amino acid analysis).
The usefulness of the SH-trapping procedure is illustrated in the following two examples.

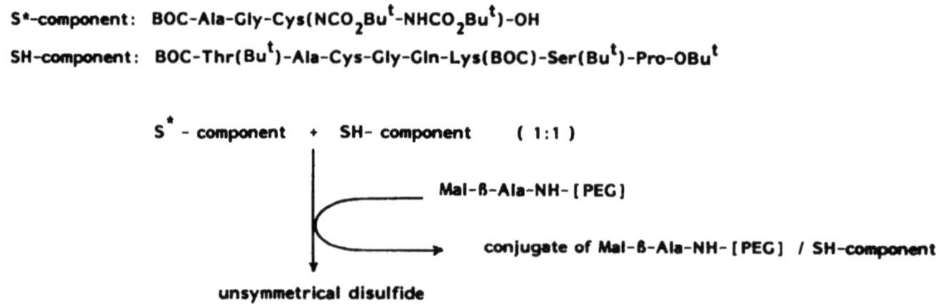

Fig. 3. Example 1: use of Mal-ß-Ala-NH-[PEG] as thiol-trapping reagent in the synthesis of unsymmetrical cystine-peptides.

In the case of example 1 reported in Fig. 3 at the maximum concentration of unsymmetrical disulfide the content of un-reacted cysteine-component -as determined by hplc- was 37%. After treatment of the reaction mixture with an excess of 3-maleimidopropionic acid amido-PEG for 30 min, the unsymmetrical cystine-peptide was isolated by precipitation with water and upon recrystallization from methanol/water it was obtained in 33% yield as homogeneous compound with analytical data identical within the limits of error with those reported above (the low yield is due to the partial water-solubility of the unsymmetrical disulfide). Amino acid analysis of the water-soluble PEG/SH-component conjugate confirmed that 37% of the cysteine-peptide was trapped by this procedure.
In example 2 (Fig. 4) again at the maximum concentration of unsymmetrical disulfide the content of unconsumed cysteine peptide was determined by hplc (37%). The reaction mixture was then stirred with an excess of 3-maleimidopropionyl-L-

S* - component : BOC-Ala-Gly-Cys(NCO$_2$Bu^t-NHCO$_2$Bu^t)-OH

SH- component : BOC-Thr(Bu^t)-Ala-Cys-Gly-Gln-Lys(BOC)-Ser(Bu^t)-Pro-OBu^t

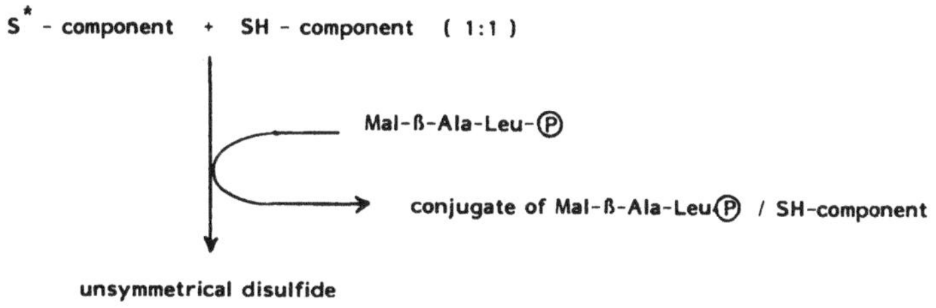

Fig. 4 Example 2: use of Mal-ß-Ala-Leu-amidomethylpolystyrene for the thiol-capture in the synthesis of unsymmetrical cystine-peptides.

leucyl-oxymethylpolystyrene and the disappearance of the thiol-component was monitored, both, by Ellman's test and hplc. The reaction was found to proceed at a remarkably lower rate than with the PEG-reagent. A quantitative removal of the cysteine-peptide was achieved only in 12 h. The resin was filtered off and the unsymmetrical cystine-peptide isolated upon addition of ether in 62% yield. Quantitative amino acid analysis of the resin/SH-component conjugate yielded 35% bound cysteine-peptide.

To summarize, above experimental models have clearly shown the usefulness of the sulfenohydrazide method for the synthesis of unsymmetrical cystine peptides.

The authors are grateful to the Deutsche Forschungsgemeinschaft for the generous financial support (grant Wu 20/12-I).

References

1. Wünsch, E.: in Houben Weyl, Methoden der Organischen Chemie, Vol. 15, Teil I, pp. 822-835, Thieme Verlag, Stuttgart (1974).

2. Sieber, P., Kamber, B., Hartmann, A., Jöhl, A., Riniker, B.,

Rittel, W.: Helv. Chim. Acta 57, 2617-2621 (1974).

3. Kamber, B.: Helv. Chim. Acta 56, 1370-1381 (1974).

4. Anderson, J.C., Barton, M.A., Hardy, P.M., Kenner, G.W., MacLeod, J.K., Preston, J., Sheppard, R.C.: Acta Chim. Acad. Sci. Hung. 44, 187-195 (1965).

5. Wünsch, E., Drees, F.: Chem. Ber. 100, 816-819 (1967).

6. Wünsch, E., Fontana, A., Dress, F.: Z. Naturforsch. 22b, 607-609 (1967).

7. Wünsch, E., Romani, S.: Hoppe-Seyler's Z. Physiol. Chem. 363, 449-453 (1982).

8. Keller, O., Rudinger, J.: Helv. Chim. Acta 58, 531-541 (1975).

SYNTHESIS OF MID REGIONAL PARATHYRIN SEGMENTS

Monika Casaretto, Helmut Zahn
Deutsches Wollforschungsinstitut an der Rheinisch-Westfälischen Technischen Hochschule Aachen
D-5100 Aachen, Germany

Waleed Danho
Chemical Research Department Hoffmann-La-Roche Inc.
07110 Nutley, New Jersey U.S.A.

Rolf-Dieter Hesch
Medizinische Hochschule Hannover
D-3000 Hannover 61, Germany

In 1978 Keutmann et al. (1) determined the complete sequence, 84 amino acids, of human parathyrin. The total synthesis was achieved by Kimura et al. (2) in 1981.

Radioimmunochemical measurement of parathyrin in plasma is complicated by heterogeneity of the circulating hormone. Fragments are found, differing in size, biological activity, immunochemical reactivity and stability. The C-terminal and mid regional fragments have prolonged half lives and are enriched in plasma, in contrast to the biologically active N-terminal sequence (1-31).

To diagnose abnormal parathyrin metabolism, specific RIA for these regions are employed. Rosenblatt et al. (3,4) have prepared the segments (44-68) and (53-84) by solid phase synthesis. We report the synthesis of the mid regional sequences (32-43) Tyr-(43-55) (7) and Tyr-(62-68).

Chemistry of Peptides and Proteins, Vol. 2

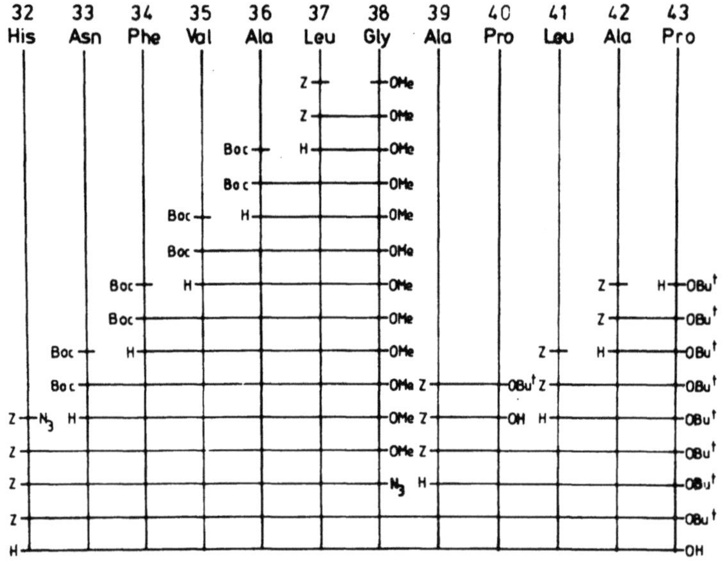

Scheme 1. Synthesis of parathyrin sequence (32-43);
Amino acid analysis: 1,01 (1) His, 1,03 (1) Asp, 0,96 (1) Phe, 0,97 (1) Val, 2,94 (3) Ala, 1,90 (2) Leu, 1,10 (1) Gly, 2,10 (2) Pro.

Synthesis of Parathyrin Sequence (32-43)

This sequence was constructed as shown in scheme 1 from the segments (32-38) and (39-43).

Segment (32-38) : Z-Leu-Gly-OMe was hydrogenated (H_2/Pd) and elongated by stepwise mixed anhydride couplings (M.A.) with Boc-derivatives of Ala, Val and Phe after successive deprotection with trifluoroacetic acid (TFA). Coupling of Boc-Asn with DCC/HOBt and Z-His-N_3 completed this segment.

Segment (39-43) : All condensations were performed by M.A. methods using Z amino acids. Z-Ala-Pro-OBu^t was deprotected with H_2/Pd and coupled with Z-Leu to yield Z-(41-43)-OBu^t. After hydrogenation this was reacted with Z-Ala-Pro-OH to give Z-(39-43)-OBu^t. This was hydrogenated and purified by counter current distribution (CCD).

Sequence (32-43) : Z-(32-38)-OMe was converted to the hydrazide activated to the azide and reacted with H-(39-43)-OBu^t. Treatment with hydrobromic acid in acetic acid using ethanethiol as scavenger allowed the cleavage of all protecting groups in one step.
Purification was achieved by gel filtration on G 25 sf in 30% acetic acid and ion exchange on CM Cellulose in isopropanol/ water 4:6 (v/v) with a pH gradient.

Synthesis of Parathyrin Sequence Tyr-(43-55)

The protected segments Tyr-43, (44-47), (48-51) and (52-55)were prepared, assembled to Tyr-(43-47) and (48-55)and finally to the sequence Tyr-(43-55) (Scheme 2).

Segment Tyr-43 : was prepared by salt coupling of Boc-Tyr-ONSu with Pro and was converted to the ONSu-ester

Segment (44-47) : was synthesized by M.A. coupling of Z-Asp(Bu^t) to TFA treated Boc-Ala-Gly-OBzl.Catalytic hydrogenation of Z-(45-47)-OBzl resulted C-and N-terminal deprotection and was followed by M.A. elongation with Z-Arg$(Z)_2$ to Z-(44-47)-OH.

Segment Tyr-(43-47) : was assembled by salt coupling of Tyr-43 with H-(44-47)-OH. The resulting hexapeptide was purified by CCD.

Segment (49-51) : was constructed of two dipeptides.Salt coupling of Z-Ser(Bu^t)-ONSu with Gln yielded Z-(48-49)-OH . Condensation of Z-Arg$(Z)_2$ with Pro-OBu^t was effected by DCC/HOBt. The resulting Z-(50-51)-OBu^t was deprotected with TFA and H_2/Pd treatments and purified by CCD.

Segment (53-55) : Starting with Glu(OBu^t)-OBu^t this synthesis was achieved by stepwise M.A. couplings of Z-protected amino acid derivatives of Lys,Lys and Arg. For final purification CCD of the hydrogenated product was employed.

Segment (48-55) : The assembly of Z-(48-51)-OH and H-(52-55)-OBu^t was performed by DCC/HOBt. Gel filtration on Sephadex LH 20 in

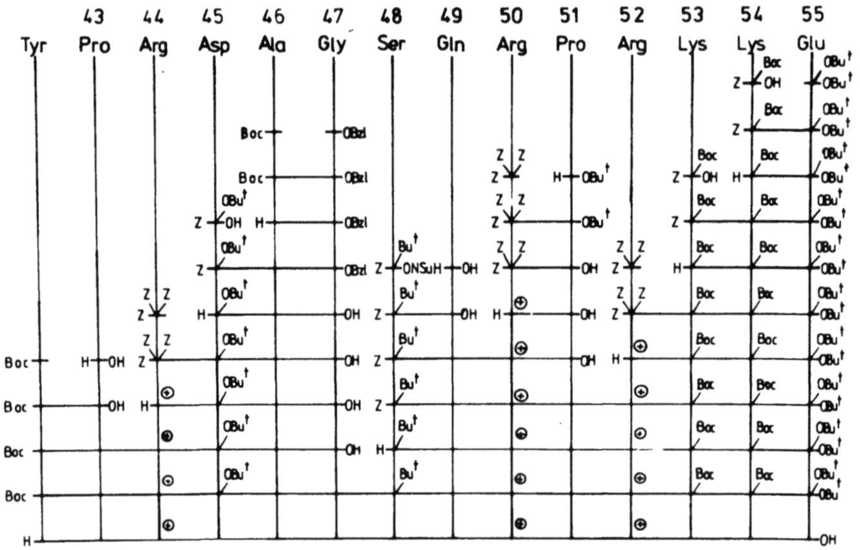

Scheme 2. Synthesis of Parathyrin Sequence Tyr-(43-55);
Amino acid analysis: o,87 (1) Tyr, 1,90 (2) Pro, 3,00 (3) Arg, 0,96 (1) Asp, 0,98 (1) Ala, 0,97 (1) Gly, 0,78 (1) Ser, 2,02 (2) Glu, 2,00 (2) Lys.

methanol led to the isolation of pure Z-(48-55)-OBu^t.

Sequence Tyr-(43-55) : was completed by condensation of Boc-Tyr (43-47)-OH with H-(48-55)-OBu^t using M.A. procedure. Boc-Tyr-(43-55)-OBu^t was completely deprotected with TFA and anisol as scavenger. For purification gel chromatography on Sephadex G 25 in 1m acetic acid and ion exchange at pH 5,5 were employed.

Synthesis of Parathyrin Sequence Tyr-(62-68)

Segment (64-68) : starting with Gly-OMe this peptide was condensed stepwise by M.A. procedure using Z-protected amino acid derivatives of Leu, Ser, Lys and Glu.

Sequence Tyr-(62-68) : Z-(62-63)-N_2H_3 was activated to the azide and coupled to H-(64-68)-OMe. Crude Z-(62-68)-OMe was purified by gel chromatography on Sephadex LH 2o in methanol. Cleavage

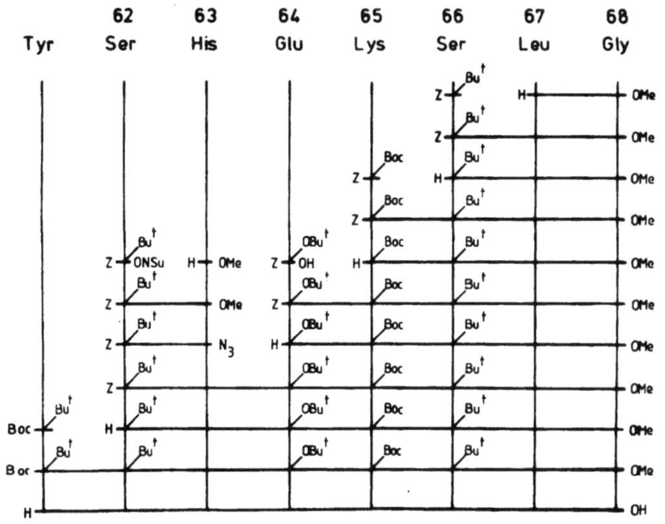

Scheme 3. Synthesis of Parathyrin Sequence Tyr-(62-68);
Amino acid analysis: 0,81 (1) Tyr, 1,68 (2) Ser, 1,00 (1) His, 1,00 (1) Glu, 1,00 (1) Lys, 0,99 (1) Leu, 1,02 (1) Gly.

of the Z-group was followed by M.A. coupling of Boc-Tyr. The removal of all protecting groups was achieved by saponification and TFA treatment. The crude product was purified on Sephadex G 15 in 10^{-3}m HCl.

Alternative pathways to prepare the described parathyrin sequences were attempted without success. Insoluble intermediates and products were isolated when (35-38) was synthesized using Z-protected amino acids.
Furthermore the yield, after deprotection and purification, of DCC/HOBt mediated coupling to sequence Tyr-(43-55) was only 8% in contrast to 40% using M.A. method.
Antibodies have been prepared against the synthetic segments (32-43) and Tyr-(43-55),which have also been used to help in characterization of various plasma fragments.

After radioiodination the synthetic parathyrin sequences have also been used for RIA and antibody characterization (5,6).

We thank the Stiftung Volkswagenwerk for financial support.

References

1. Keutmann, H.T., Sauer, M.M., Hendy, G.N., O'Riordan, J.L.H. Potts, J.T.Jr.: Biochemistry 17, 5723-5729 (1978).

2. Kimura, T., Takai, T., Masui, Y., Morikawa, T., Sakakibara, S.: Biopolymers 20, 1823-1832 (1981).

3. Rosenblatt, M., Segre, G.V., Tregear, G.W., Potts, J.T.Jr.: J.Med.Chem. 20, 1452-1456 (1977).

4. Rosenblatt, M., Segre, G.V., Tregear, G.W., Shepard, G.L., Tyler, G., Potts, J.T.Jr.: Endocrinology 103, 978-984 (1978).

5. Atkinson, M.J., Niepel, B., Jüppner, H., Casaretto, M., Zahn, H., Hehrmann, R.: Acta Endocrinol. (Kbh) 96, Suppl.24o, 42-43 (1981).

6. Wood, W.G., Butz, R., Casaretto, M., Hehrmann, R., Jüppner, H., Marschner, I., Wachter, C., Zahn, H., Hesch, R.D.: J. Clin. Chem. Clin. Biochem. 18. 789-795 (1980).

7. Casaretto, M., Danho, W., Hesch, R.D., Zahn, H.: Hoppe-Seyler's Z. Physiol. Chem. 363, 407-423 (1982).

APPLICATION OF MODERN METHODS IN THE SYNTHESIS OF CYCLIC ANALOGUES OF BRADYKININ

Felikss Mutulis, Ilze Mutule, Gunars Chipens,
Vera Grigoryeva, Mikhail Breslav
Institute of Organic Synthesis, Latvian SSR Academy of Sciences, 226006, Riga, USSR

Introduction

Some of the most stable bradykinin conformers have been shown to possess a cyclic structure (1). In an attempt to simulate the quasi-cyclic structure of the natural hormone, we have synthesized a number of bradykinin cycloanalogues. It has been found that these cyclopeptides exhibit a long-lasting hypotensive action (2,3) and therefore present interest both from theoretical and practical standpoint. However, limited availability of cyclic analogues due to low cyclization yields in our first synthetic attempts was the main obstacle on the way to their detailed examination and hence dictated the need to improve the preparation procedure.

Synthetic Procedure

A new synthesis of cycloanalogues was carried out using the classical methods of peptide chemistry. The synthetic procedure is illustrated by the preparation of cyclo-[δ-(Lys^1,Gly^6)-bradykinin] depicted in the Scheme. Dipeptide I was obtained using Woodward's reagent K, its removal by catalytic hydrogenation yields compound II. Similarly, addition of a second proline residue was achieved. Another dipeptide V was prepared by the mixed anhydride method. Coupling of segments IV and V

Chemistry of Peptides and Proteins, Vol. 2

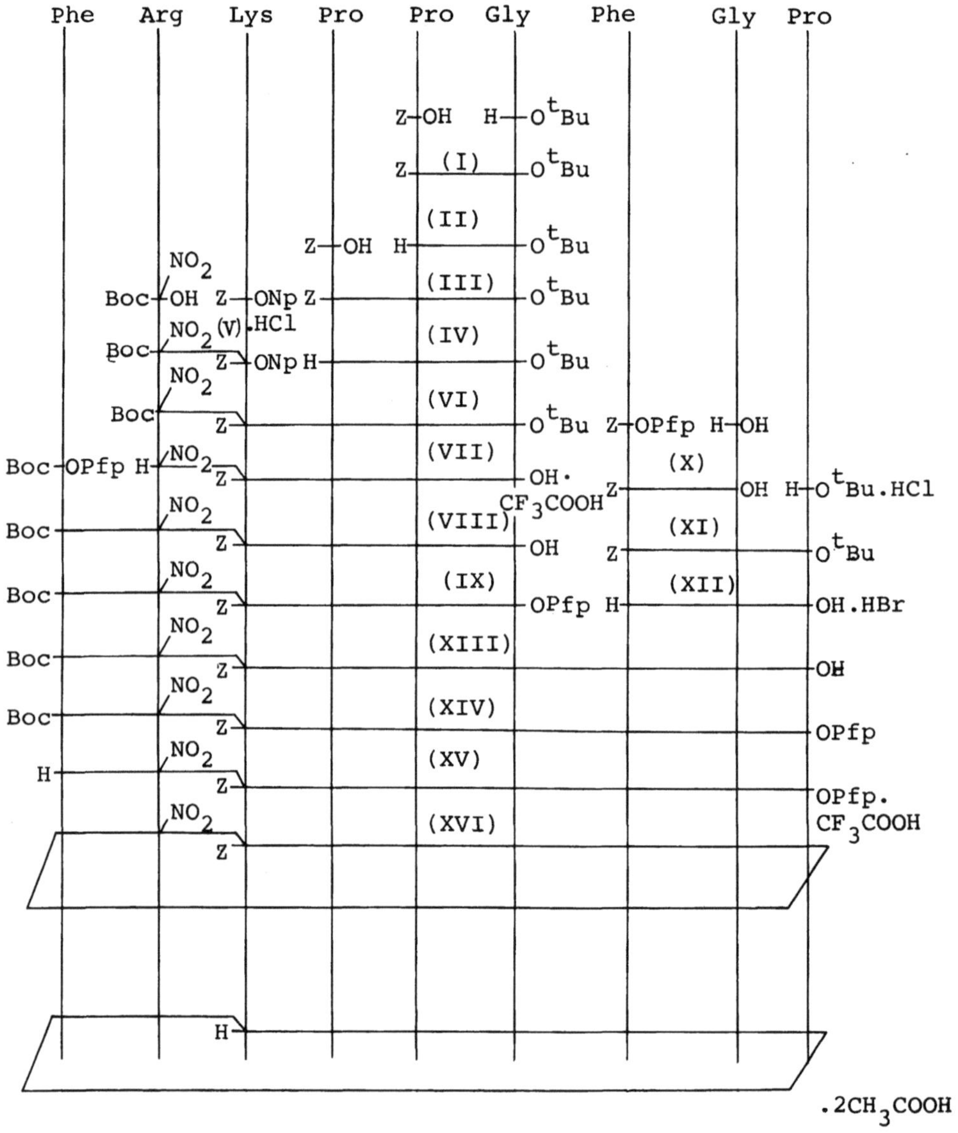

Scheme. Synthesis of cyclo-[ε-(Lys^1,Gly^6)-bradykinin].

afforded pentapeptide VI which was purified chromatographically. Removal with trifluoroacetic acid, treatment with Boc-Phe-OPfp and then with DCC-pentafluorophenol complex gave compounds VII, VIII and IX, respectively. Preparation of tripeptide XII was attained by the pentafluorophenyl ester method followed by the reaction with Woodward's reagent and subsequent cleavage of protective groups with HBr. Coupling of segments IX and XII yielded nonapeptide XIII which could be converted to pentafluorophenyl ester XIV. Cleavage of the Boc group gave compound XV which was treated with diisopropylethylamine in dilute dioxan solution. The protected cyclopeptide XVI formed was obtained in a 60% yield after chromatographic purification. The improved cyclization yield is evidently due to the use of dioxan as solvent instead of DMF which might have contained impurities such as dimethylamine preventing the formation of cyclopeptide. It is possible, too, that the closure of cycle via Gly-Phe bond in the previous synthesis of XVI was made difficult because of unsuitable conformation of the corresponding precursor. At the end of procedure, the protecting side groups were removed by anhydrous hydrogen fluoride in the presence of anisole and after purification a homogeneous end product was obtained. Similarly, 6 cyclic analogues in amounts equal to several hundred milligrams were obtained.

Purification of Peptides

Purification of protected peptides was successfully achieved using a "Jobin Yvon Chromatospac Prep" preparative chromatograph providing axial compression of sorbent. Column packing was carried out prior to use, therefore it was possible to have a new portion of silica gel for each separation. The main component of the eluents used was chloroform, other constituents being alcohols, ethyl acetate, acetic acid, water.

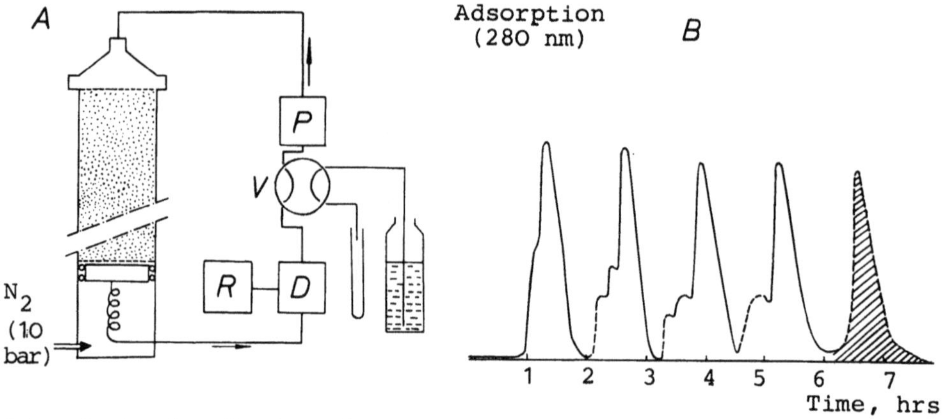

Fig. 1. Recirculation chromatography. A. Schematic picture of a modified "Jobin Yvon Chromatospac Prep 100" chromatograph Designations: P - pump, D - detector, R - recorder, V - 4-way valve. The circulation ←→ elution transition is achieved by turning V by 90°. B. Chromatogram of Boc-Arg(NO_2)-Pro_2-Gly-Phe-Ser(Bzl)-Pro-Phe-Arg(NO_2)-NH$(CH_2)_{11}$COOH purification using recirculation. Conditions: 1.1 g of initial mixture, 1 kg silica gel, eluent $CHCl_3$-EtOH-n·BuOH-AcOEt-H_2O (10:6:4:3:1). Circulation is shown by solid line, elution by broken line. Evaporation of hatched fraction gave 0.56 g of pure peptide.

When necessary, stepwise gradient elution was applied. Used solvent systems were wholly distilled and reutilized for subsequent separations. Proper handling of the instrument made it possible in most cases to get rid of all by-products detectable by analytical chromatography. Attachment of a piston pump and a four-way valve also permitted to perform recirculation chromatography (Fig. 1). During this procedure we observed recorder diagram and fractions containing by-products only were disposed of.. At the same time fresh eluent was introduced into the system. This was necessary to avoid overlapping of the beginning and end within the peak series.

Preparative reversed-phase HPLC enabled us to perform fast and effective purifications of free peptides. For such purposes, a "Du Pont" instrument with a 2.1 x 25 cm column was used, eluent-aqueous ethanol or methanol with addition of ammonium acetate.

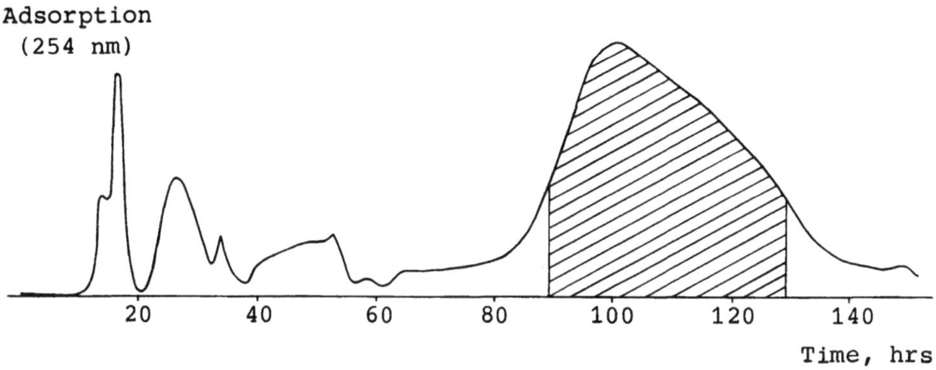

Fig. 2. Chromatogram of cyclo-(ArgProProGlyPheGlyProLeuLys). 2 AcOH, purification by droplet countercurrent chromatography. Conditions: 150 tubes; 190 mg of initial mixture; solvent system - n.BuOH-AcOH-water (4:1:5), moving phase - upper layer; flow-rate - 3.8 ml / hr. Evaporation of hatched fraction gave 125 mg of pure peptide.

However, a significant portion of peptide was lost, obviously because of irreversible adsorption. For example, chromatography of crude cyclo-ε-kallidin (100 mg containing 72 per cent of the main product as determined by quantitative HPLC) gave only 57 mg of pure peptide, i.e. 15 mg of cyclopeptide was lost. Purification of end product by droplet countercurrent chromatography on a "DCC-A" device was also successful. The results of purification of cyclo-(Gly^6,Leu^8,Lys^9-bradykinin) are shown in Fig. 2. This method is preferable because of quantitative yield of substances, its main disadvantage being the long duration of process.

Peptide Identification

In order to make sure that the assumed protected cyclopeptides are not dimers we carried out microscale molecular mass determination using urea melt by analogy with the Rast's method. In comparison with the commonly used camphor, urea melt dissolves peptides better and has a lower temperature (m.p. 132°C for urea, as compared to 178°C for camphor). This preserves peptides from decomposition. The observed discrepancy between the found and calculated values for 9 cyclopeptides did not exceed 20 per cent.

Satisfactory results of amino acid and elemental analyses were obtained for the end products. To prove the cyclic structure of these compounds they were submitted to enzymatic digestion. The course of digestion observed by paper electrophoresis confirmed the supposed structure. Usually, a single product was formed first thus demonstrating the cyclic structure of the substrate molecule. Long-term incubation often led to products of further degradation. For example, tryptic digestion of cyclo-ε-kallidin yielded kallidin first, and only then lysine and bradykinin.

Racemization test with L-amino acid oxidase performed for 3 cyclopeptides demonstrated that practically the entire amount of arginine, lysine, phenylalanine and leucine is in the L-configuration.

References

1. Ivanov V.T., Filatova M.P., Reissmann S., Reutova T.O., Efremov E.S., Pashkov V.S., Galaktionov S.G., Grigorian G.C., Ovchinnikov Yu.A. Peptides. Chemistry, Structure & Biology. Proc. 4-th Amer.Pept.Symp., Ann Arbor Sci., Ann Arbor, 1975, p.151-157.

2. Chipens G.I., Mutulis F.K., Katayev B.S., Klusha V.E., Misina I.P., Myshlyakova N.V.: Int.J.Pept.Prot.Res. 18, 302-311 (1981).

3. Chipens G., Mutulis F., Mishlyakova N.: Chemistry of Peptides and Proteins, Proc. 3-rd USSR-FRG Symposium: Voelter,W., Wünsch,E., Ovchinnikov,Y., Ivanov,V. (eds) W. de Gruyter-Verlag, Berlin, New York, 1982, p. 269-273.

TOTAL SYNTHESIS OF NEUROTOXIN II FROM NAJA NAJA OXIANA COBRA VENOM

Vadim T.Ivanov, Vladslav I.Deigin, Victor V.Ulyashin, Inessa I.Mikhaleva

Shemyakin Institute of Bioorganic Chemistry, USSR Academy of Sciences, 117988, Moscow, USSR

As a continuation of synthetic studies of polypeptide neurotoxins (α-bungarotoxin (1-6), apamin (2,7-9)) the synthesis of the title toxin (NT II) has been carried out by fragment coupling in solution. The general approach to synthesis, and the choice of protecting groups were discussed at the previous meeting (10). A number of novel chromatographic techniques was employed for purification of protected intermediates (10, 11). The optimal synthetic routes to protected "quarters" 1-19, 20-33, 34-47 and 48-61 are presented in Figs 1-4, and the further coupling steps - in Fig. 5. Description of the final stages of the synthesis will be presented below.

The fully protected 61-membered neurotoxin II, shown in Fig.6 was obtained from the respective "halfs" by "F" - complex activation in 59% yield after purification on Zorbax RP-8 column (Fig. 7). Liberation of C-terminal carboxyl in the 1-33 segment was afforded by 50% TFA/CH_2Cl_2 in the presence of thioanisole and N-acetyltryptamine as scavengers preventing alkylation of the Trp-27,28. Protected neurotoxin was deblocked by liquid HF in conditions described in (2), i.e. 1.5hr, 0°, in the presence of 10% anisole and 50 equiv. of N-acetyltryptamine. After HF evaporation the reaction mixture was treated with 8M guanidinium chloride in 0,1M Tris buffer (pH 8.5) and ether. Some insoluble material and the ether layer were separated by centrifugation. 2-Mercaptoethanol was added to the aqueous phase which after 2,5 hr at room temperature was applied to Sephadex G-50 column and eluted with 0,1 M NH_4OAc buffer (pH 7.5)

Chemistry of Peptides and Proteins, Vol. 2

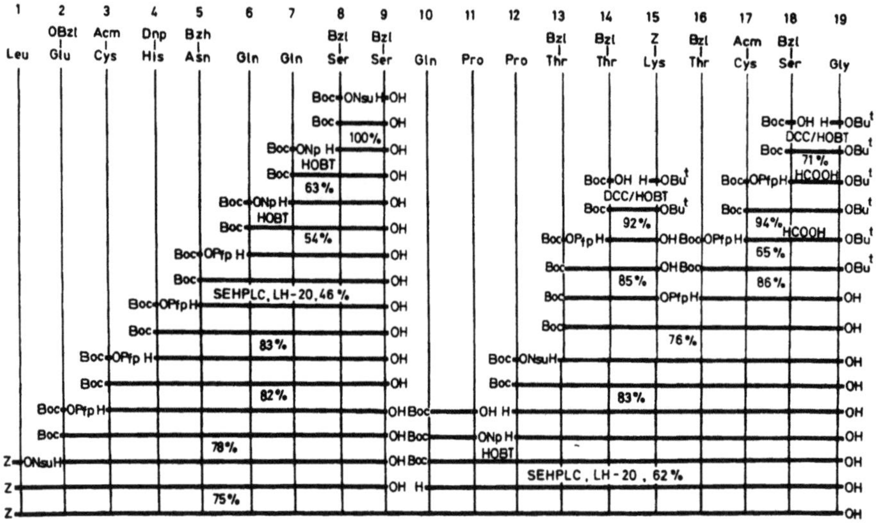

Fig. 1. Preparation of the protected segment 1-19.

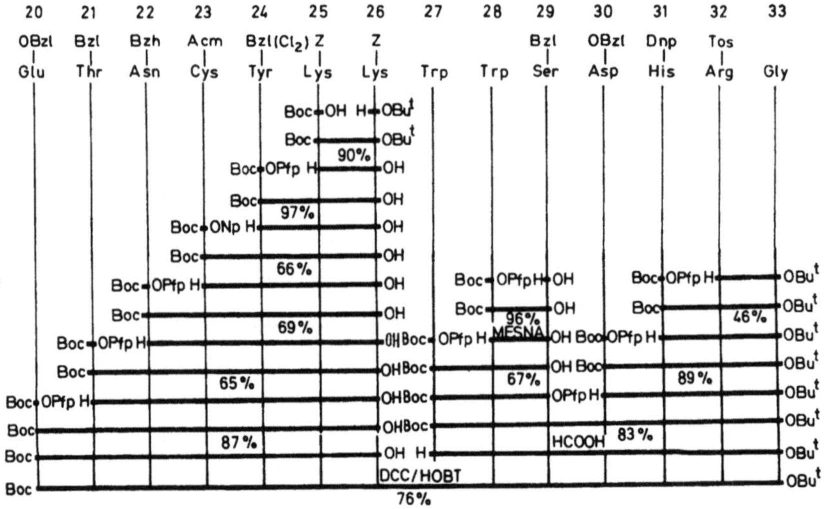

Fig. 2. Preparation of the protected segment 20-33.

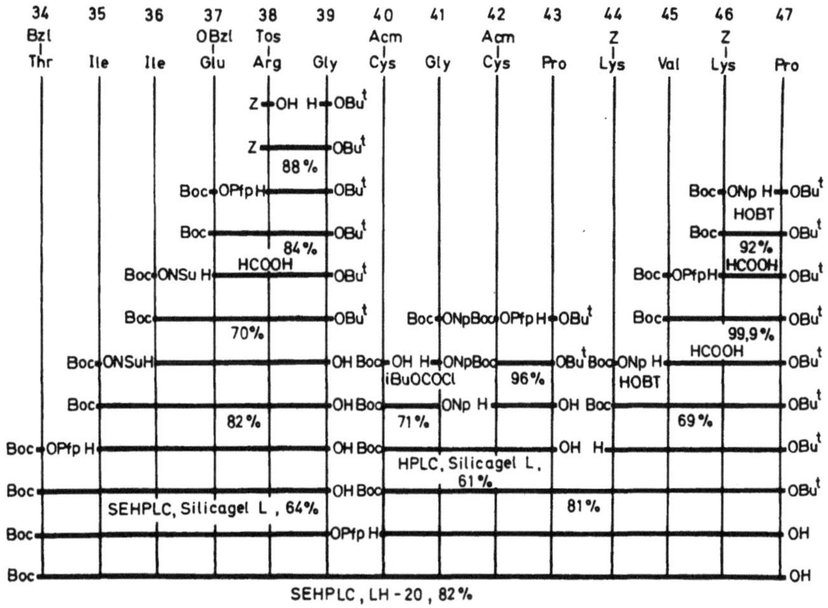

Fig. 3. Preparation of the protected segment 34-47.

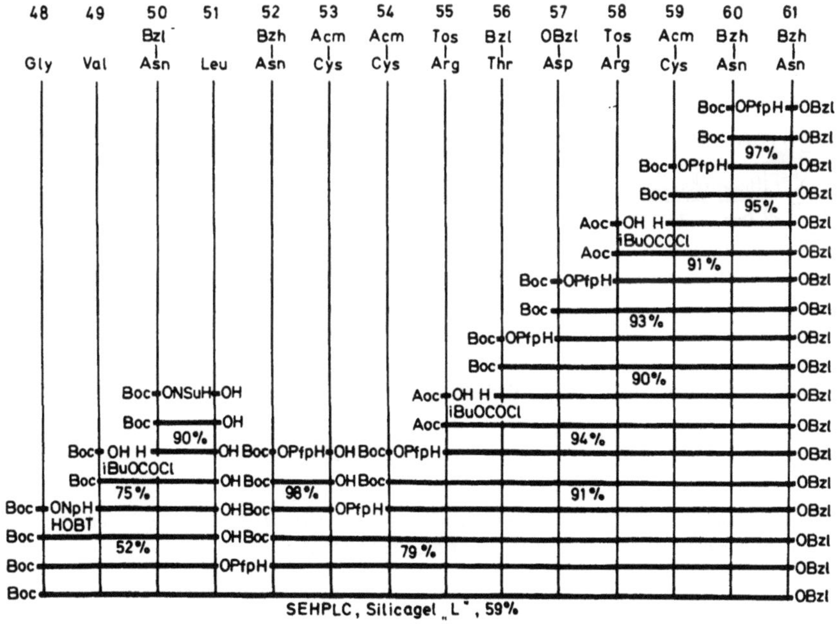

Fig. 4. Preparation of the protected segment 48-61.

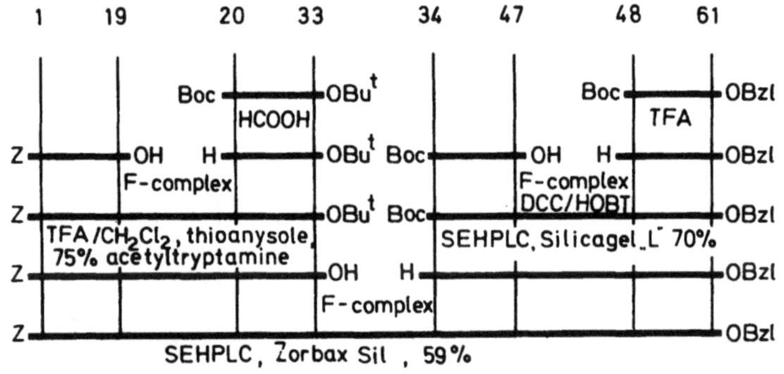

Fig. 5. Preparation scheme of protected NT II.

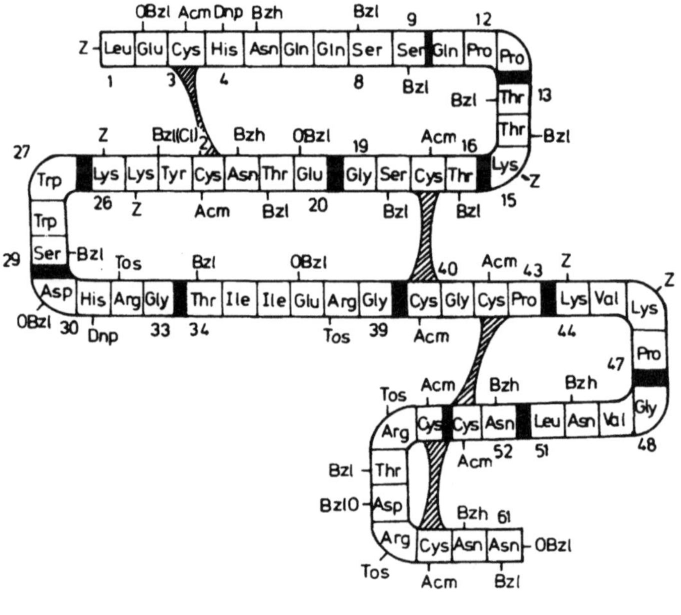

Fig. 6. Protected NT II. Hatched areas show the positions of disulfide bonds in the native protein.

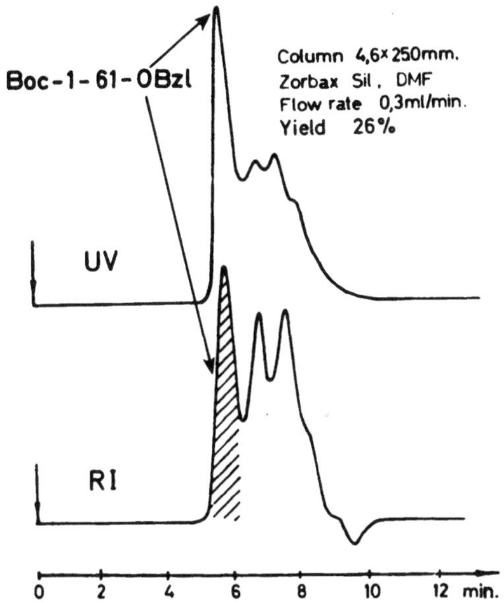

Fig. 7. SEHPLC of protected NT II.

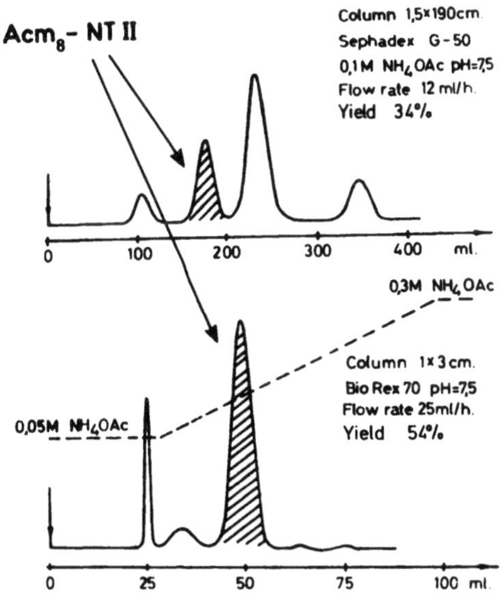

Fig. 8. Isolation of Acm$_8$ - NT II.

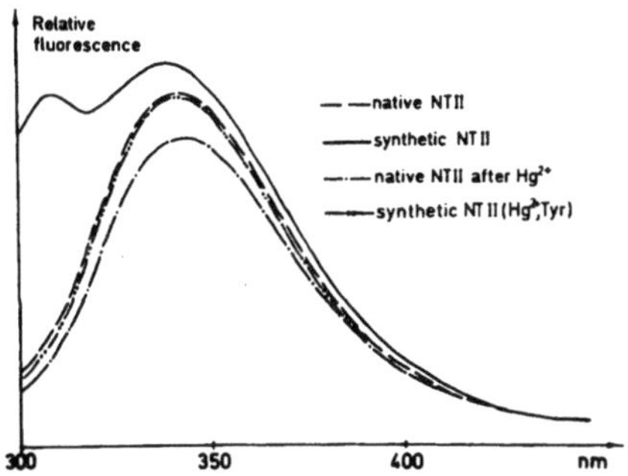

Fig. 9. Fluorescence spectra of the native and synthetic NT II.

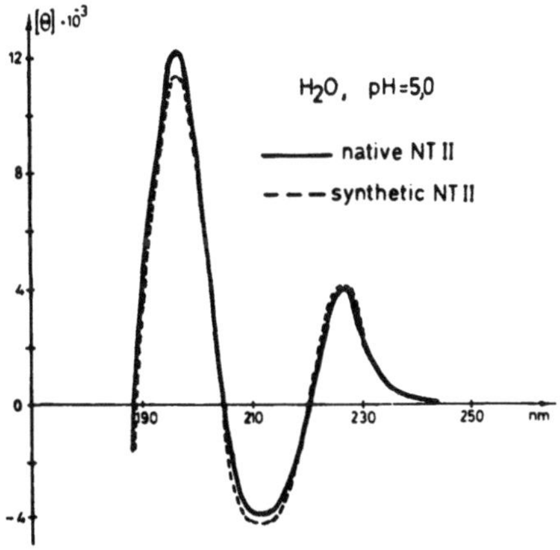

Fig. 10. CD spectra of the native and synthetic NT II.

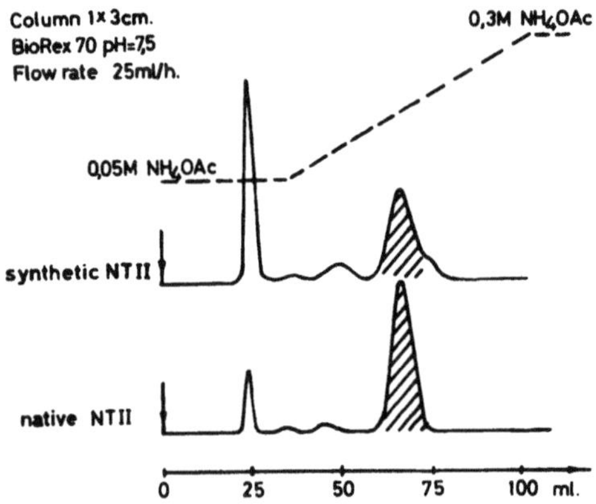

Fig. 11. Isolation of synthetic NT II.

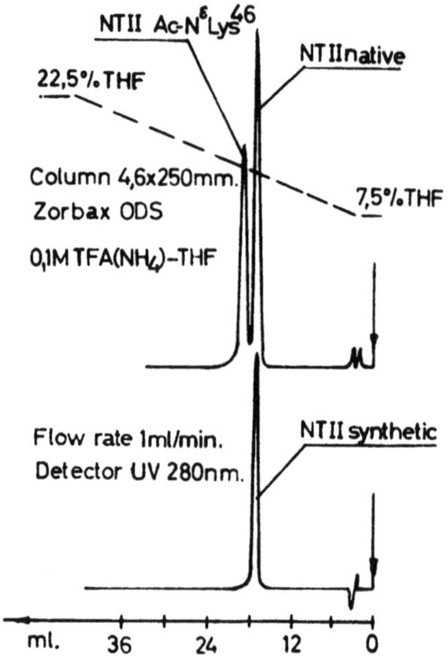

Fig. 12. Reverse phase HPLC of synthetic NT II and the mixture of native NT II with its Ac-N^{ε}Lys^{46} derivative.

and further purified on Bio-Rex-70 column (Fig. 8). The resultant material had correct amino acid and N-terminal analyses as well as the UV spectrum.

Next, an attempt to remove the Acm moieties was made by Hg^{2+} ions at standard conditions of (13). Subsequent air oxidation (14) followed by purification on Bio-Rex-70 afforded a product with satisfactory amino acid analysis, UV and CD spectra. At the same time the toxicity of that product was somewhat lower than with the native toxin (by ca 30%) and the fluorescence spectrum in the region of Tyr fluorescence also differed from the spectrum of neurotoxin II (Fig. 9).

Mercuration of the single Tyr-24 residue in the course of Acm removal from the neighbouring Cys-25 was suggested as a plausible explanation of these discrepances. Therefore 50 eguivalents of tyrosine were added to the reaction mixture at further Acm removal experiments (10% molar excess of $Hg(OAc)_2$, pH 4,25°, 90 min.). Chromatographic purification (Fig. 10) afforded a product undistinguishable from the native toxin in UV-, CD- (Fig 10) and fluorescence spectra (Fig. 9). The synthetic toxin was as active as the native toxin in replacing Dns-Lys^{46}-neurotoxin II from its complex with the acetylcholine receptor in conditions described in (15) . Reverse phase HPLC in conditions securing separation of NT II from the closely related derivatives confirmed the homogeneity of the synthetic material (Fig.12). The synthetic product showed LD_{50} of 45 µg while with the native toxin LD_{50} was 40 µg(intraperitoneal injection to mice). The difference between these values is well within the experimental error. Further characterization of the synthetic product is now being carried out including racemization and NMR studies. None the less we believe there are all reasons to assume that the present paper represents one of the very few clean total syntheses of proteins. The work might have a general methodological interest since the neurotoxin II molecule contains all the most troublesome, from synthetic viewpoint, amino acid residues - tyrosine, two tryptophans, two histidines and of course eight cysteines.

References

1. Ivanov, V.T., Mikhaleva, I.I., Volpina, O.M., Myagkova, M.A., Deigin, V.I.: Peptides 1976 (A.Loffet, ed.), Edit.Univ. Bruxelles, Bruxelles, 219-229 (1976).

2. Ivanov, V.T., Tsetlin, V.I., Mikhaleva, I.I., Volpina, O.M., Nuriddinov, A.R., Utkin, Yu.N., Arseniev, A.S., Pashkov, V.S., Karlsson, E., Surin, A.M., Bystrov, V.F.: Peptides 1978 (I.Z. Siemion and G.Kupryszewski, eds.), Wrocław Univ.Press, Wrocław, 41-62 (1979).

3. Mikhaleva, I.I., Myagkova, M.A., Zhukova, G.F., Ivanov, V.T.: Bioorg.Khim. 6, 982-1007 (1980).

4. Volpina, O.M., Deigin, V.I., Mikhaleva, I.I., Ivanov, V.T.: Bioorg. Khim. 6, 1133-1154 (1980).

5. Volpina, O.M., Deshko, T.N., Mikhaleva, I.I., Ivanov, V.T.: Bioorg.Khim. 6. 1155-1162 (1980).

6. Michaleva, I.I., Volpina, O.M., Utkin, Yu.N., Ivanov, V.T.: Bioorg.Khim. 7, 805-818 (1981).

7. Nuriddinov, A.R., Zhukova, G.F., Tsetlin, V.I., Ivanov, V.T.: Bioorg.Khim 4, 1322-1333 (1978).

8. Nuriddinov, A.R., Okhanov, V.V., Tsetlin, V.I., Ivanov, V.T.: Bioorg.Khim. 7, 5-15 (1981).

9. Nuriddiniv, A.R., Tsetlin, V.I., Ivanov, V.T.: Bioorg.Khim. 7, 165-178 (1981).

10. Deigin, V.I., Ulyashin, V.V., Michaleva, I.I., Ivanov, V.T.: Proc. III USSR-FRG Symp. Chemistry of Peptides and Proteins, Voelter,W., Wünsch,E., Ovchinnikov.Y., Ivanov,V. (eds), Walter de Gruyter, Berlin-New York, 229-237 (1982).

11. Ulyashin, V.V., Deigin, V.I., Ivanov, V.T., Ovchinnikov, Yu. A.: J.Chromatogr. 215, 263-277 (1981).

12. Tan, N.H., Kaiser, E.T.: J.Org.Chem. 41, 2787-2793 (1976).

13. Veber, D.F., Milkowski, J.D., Varga, S.,Denkewalter, R.G., Hirschmann, R.: J. Am. Chem. Soc., 94, 5456-5461 (1972).

14. Yang, C.C.: Biochim. Biohpys. Acta 133, 346-355 (1967).

15. Tsetlin, V.I., Karlsson, E., Utkin, Yu.N., Pluzhnikov, K.A., Arseniev, A.S., Surin, A.M., Kondakov, V.I., Bystrov, V.F., Ivanov, V.T., Ovchinnikov, Yu.A.: Toxicon 20, 83-93 (1982).

THE SYNTHESIS OF THE N-TERMINAL AMINO ACIDS OF THE PEPTIDE-NUCLEOSIDE ANTIBIOTIC NIKKOMYCIN AND THEIR USE IN PEPTIDE SYNTHESIS

Wilfried A. König, Werner Hass, Holger Faasch

Institut für Organische Chemie und Biochemie der Universität Hamburg, D-2000 Hamburg 13

Introduction

The nikkomycins, a group of peptide-nucleoside antibiotics, were isolated by Zähner and his associates from *Streptomyces tendae* (1) and have fungicidal and insecticidal properties. The structure and stereochemical properties were elucidated by chemical degradation and spectroscopic techniques (2 - 6) and show close relationship to the well-known polyoxins (7). The major components of the nikkomycins consist of an α-amino-γ-hydroxy butyric acid derivative with an aromatic or pyridine substituent in γ-position, a 5-aminohexuronic acid residue and a heterocyclic base, which may be either a uracil or a 4-formyl-4-imidazolin-2-one unit (figure 1).
The N-terminal amino acids contain 3 asymmetric centers. Our synthetic studies intended to prove the structure and stereochemistry of these amino acids. For the investigation of structure-activity relationships the influence of stereochemistry and the chemistry of protected derivatives of the amino acids for use in peptide synthesis had to be studied.

Results

For the synthesis of the amino acids of nikkomycins several approaches are possible. A convenient way seemed to be an alkylation of acetylaminomalonic acid ester, followed by

Chemistry of Peptides and Proteins, Vol. 2

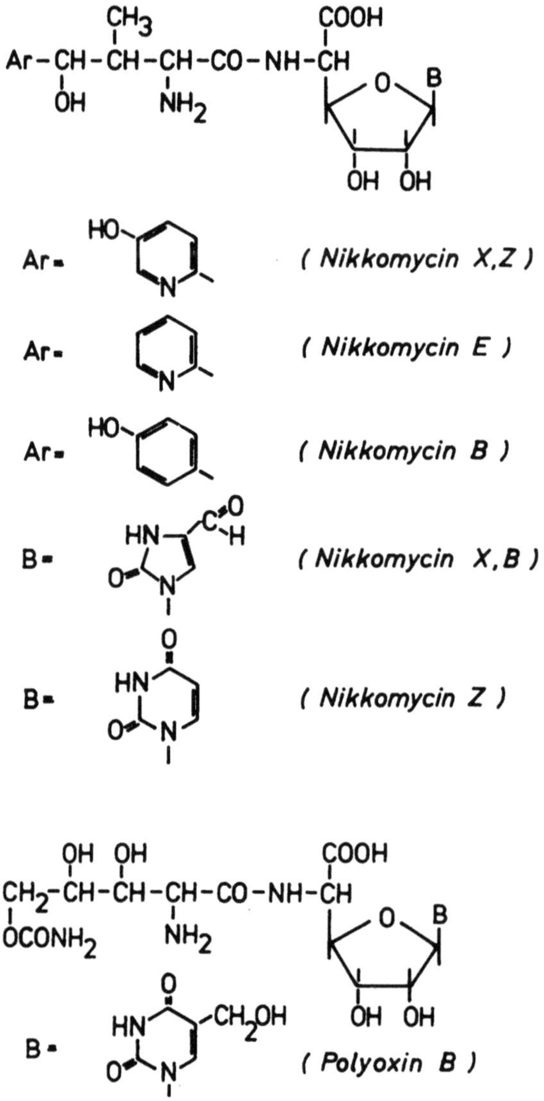

Figure 1.
Structures of nikkomycins and comparison with polyoxin G.

hydrolysis, decarboxylation and reduction of the carbonyl group as shown in figure 2.

Unfortunately, the product could only be obtained in low yields and, of course, as a mixture of eight stereoisomers.

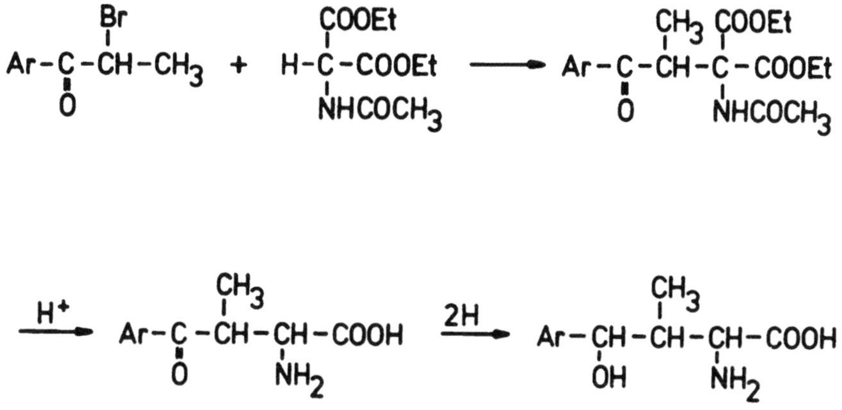

Figure 2.
Synthetic pathway to α-amino-γ-hydroxy acids by alkylation of acetylamino malonic ester.

A more elegant synthesis is possible by 1,3-dipolar cycloaddition of a nitrile-oxide to an olefine with defined configuration (4, 8, 9). As an intermediate an isoxaline carboxylic acid ester is formed. The α-amino-γ-hydroxy acid can be obtained after reductive ring opening and hydrolysis as demonstrated in figure 3.

Since the cycloaddition proceeds stereospecifically as a cis addition, only two stereoisomers are formed, when starting from a trans olefine. From the two possible regioisomers fortunately the correct one is favored by sterical reasons and is obtained in a large excess (figure 4).
The cyclic intermediate is reduced with zinc, after activation with copper, in acetic acid anhydride.

A separation of the enantiomers can be achieved on the stage of the isoxazoline derivative by formation of the diastereoisomeric α-phenylethylamides by crystallization.
The synthesis of the major amino acid constituent of the nikkomycins X and Z was performed according to the same schedule, although the corresponding olefine had to be prepared in a rather complicated series of reactions as shown in figure 5. The intermediate aldehyde 8 can be converted in a Wittig re-

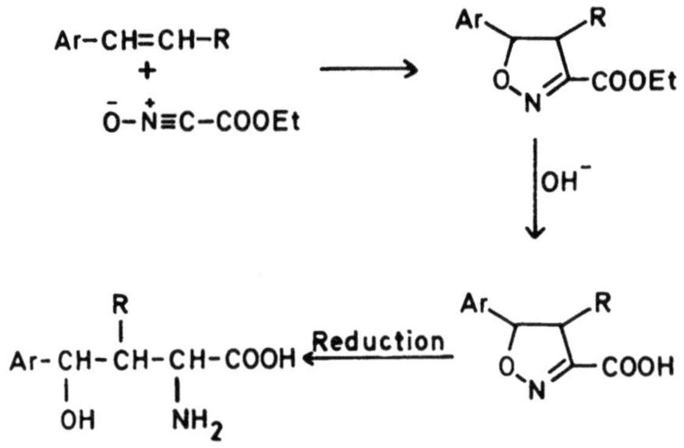

Figure 3.
Synthetic pathway to α-amino-γ-hydroxy acids by 1,3-dipolar cycloaddition.

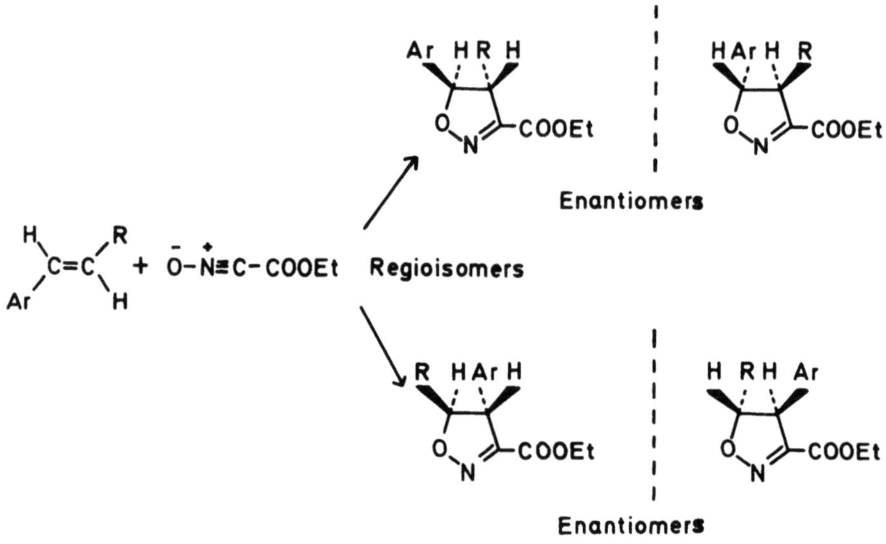

Figure 4.
Stereospecific cycloaddition of a nitrile-oxide to a trans olefine.

action with ethylidene-triphenylphosphorane to the trans olefine which is formed in a 2.6 fold excess over the corresponding cis olefine (9).

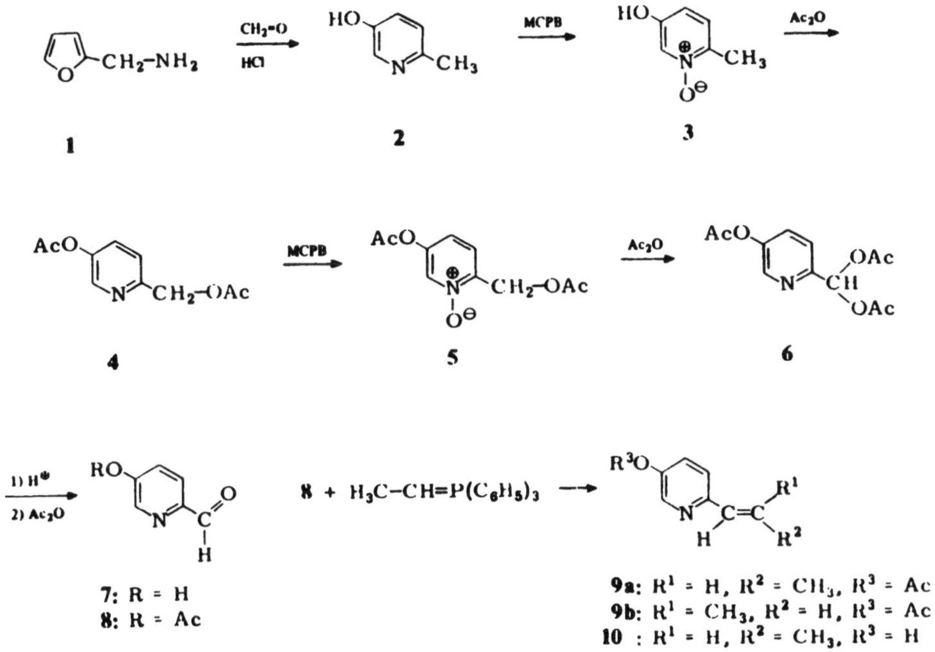

Figure 5.
Preparation of 5-hydroxy-2-propenylpyridine (MCPB = meta-chloro-perbenzoic acid).

By reductive ring opening of the isoxazoline derivative a new chiral centre in position 2 of the molecule is formed. The resulting diastereoisomers can be separated by chromatography on silica gel. The identity of the natural and synthetic amino acids was proved by NMR, mass spectrometry and gas chromatography.

It is known from systematic studies of Zähner et al. (10, 11) that biologically active amino acids are using a tripeptide transport mechanism to cross the cell membrane barrier. In order to find out if our synthetic amino acids have any biological activities they had to be connected to a dipeptide. We chose a somewhat simplified model amino acid and two L-alanine residues. The tripeptide was prepared by coupling the isoxazoline carboxylic acid azide to L-Ala-L-Ala-OMe. In this procedure no special protecting groups are necessary and the reductive ring opening could be performed after peptide formation (figure 6).

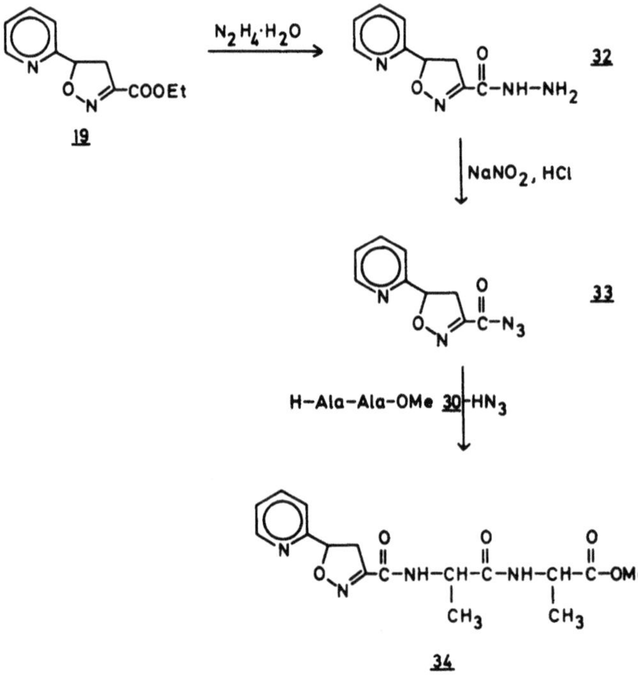

Figure 6.
Peptide synthesis with α-amino-γ-hydroxy acids.

The biological test of two stereoisomers of the tripeptide showed inhibition of *Bacillus subtilis* (12). It could also be shown that the biological activity is ineffective in the presence of certain natural amino acids (Asp, Asn, Orn, Arg, Pro). From this observation it can be concluded that the synthetic amino acids are amino acid antagonists.

The semi-synthetic preparation of modified nikkomycins by coupling of synthetic amino acid derivatives to the natural nucleoside moiety is now in progress.

References

1. Dähn, U., Hagenmaier, H., Höhne, H., König, W.A., Wolf, G., Zähner, H.: Arch. Microbiol. 107, 143-160 (1976).

2. König, W.A.: in "Advances in Mass Spectrometry", N.R. Daly, Ed., Vol. 7, Heyden & Son Ltd., 1978, p. 1530-1533.

3. Hagenmaier, H., Keckeisen, A., Zähner, H., König, W.A.: Liebigs Ann. Chem. 1979, 1494-1502.

4. König, W.A., Hass, W., Dehler, W., Fiedler, H.-P., Zähner, H.: Liebigs Ann. Chem. 1980, 622-628.

5. König, W.A., Pfaff, K.-P., Bartsch, H.-H., Schmalle, H., Hagenmaier, H.: Liebigs Ann. Chem. 1980, 1728-1735.

6. Hagenmaier, H., Keckeisen, A., Dehler, W., Fiedler, H.-P., Zähner, H., König, W.A.: Liebigs Ann. Chem. 1981, 1018-1024.

7. Isono, K., Asahi, K., Suzuki, S.: J. Amer. Chem. Soc. 91, 7490-7505 (1969).

8. Drefahl, G., Hörhold, H.H.: Chem. Ber. 97, 159-164 (1964).

9. Hass, W., König, W.A.: Liebigs Ann. Chem. 1982, 1615-1622.

10. Diddens, H., Zähner, H., Kraas, E., Göhring, W., Jung, G.: Eur. J. Biochem. 66, 11-23 (1976).

11. Diddens, H., Dorgerloh, M., Zähner, H.: J. Antibiotics 32, 87-90 (1979).

12. Zähner, H.: personal communication (1982).

SYNTHESIS AND ACTIVITY OF PEPTIDES ANALOGOUS TO THE TOXIC PEPTIDES ISOLATED FROM AMANITA VIROSA MUSHROOMS

Jens-Uwe Kahl, Tamiko Miura, Theodor Wieland
Max-Planck-Institut für medizinische Forschung,Abteilung Naturstoffchemie,Jahnstraße 29, 6900 Heidelberg

Introduction

Extracts of the deadly poisonous Amanita mushrooms have been objects of investigation for more than a century. Analysis and synthesis as well as the biological actions of the main toxins, the amatoxins and the phallotoxins have been extensively investigated(for a review see (1)).
So it was of great interest that among the extracts of the white species,Amanita virosa,a new group of cyclic peptides, the virotoxins was found (2).In contrast to the known bicyclic toxins,the virotoxins turned out to be monocyclic heptapeptides. Variations in the sequence occured only in positions 1 and 7, analogous to the phallotoxins.
Monocyclic phallotoxins are nontoxic (3), whereas the peptides reported here are of the same toxicity as e.g. phalloidin. And,the biological activity of viroisin is comparable to that of the phallotoxin,e.g., with a dose of 2.5 mg of viroisin per kg white mouse, 50% of the animals die within 2 - 5 hours by hemorrhagia of the liver.
On the molecular level, virotoxins behave similar to the phallotoxins.Thus,both bind very strongly to the liver receptor protein F-actin.The phallotoxins possess a rigid binding site caused by the bicyclic structure,with any changes decreasing the toxicity rapidly.So, the very flexible monocyclic structure of the virotoxins may adopt the same biologocally active conformation by an induced fit mechanism upon contact with actin.This conformation is gained and stabilized by

Chemistry of Peptides and Proteins, Vol. 2

further bonds in comparison to the phallotoxins. These further structural parameters are the hydroxyl groups of serine and dihydroxyproline, the methylsulfonyl group of tryptophan and the inverse side chain of serine.

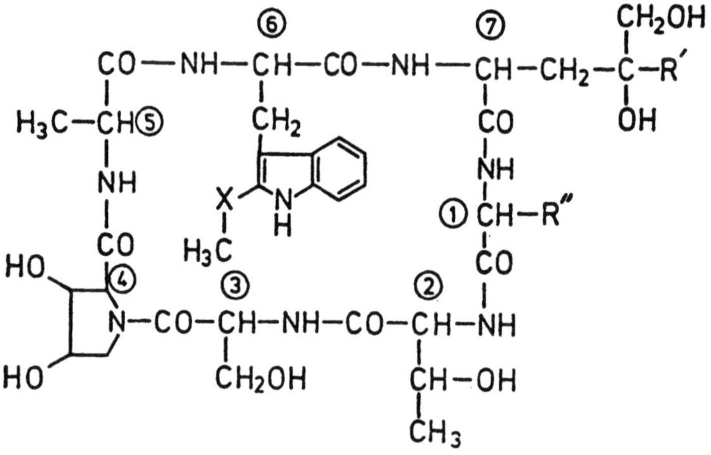

Fig.1. Structures of isolated virotoxins.

Compound	X	R'	R''
Viroidin	SO_2	CH_3	$CH(CH_3)_2$
Desoxo-viroidin	SO	CH_3	$CH(CH_3)_2$
Ala^1-virodin	SO_2	CH_3	CH_3
Ala^1-desoxo-viroidin	SO	CH_3	CH_3
Viroisin	SO_2	CH_2OH	$CH(CH_3)_2$
Desoxo-viroisin	SO	CH_2OH	$CH(CH_3)_2$

Synthesis

The synthesis of the virotoxin analogous peptides is limited to derivatives of monohydroxylated leucine instead of naturally occuring di - or trihydroxylated leucine. Hence, one main problem in the synthesis is the very easy opening of one peptide bond under soft acidic conditions due to the neighbouring effect of the γ- hydroxyl group in leucine forming a stable γ-lactone.

Vice versa, once the lactone is opened to the free acid, the conditions of peptide synthesis are conducive to then formation of lactone, thus reducing the yield of the synthesis to a very low level (4 , 5).

However, several peptides have been successfully synthesized. Four peptides containing allo-hydroxyproline instead of dihydroxyproline were synthesized together with one peptide that contains dihydroxyproline and leucine. The synthesis for dihydroxyproline on a preparative scale, necessary for the synthesis of the natural product, has been previously described(6).

So far the five peptides are:

1) Val - D - Thr - D - Ser - Hypro - Ala - Trp - Leu
with SO_2-CH_3 on Trp

2) Val - D - Thr - D - Ser - Hypro -Ala - Trp - Hyleu
with SO_2-CH_3 on Trp

3) Val - D -Thr - D - Ser - Hypro - Ala - Trp - Hyleu
with $S-CH_3$ on Trp

4) Val - D - Thr - Ala - Hypro - Ala - Trp - Hyleu
with SO_2-CH_3 on Trp

5) Val - D - Thr - D - Ser - Dipro - Ala - Trp - Leu
with SO_2-CH_3 on Trp

The synthesis of the linear heptapeptides itself was carried out, both, stepwise and by condensation of fragments according to the classical method in solution, using single protected amino acids. As coupling reagents mixed anhydrides were used as well as the DCC method. The synthesis of the linear peptides proceeded without problems, with an overall yield of about 40 %. All peptides were synthesized in the same manner, see following schemes.

The preparation of 2'-(methylsulfonyl)tryptophan is tedious. This compound, first isolated by W.E. Savige (7), is produced from tryptophan by oxidation with 40 % peracetic acid.

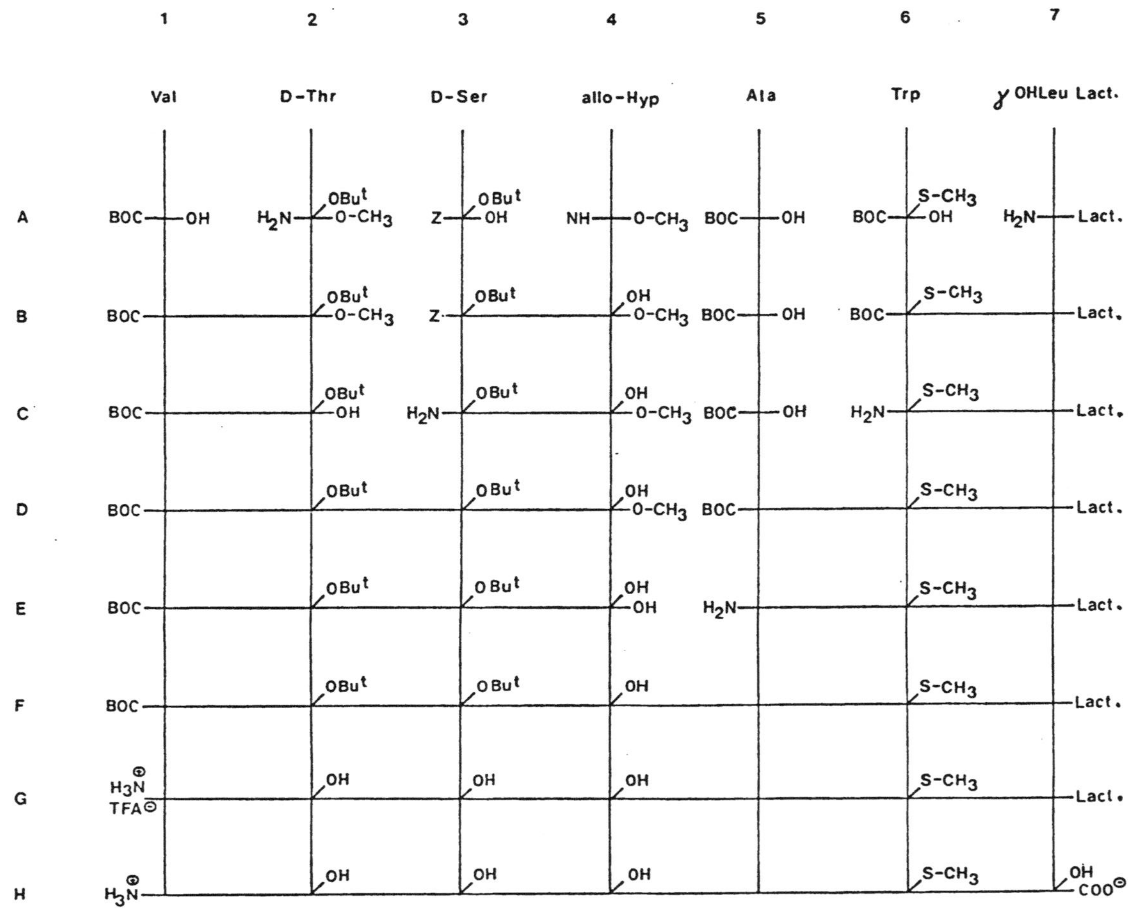

Fig.2.Fragment condensation method used for the syntheses of virotoxin analogues.

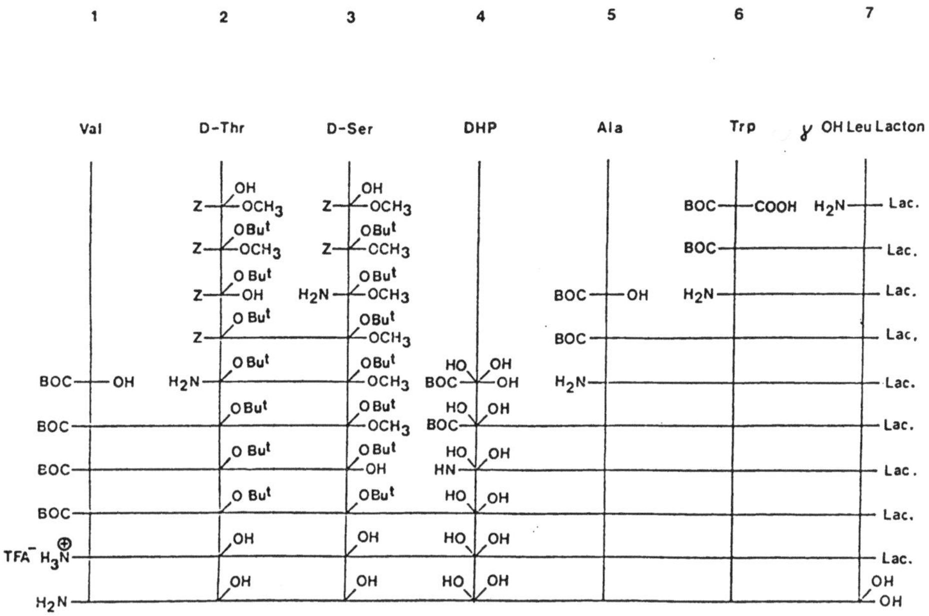

Fig.3. Stepwise synthesis of virotoxin analogues.

An easier and more convient way is the conversion of the tripeptide Boc-Ala-Trp-Hyleulac with methylsulfenyl chloride which can be easily obtained by reacting dimethyl disulfide with sulfuryl chloride(8). The reaction is accomplished in acetic acid in about 5 min, the products are worked up by column chromatography on silica gel. The thioether has to be oxidized with peracetic acid to give the desired product(9). Becaue of the problems associated with the hydroxyleucine unit, the amino acid was introduced as a lactone over the whole synthesis to avoid the possible cleavage of the peptide by treatment with acids. Thus, before cyclisation of the peptide the lactone has to be opened. This is done after removing all protecting groups by column chromatography using Sephadex LH 20 and 0,04 N ammonia as eluent(10). Under these very mild conditions, the carboxylic generated, is ready for cyclisation.

The cyclisation ,however,is performed in very dilute solution. Therefore, the carboxylic acid is first protonated with one equivalent trifluoracetic acid and then activated with isobutyl chloroformate for two minutes. Then, N-methylmorpholine is added and the reaction mixture is kept at 0°C for three hours and at room temperature for two days. After Sephadex column chromatography the cyclic product can be isolated in a yield below 5 %.
Peptides with the leucine unit were cyclized in the classical way using 5-fold excess of hydroxysuccinimide and DCC.

Results

So far, four peptides show biological activity at equilibrium dialysis with actin. For determination of the apparent equilibrium dissociation constant the ^{3}H - demethylphalloin-actin complex was prepared, and the labeled toxin replaced competitively the peptide in question during equilibrium dialysis. As shown in the curves of figure 4, the tested peptides were a little less effective than unlabeled demethylphalloin, used in a standard experiment. Only one peptide showed absolutly no biological activity.

Discussion

To conclude, the obtained virotoxin analogs exhibit a clear binding to actin. The binding is less effective than the standard material in comparison, demethylphalloin. For the binding to actin the hydroxyl group of serine and the inverse side chain are here of essential significance. The same is valid for the second hydroxyl group in proline and for the methylsulfonyl group in tryptophan. Surprisingly, the 2'-(methylthio)- tryptophan derivative shows pronounced biological activity.

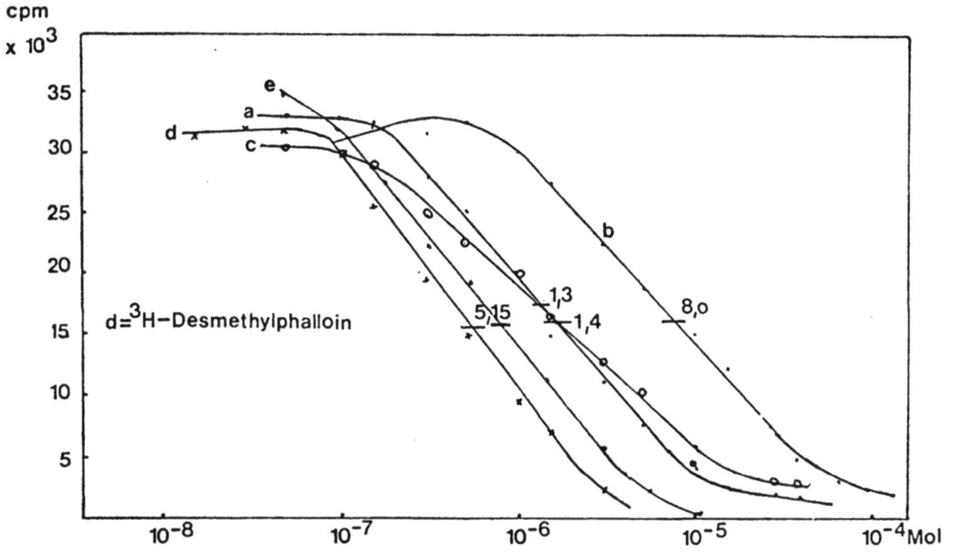

Fig.4. Activity constants,determined by equilibrium dialysis.

c) Val- D - Thr - D - Ser - Hypro - Ala - Trp - Leu (cyclic; Trp carries SO_2-CH_3)

b) Val- D - Thr - D - Ser - Hypro - Ala - Trp - Hyleu (cyclic; Trp carries SO_2-CH_3)

a) Val - D - Thr - D - Ser - Hypro - Ala - Trp - Hyleu (cyclic; Trp carries SO_2-CH_3)

x) Val - D - Thr - Ala - Hypro - Ala - Trp - Hyleu (cyclic; Trp carries SO_2-CH_3)

e) Val - D - Thr - D - Ser - Dipro - Ala - Trp - Leu (cyclic; Trp carries SO_2-CH_3)

The role of the hydroxyl groups in leucine are not yet clear. According to the phallotoxins, norphalloin is even more toxic than phalloidin. In the class of the virotoxins the derivative with dihydroxyproline and leucine was expected to be of same toxicity, but this is not the case. It shows a good binding to actin,but it is nontoxic. A possible explanation is a side reaction during cyclisation procedure,leading to a different conformation. Another explanation would be that the hydroxyl groups in leucine are stronger involved in the binding to actin than it is the case in the class of phallotoxins. This problem is still under investigation.

References

1. Wieland, Th., Faulstich, H.: C. R. C. Crit. Rev. Biochem. 5, 184 - 260 (1978).
2. Faulstich, H., Buku, A., Bodenmüller, H., Wieland, Th.: Biochemistry 19, 3334 - 3343 (1980).
3. Wieland, Th., Schmidt, G.: Liebigs Ann.Chem. 577, 215.
4. Faulstich, H., Trischmann, H.: Liebigs Ann.Chem. 741, 55-63 (1970).
5. Munekata, E., Faulstich, H., Wieland, Th.: Angew.Chem. 89, 274, (1977).
6. Kahl, J.U., Wieland, Th.: Liebigs Ann.Chem. 1445-1450, (1981).
7. W.E.Savige, Austr.Journ.Chem., 2275 - 2287 (1975).
8. Wieland, Th., Jochum, C., Faulstich, H.: Liebigs Ann. Chem. 727, 138 (1969).
9. Faulstich, H., Wieland, Th.: Liebigs Ann.Chem. 713, 186 (1968).

SYNTHESIS OF A NEW PHOTOLABILE TRITIATED ANTAMANIDE-ANALOGUE

Michael Nassal, Pedro Buć, Theodor Wieland

Max-Planck-Institut für medizinische Forschung, Heidelberg
Abt. Naturstoff-Chemie
D-6900 Heidelberg, FRG

Introduction

Besides the toxic peptides of the amatoxin and phallotoxin families the deadly poisonous mushroom Amanita phalloides contains a cyclic decapeptide of the following structure:

$$\begin{array}{l} \mathrm{Pro^{8}\text{-}Phe^{9}\text{-}Phe^{10}\text{-}Val^{1}\text{-}Pro^{2}} \\ | \qquad\qquad\qquad\qquad\quad | \\ \mathrm{Pro^{7}\text{-}Phe^{6}\text{-}Phe^{5}\text{-}Ala^{4}\text{-}Pro^{3}} \end{array} .$$

This compound, called antamanide (1), in a dose of 0.5 mg/kg of bodyweight, is able to antagonize the lethal effect of 5 mg/kg phalloidine on the white mouse, provided it is given before or at least at the same time as the toxin. Likewise the alterations caused by phalloidine in the perfused rat liver and in isolated rat hepatocytes can be prevented by antamanide.
The peptide is not a competitive inhibitor of the phallotoxins; it rather acts as a permeation inhibitor. The properties of the cell membrane are changed in such a way, that critical toxin concentrations are not reached inside the cell (2).

Photoaffinity labeling (3,4) seems to be a suitable technique for finding out more about the interaction of antamanide with the cell membrane. All the photolabile antamanide analogues used so far, however, had more or less serious disadvantges: either the derivative could not be synthesized conveniently with high specific radioactivity (O-[azidoacetyl]-Tyr^{6}-AA) or like O-[(4-azido-2-nitrophenyl)-[^{3}H]-γ-aminobutyryl-aminoethyl]-Tyr^{6}-

Chemistry of Peptides and Proteins, Vol. 2

AA, it led predominantly to unspecific labeling, a result described also by others using aromatic azides as nitrene generating photolabels (5). O-[3,3,3-trifluoro-2-diazopropionyl-N-methyl-[^{3}H]-γ-aminobutyryl-aminoethyl]-Tyr^{6}-AA upon irradiation produces a carbene instead of a nitrene, and photolysis in the presence of isolated hepatocytes led to labeling of only a few proteins. However, irradiation times of more than one and a half hour are necessary to photolyse at least half of the diazo compound (6). During that time a considerable portion of the liver cells dies.

Thus, there is a need for a photolabile antamanide-analogue with the following properties:

i) generation of a carbene rather than a nitrene by irradiation,

ii) complete photolysis with wavelengths longer than 300 nm within several minutes or less time,

iii) specific radioactivity of at least 10 Ci/mmol, easily introducable into a suitable precursor

N_{ε}-[4-(2,2,2-trifluoro-1,1-azoethyl)benzoyl]-[4,5-^{3}H]-Lys^{6}-AA, the synthesis of which is described here, is likely to meet all these prerequisites.

Experimental Procedures

Synthesis of 4-(2,2,2-trifluoro-1,1-azoethyl)benzoic acid

In 1980, Brunner et al. (7) published the synthesis of 3-trifluoromethyl-3-phenyldiazirine (I) and its derivative (II), which should permit the attachment of the photolabile residue to other molecules. Since the diazirine is photolysed within seconds at 350 nm yielding a carbene, it seems very promising as photolabel, although there is a partial rearrangement of the 3-membered ring to the isomeric diazo compound, which is photolysed only slowly with a half-life time of about 22 minutes. We intended to synthesize (II) and to couple it to an antamanide

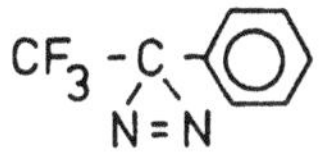

(I)

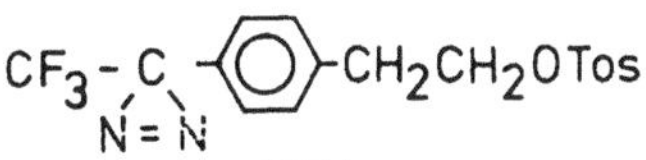

(II)

analogue containing a free amino function. During the preparation we followed the described procedures with one major variation: introduction of the trifluoroacetyl group to give (IV) was not performed via the Grignard reagent, but via the organolithium compound prepared by treating the silylated 4-bromophenylethanol (III) with n-butyl lithium followed by trifluoroacetylpiperidine. Hereby the yield of (IV) with respect to (II) increased from 18% to 70%.

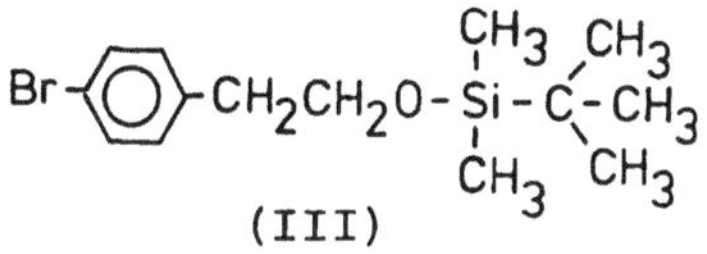

(III)

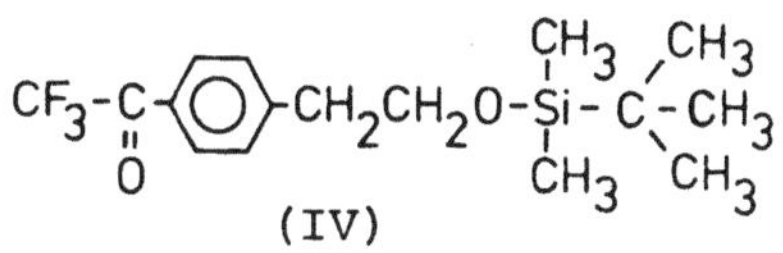

(IV)

Attempts to attach the tosylate (II) to O-[γ-aminobutyryl-aminoethyl]-Tyr^6-AA did not yield the expected product. In addition to unreacted antamanide derivative, a low molecular weight compound was isolated, not identical with (II). By nmr spectroscopy the molecule was identified as 4-(2,2,2-trifluoro-1,1-azoethyl)-styrene, resulting from elimination of p-toluenesulfonic acid. From the structure of the tosylate it is evident that the molecule must be prone to β-elimination, because this results in a conjugated system.

Thus, we decided to shorten the alkyl chain by one CH_2-group. The first synthesis of the benzylic alcohol derivative (V) followed the above route. We isolated (V), but problems arose tosylating the alcohol. The main product isolated from a reaction at room temperature probably was the pyridinium tosylate (VI). At 0^oC

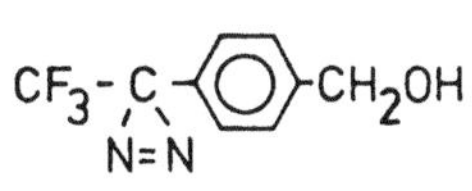

(V)

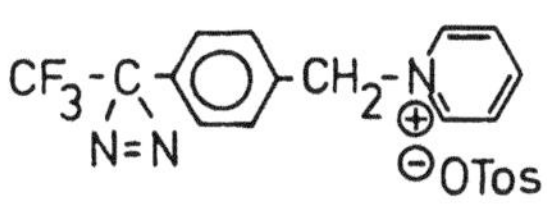

(VI)

a product was obtained in rather poor yield, which, according to nmr data had the correct structure.
Because of the low yield and the principal difficulties in monoalkylating amines, the benzylic alcohol was oxidized by means of $KMnO_4$ in dioxane/0.2 n KOH to give 70% of 4-(2,2,2-trifluoro-1,1-azoethyl)benzoic acid (VII).

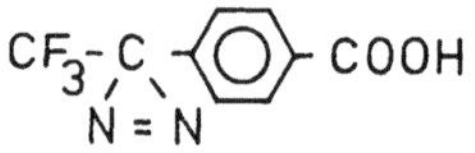

(VII)

Since the photochemical properties of the compound seemed very promising with respect to photoaffinity labeling (see below), we developed a more economic synthesis avoiding the use of the expensive t-butyl-dimethyl-silyl group for the OH protection, according to the following scheme (8):

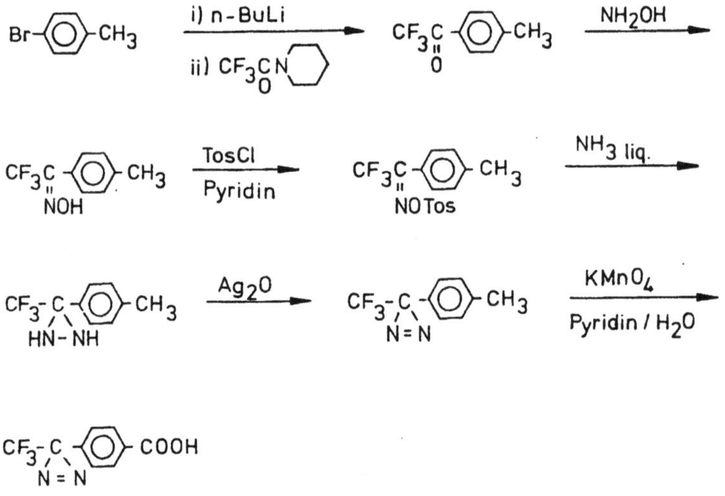

Photochemical properties of 4-(2,2,2-trifluoro-1,1-azoethyl)-benzoic acid

The photochemical behaviour of the new benzoic acid derivative was tested by irradiating an ethanolic solution with a Philips HPK 125 W/L mercury lamp in Pyrex glass test tubes as filter for wavelengths shorter than 300 nm and recording the uv spectra

at different times of irradiation. The diazirine-absorption with its maximum at 348 nm (ε=390) decreases with a half-life time of about 25 seconds. During the first half minute there is a strong increase of the absorption around 280 nm, which is probably due to the rearrangement of the diazirine to the diazo compound. Absorption then decreases to a constant value which is reached after about 5 minutes. Irradiation for another 90 minutes does not change the spectrum anymore. The half-life of the diazo compound can be estimated to be in the range between 1 and 2 minutes. For comparison the tosylate (II) was photolysed under the same conditions. Here the maximum of the diazo-absorption is reached after about 2 minutes, then decreasing with an estimated half-life time of about 24 minutes. This value is very near to the one of 22 minutes originally published (7).
The benzoic acid derivative was coupled to O-[γ-aminobutyryl-aminoethyl]-Tyr^6-AA and irradiated under the above conditions. The diazirine-absorption disappears with a half-life time of also 25 seconds. The initial increase at shorter wavelengths, due to the diazo rearrangement, and the subsequent decrease go on somewhat slower than with the free benzoic acid; the half-life time for the photolysis of the diazo compound is between 2 and 3 minutes. Nevertheless, the new photolabel should be applicable also in experiments with living cells.

Synthesis of L-2-(α,α-dimethyl-3,5-dimethoxybenzyloxycarbonyl)-amino-6-t-butyloxycarbonylamino-hexynoic acid [2-Ddz-6-Boc-L-Dah]

In a first approach L-2,6-diamino-4-hexynoic acid was prepared according to reference (9). The attempt to protect selectivly the 6-amino function with the Boc group by the copper complex method gave the desired product in only very poor yield. In a recent publication Sasaki & Bricas (10) described the same difficulties and developed another scheme of synthesis. We followed that route with only one modification during the separation of the reaction mixture resulting from the enzymatic racemate resolution. The described chromatography with silicagel and 96% EtOH as solvent proved rather poor in our hands. We obtained much better results using a reversed phase Lobar RP-8 column with water/MeOH as eluant. Reaction of L-2-amino-6-t-butyloxycarbonylamino-4-hexynoic acid with Ddz-azide (11) gave 2-Ddz-6-Boc-L-Dah, a derivative, suitable for the use in peptide synthesis with t-butyl side chain protection.

Synthesis of cyclo-[Dah(Boc)-Pro_2-Phe_2-Val-Pro_2-Ala-Phe]

The unsaturated lysine analogue was coupled by the mixed anhydride method to the linear nonapeptide H-Pro_2-Phe_2-Val-Pro_2-Ala-Phe-OMe, the synthesis of which is given in fig.1:

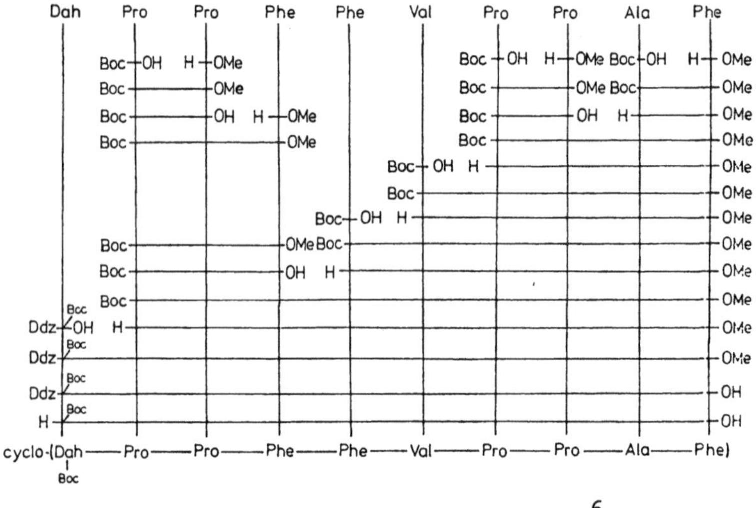

Fig.1 Scheme of synthesis of N_ε-Boc-Dah^6-AA.

After saponification of the methyl ester, the Ddz group was removed by 5% TFA in CH_2Cl_2. The resulting free peptide was cyclized with dicyclohexylcarbodiimide/N-hydroxysuccinimide. After separation of the reaction mixture on Sephadex LH 20/MeOH the cyclic peptide could be crystallized from acetone/water.

Hydrogenation of the unsaturated antamanide-analogue

Catalytic hydrogenation with Pd metal went rather slowly in either MeOH or dioxane. A new spot with slightly higher r_f-value than the starting material appeared on TLC, but even after 16 h only minor amounts of a substance identical in r_f-value and ninhydrine colour with an authentic sample of N_ε-Boc-Lys^6-AA could be detected. Better results were obtained with Pd on charcoal as catalyst. In MeOH hydrogenation was complete after 3 h, in dioxane it took 5 h. In both reaction mixtures after several minutes the above mentioned further migrating compound could be observed by TLC. Since during the hydrogenation it disappeared, whereas the amount of the desired saturated product increased, it obviously was the peptide containing 2,6-diamino-4-hexenoic acid which resulted from partial hydrogenation of the original triple bond.
Hydrogenation with tritium gas was performed in the lab of Dr.Morgat at C.E.N., Saclay, France, and yielded a product with a specific radioactivity of about 20 Ci/mmol. After cleavage of the Boc group the cyclopeptide was coupled to 4-(2,2,2-trifluoro-1,1-azoethyl)benzoic acid. The resulting derivative did not differ markedly in its photochemical properties from those desribed for the photolabile Tyr-AA-analogue. A nonradioactive sample of N_ε-[4-(2,2,2-trifluoro-1,1-azoethyl)benzoyl]-Lys^6-AA was tested for biological activity and proved to be fully active with less than 1.25 mg/kg against 5 mg/kg of phalloidine.

Conclusions

4-(2,2,2-trifluoro-1,1-azoethyl)benzoic acid can be prepared conveniently by the described procedure starting with 4-bromotoluene. The stable crystalline compound is photolysed completely in less than 5 minutes with wavelengths longer than 300 nm, although there is a partial rearrangement of the diazirine to the diazo compound. Additionally the benzoic acid derivative is easily coupled to other molecules containing NH_2 or OH functions. Linking the photolabile residue to suitable antamanide analogues does not change the photochemical properties dramatically. The half-life time for the photolysis of the diazo compound - resulting from photochemical rearrangement of the diazirine - is somewhat longer, but even here photolysis is complete in less than 10 minutes, a time short enough for living cells to survive irradiation. After it has been shown, that N_{ε}-[4-(2,2,2-trifluoro-1,1-azoethyl)benzoyl]-Lys^6-AA is biologically active, the tritiated analogue with its high specific radioactivity should be applicable to photolabeling experiments.

References

1. Wieland, Th., Lüben, G., Ottenheym, H., Faesel, J., Konz, W., de Vries, J.X., Prox, A., Schmid, J.: Angew.Chem. 80, 209-213 (1968).
2. For a recent review see: Wieland, Th., Faulstich, H.: C.R.C. Crit.Rev.Biochem. 5, 184-260 (1978).
3. Bayley, H., Knowles, J.R.: Meth.Enzym. 46, 69-114.
4. Chowdry, V., Westheimer, F.H.: Ann.Rev.Biochem. 48, 293-325 (1979).
5. Bayley, H., Knowles, J.R.: Biochem. 17, 2420-2423 (1978).
6. Nassal, M.: Thesis, University of Heidelberg (1980).
7. Brunner, J., Senn, H., Richards, F.M.: J.Biol.Chem. 255, 3313-3318 (1980).
8. Nassal, M., Wieland, Th.: manuscript in preparation.
9. Jansen, A.C.A., Kerling, K.E.T., Havinga, E.: Rec.Trav. Chim. 89, 861-864 (1970).
10. Sasaki, A.N., Bricas, E.: Tetr.Lett. 21, 4263-4264 (1980).
11. Birr, Chr., Lochinger, W., Stahnke, G., Lang, P.: Ann. 763, 162-172 (1972).

SYNTHESIS AND PROPERTIES OF ($NLe^{56,60}$) 34-65 BACTERIORHODOPSIN FRAGMENT

A.T. Kozhich, B.V.Vaskovsky, I.I.Mikhaleva, V.T.Ivanov
Shemyakin Institute of Bioorganic Chemistry, USSR Academy of Sciences, Moscow, USSR

Bacteriorhodopsin is the main structural element of the purple membrane of halofilic bacteria Halobacterium halobium. Together with membrane lipids it forms a unique system, performing photoinduced proton transport through the membrane against the pH gradient (1). A study of conformational properties of bacteriorhodopsin fragments and of their interaction with lipids could help to understand a structure activity relationship. With this purpose in mind we synthesized a series of peptides of to the second α-helical rod of bacteriorhodopsin and studied their conformational properties. The 34-65 segment was chosen as first target of this study, with Met^{56} and Met^{60} replaced by the isosteric NLe to simplify the synthesis (2).

A strategy of solution synthesis with maximal side chain protection with tert.butyl-derived protecting groups was used. The Z group was used for temporary protection (Figs 1 und 2).

The tetrapeptide 34-37 was prepared starting from proline methyl ester. The carbodiimide method with addition of pentafluorophenol gave the best yield of the dipeptide 36-37, serine and valine were coupled with the aid of N-hydroxysuccinimide esters.

The heptapeptide 38-44 was synthesized in two ways, by 3+4 condensation (not shown in the scheme) and by stepwise elongation. The latter gave better overall yields and was considerable less time-consuming.

Chemistry of Peptides and Proteins, Vol. 2

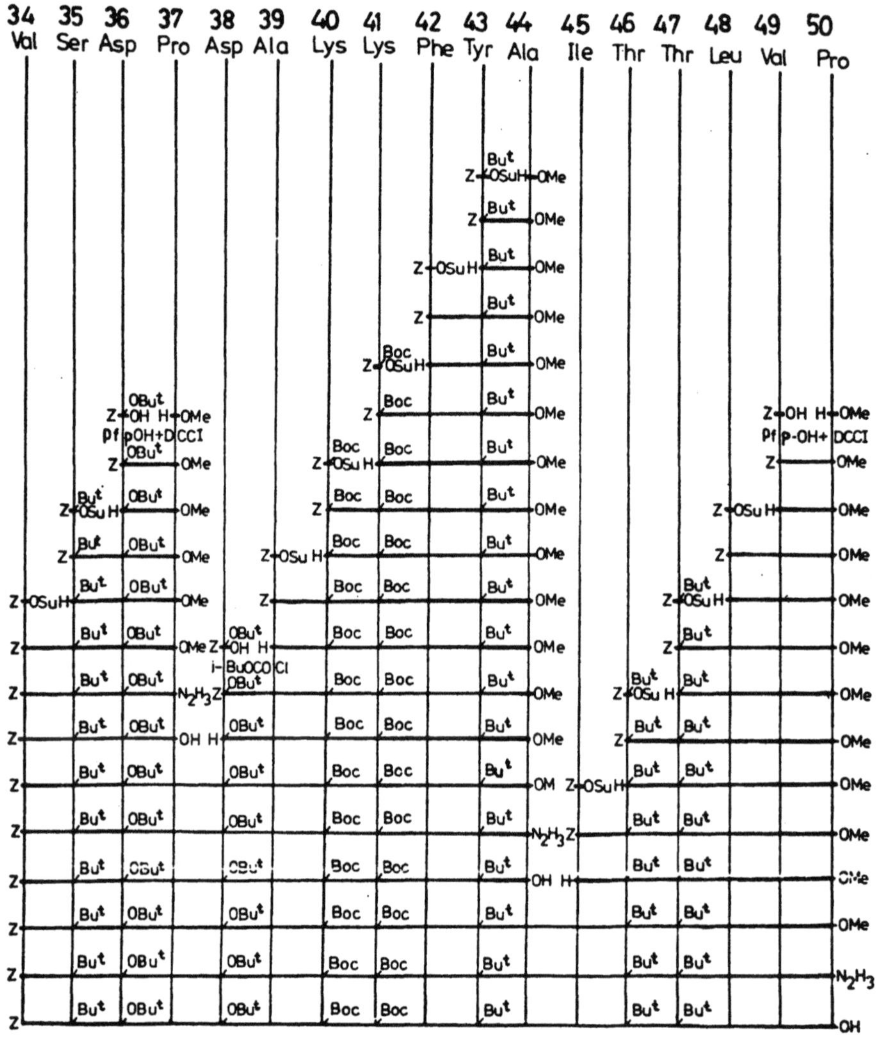

Fig.1. Synthetic scheme of the protected segment 34-50 .

A similar approach was used for the preparation of the protected segment 45-50. Azide coupling with isoamyl nitrite activation of the tripeptides 45-47 and 48-50 showed low (15 %) yield. Attempts of azide (34-37) + (38-44) and (38-44) + (45-50) couplings also resulted in unsatisfactory yields

(8 % and 10 % respectively). We therefore converted both hydrazides, shown in the scheme, to respective carboxylic acids by N-bromosuccinimide oxidation (3) and coupled those with the aid of DCCI, Pfp-OH and HONSu techniques. In all cases the yields were in the 60-70% range.

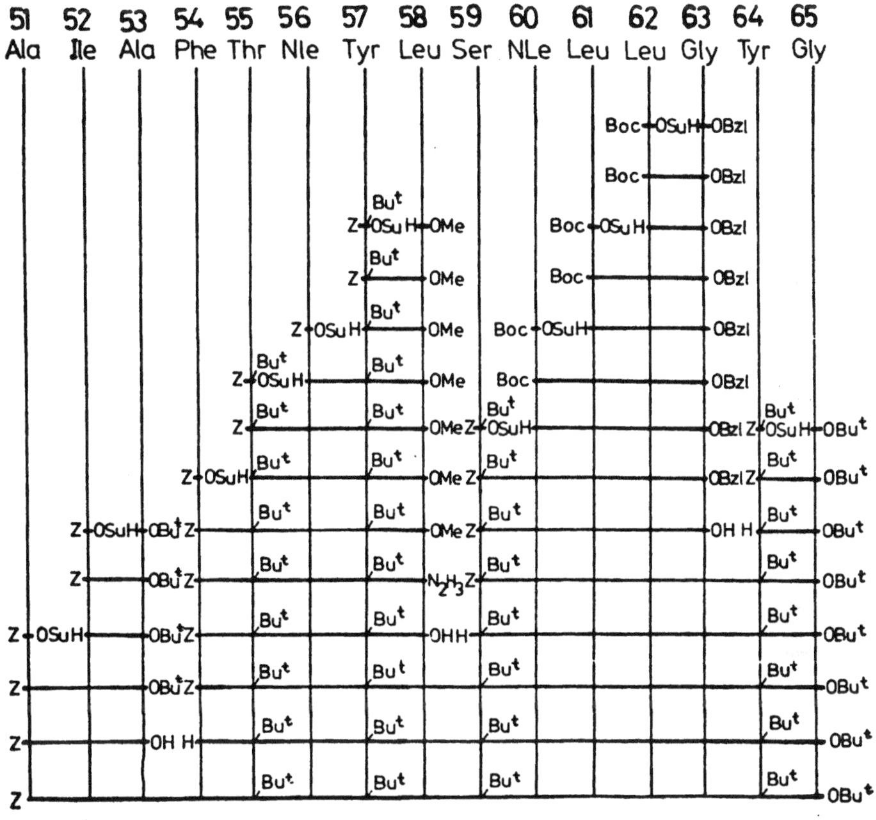

Fig.2. Synthetic scheme of the protected segment 51-65.

The protected segments 51-53, 54-58, 59-63 and 64-65 were obtained by stepwise N-hydroxysuccinimide reaction. Thus, the couplings (51-53) + (54-58) and (59-63) + (64-65) (activation with the F complex)were performed. One of the resulting products Z-51-58-OMe showed extremely poor solubility in all solvents studied (DMF, TFE, n-BuOH, DMSO,tetramethylurea), precluding further utilization of this intermediate. An alternative route, shown in Fig.2, was used for the preparation of the 51-65-protected segment.

The fully protected peptide 34-65 was obtained in 31 % yield by coupling of the peptides 34-50 and 51-65 using the carbodiimide procedure with pentafluorophenol as additive.

The purity of the peptides obtained was monitored by TLC on Merck silica gel precoated plates, HPLC (gel permeation, reverse phase and adsorption mode), mass spectrometry and amino acid analysis. Open column and preparative high performance adsorption chromatography was widely used throughout the synthesis for purification of the intermediates possesing good solubility in methylene chloride, containing 3-10 % of polar solvents like MeOH, Me_2CO, AcOEt. The protected peptides 59-65, 54-65, 34-50, 51-65 and 34-65, sparingly soluble in organic solvents, were purified by preparative size exclusion high performance liquid chromatography (SEHPLC) in dimethylformamide, using a LH-20 (2,5 x 45 cm) HPLC or a Zorbax Sil (2,5 x 25 cm) column and direct pump injection. High performance reverse phase chromatography was used mainly to monitor the homogeneity after preparative adsorption or SEHPLC. SEHPLC on a silica gel Zorbax PSM-60 column was used to monitor the kinetics of the majority of the fragment couplings.

Removal of the Boc and Bu^t groups from the protected peptides (38-44, 34-44, 34-50, 34-65) was achieved by exposure to trifluoraoacetic acid at room temperature for one hour. Purification of the resultant Z peptides was performed in one step by preparative reverse phase chromatography on a Lichrosorb RP-18 column. Conventional methods, such as ion-exchange or aqueous gel filtration chromatography were unapplicable in this case, due to the poor solubility of all these peptides in water.

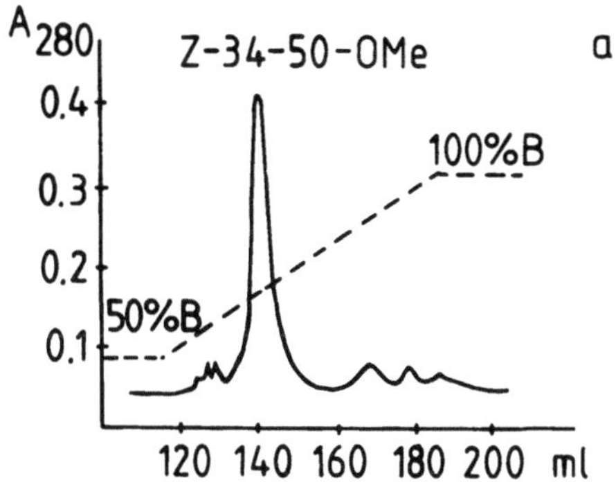

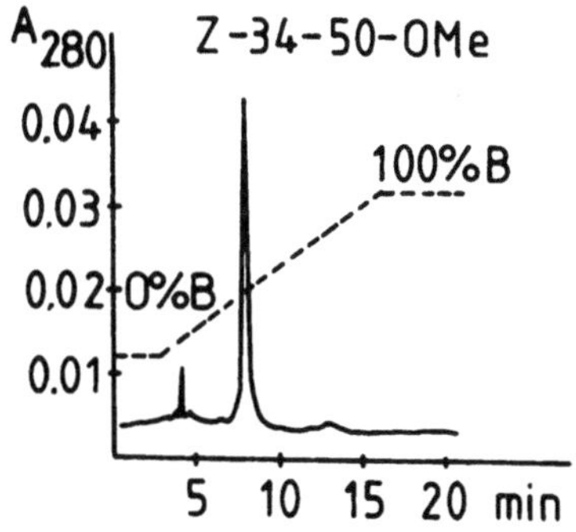

Fig.3. Chromatograms of the partially deblocked bacteriorhodopsin segment Z-34-50-OMe , a: column: 22.7 x 250 mm, LiChrosorb RP-18, flow rate:10 ml/min, A: 0.1 % CF_3COOH in H_2O, B: 0.1 % CF_3COOH in MeOH and the purified product, b: column: 4.6 x 250 mm, Nucleosil 7C-8, flow rate: 5 ml/min, A: 27 % AcN + 70 % H_2O + 3 % HCOOH, B: 57 % AcN + 40 % IPA + 3 % HCOOH.

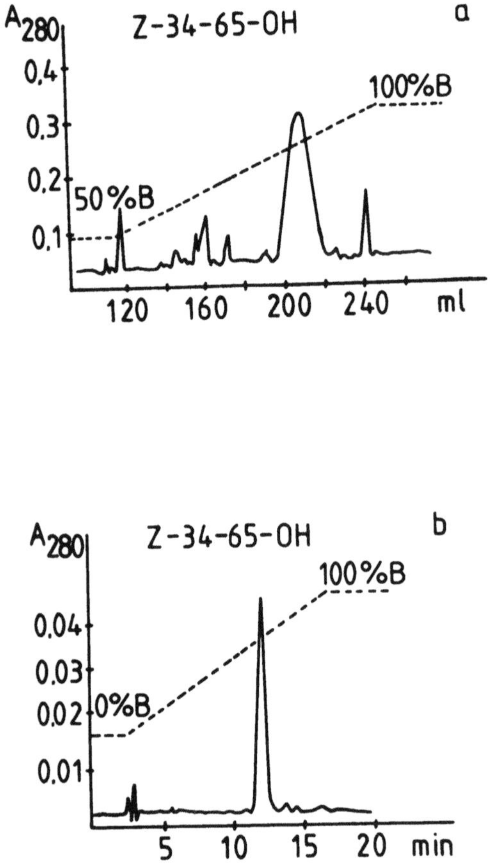

Fig.4. Chromatograms of the partially deblocked bacteriorrhodopsin segment Z-34-65-OH, a: column: 22.7 x 250 mm, LiChrosorb RP-18, flow rate: 10 ml/min, A: 0.1 % CF_3COOH in H_2O, B: 0.1 % CF_3COOH in MeOH and of the purified product, b: column: 4.6 x 250 mm Nucleosil 7C-8, flow rate: 1.5 ml/min, A: 27 % AcN + 70 % H_2O + 3 % HCOOH, B: 57 % AcN + 40 % IPA + 3 % HCOOH.

Circular dichroism spectra of the partially deblocked peptides 38-44,34-44,34-50 and 34-65 in trifluoroethanol and trifluoroethanol-water (1:1) were studied. It was rather rewarding to find the 32-membered product to form α-helical structure as follows from the typical intense minima at 220 and 207 nm.

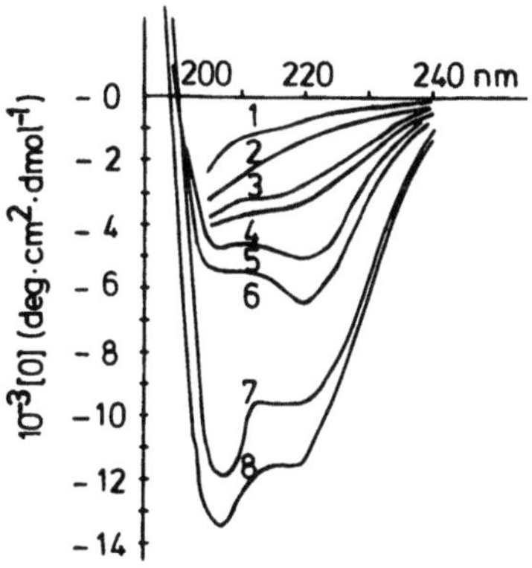

Fig.5. CD spectra of partially deblocked bacteriorhodopsin segments. (1. Z-38-44-OMe in TFE-H_2O (1:1), 2.Z-38-44-OMe in TFE, 3. Z-34-44-OMe in TFE-H_2O (1:1), 4. Z-34-44-OMe in TFE, 5. Z-34-50-OMe in TFE-H_2O (1:1), 6. Z-34-50-OMe in TFE, 7. Z-34-65-OH in TFE-H_2O (1:1), 8. Z-34-65-OH in TFE).

This result allows to suggest that α-helical structure of the bacteriorhodopsin transmembrane strands is due to inherent properties of each segment rather than to the packing effect of these strands. Further spectroscopic studies of the peptides obtained are now being carried out under various conditions, including phospholipid vesicles.

References

1. D.Oesterhelt: Angew.Chem.Int.Ed.Engl. 15, 17-24 (1976).

2. Yu.A.Ovchinnikov, N.G.Abdulaev, M.Yu.Feigina,A.V.Kiselev, N.A.Lobanov: FEBS Lett. 100, 219-224 (1979).

3. J.Meinhofer, M.Dzieduszycka, M.Smulkowski, R.C.Makofske, W.Danho: Int.J.Pept.Prot.Res. 18, 6-11 (1981).

SYNTHETIC PEPTIDE DERIVATIVES OF BACTERIAL LIPOPROTEIN AND THEIR INTERACTION WITH LYMPHOCYTE PLASMA MEMBRANES

Wolfgang G. Bessler, Karlheinz Wiesmüller, Werner Scheuer, Ronald B. Johnson

Lehrstuhl für Mikrobiologie II der Universität Tübingen, D-7400 Tübingen, Germany

Hassanein H. Fouad, Günther Jung

Institut für Organische Chemie der Universität Tübingen, D-7400 Tübingen, Germany

Introduction

Lipoprotein from the outer membrane of Escherichia coli is a mitogen towards mouse, rat, rabbit, and calf lymphocytes (1, 2, 3). It was characterized and sequenzed by V. Braun and coworkers. It is a major protein of the bacterial cell wall composed of 58 amino acids and contains at the N-terminal end one amide-linked and two ester-linked fatty acids bound to glyceryl cysteine(4).The molecular structure, responsible for the mitogenicity of the molecule, resides in its N-terminal fatty acid-containing part; lipopeptide fragments, prepared by enzymatic digestion which carried three fatty acids bound to glyceryl cysteine attached to an oligopeptide chain of 3-5 amino acids, were mitogenically active. In order to determine more precisely the minimal molecular requirements for biological activity of this B-lymphocyte mitogens, we prepared by chemical synthesis several compounds analogous to the N-terminal lipid part. In this study we demonstrate that the synthetic lipopeptide segment S-[2,3-bis(palmitoyloxy)-(2RS)-propyl]-N-palmitoyl-(R)-cysteinyl-seryl-seryl-asparaginyl-alanine ("tripalmitoylpentapeptide") (Fig. 1) is a potent B-lymphocyte mitogen. Fragments, lacking the tetrapeptide moiety Ser-Ser-Asn-Ala, are less active. Thus, for the first time defined synthetic fragments of a bacterial mitogen exhibiting marked biological activity in vitro as well as in vivo are described. In addition we show that the synthetic lipoprotein segments,

Chemistry of Peptides and Proteins, Vol. 2

after coupling them to carrier material, can be used to isolate membrane receptors from immune cells.

Results of Synthesis and Spectroscopic Characterisation

1) Chemical Synthesis of the Tripalmitoyl-pentapeptide

Fig. 1 shows our 14 step synthesis of the lipopeptide segment of bacterial lipoprotein, tripalmitoyl-pentapeptide. L-Cystine-di-tert-butyl ester was acylated with palmitic acid chloride and subsequently reduced with dithioerythritol. The resulting N-palmitoyl-L-cysteine-tert-butyl ester was reacted with 1-bromo-2,3-dihydroxy-propane. The S-glyceryl hydroxyl groups were esterified with palmitic acid using N,N-dicyclohexylcarbodiimide in the presence of dimethylaminopyridine. Finally, the tert-butyl protecting group was removed with trifluoroacetic acid to yield S-[2,3-bis(palmitoyloxy)-(2RS)-propyl]-N-palmitoyl-(R)-cysteine ("tripalmitoyl-cysteine"). Tripalmitoyl-pentapeptide was prepared via N-benzyloxycarbonyl-L-seryl(O-tert-butyl)-seryl(O-tert-butyl)-asparaginyl-alanine-tert-butyl ester which was hydrogenated and coupled to tripalmitoyl-cysteine. After purification by gel chromatography on Sephadex LH 20 (or Silica gel), the tert-butyl protection was split off by trifluoroacetic acid.

All products were pure according to thin-layer chromatography, elemental analysis, amino acid analysis, mass spectrometry and ^{13}C NMR spectroscopy.

2) Chromatography Separation of the Diastereomers of Tripalmitoyl-L cysteine-tert-butyl ester

The use of racemic 1-bromo-2,3-dihydroxy-propane yields S[2,3-bis(palmitoyloxy)-(2RS)-propyl]-N-palmitoyl-(R)-cysteine-tert-butyl ester. The diastereomers R,R and S,R could be separated on a Silica gel column (60 x 3 cm). The conditions were MM-Kieselgel 60, eluent dichloromethane ethylacetate (20/1, v/v), and samples of 400 mg/5 ml eluent were applied for each run. Thin-layer chromatography was used to test 150 fractions á 8 ml (Kieselgel Merck, eluent dichloromethane/ethylacetate = 20/1), detection by chlorine/4,4-bis(dimethylamino)diphenyl-methane (TDM) reagent. The fractions 65-76, 76-95 and 96-112 were combined and the three samples characterized by

Fig. 1. Scheme of the synthesis of tripalmitoyl-pentapeptide.

^{13}C NMR.

3) Assignment of Configuration by ^{13}C NMR

Unequivocal assignments of the ^{13}C NMR signals of tripalmitoyl-cysteine-methyl ester and the synthetic intermediates $(Pal\text{-}Cys\text{-}OMe)_2$, Pal-Cys-OMe, Pal-Cys($CH_2$-CHOH-$CH_2$)-OMe are published elsewhere (6). Even at 22,62 MHz (^{13}C) and clearly at 90,5 MHz (^{13}C) some signal doublings were detected due to the presence of diastereomers for all compounds with S-glyceryl moieties.

In order to correlate the configuration R,S and R,R with the signals, ^{13}C NMR spectra at 100,62 MHz (Bruker WM 400) were taken from the chromatographically separated diastereomers. As shown in Fig. 2 the diastereomers differ in their chemical shifts at Cys-C_β ($\Delta\ \delta$= 0.2 ppm), and at glyceryl-C-1 ($\Delta\ \delta$ = 0.1 ppm). Separated signals are also found for Cys-C_α($\Delta\ \delta$ = 0.1 ppm), and shoulders for Cys-CO and Pal(2)-CO (not shown in Fig. 2).

For the configurational assignment we prepared the diastereomers R,R via substitution of N-hexadecanoyl-L-cysteine-tert-butyl ester with (R),3-propane-diol-(1,2). The comparison of the ^{13}C NMR data of the tripalmitoyl-cysteine-tert-butyl ester with known configuration at glyceryl-C-2 afforded, that fractions 65-76 contained almost pure R,R diastereomers, fractions 76-95 a mixture of diastereomers (S,R/R,R = 3/2), and fractions 95-112 almost pure S,R diastereomer (Fig. 2).

In order to detect possible differences in the biological activities of the diastereomers the corresponding tripalmitoyl-pentapeptides are prepared at present.

Results and Discussions of Biological Assays

In previous work we have shown that pronase digestion fragments of bacterial lipoprotein were mitogenically active (5). These fragments carried 3 fatty acids bound to glyceryl cysteine attached to an oligopeptide chain of 3-5 amino acids. The synthetic lipid compound prepared by us is, with respect to the amino acid sequence, identical to these pronase fragments (4, 5), but differs in the fatty acid composition since it contains solely palmitic acids residues. This product, S-[2,3-bis(palmitoyoxy)-(2RS-oropyl]-N-palmitoyl-(R)-cysteinyl-(S)-seryl-(S)-seryl-(S)-asparaginyl-(S)-alanine, was investigated for the ability to mitogenically activate in vitro splenocytes from different mouse strains.

The results of in vitro stimulation experiments are shown in Tab. 1. Tripalmitoyl-pentapeptide was able to increase ^{3}H-thymidine incorporation in spleen cell cultures of the Balb/c mouse strain starting at concentrations around 4 µg/ml and exhibited a superior mitogenicity to native lipoprotein.

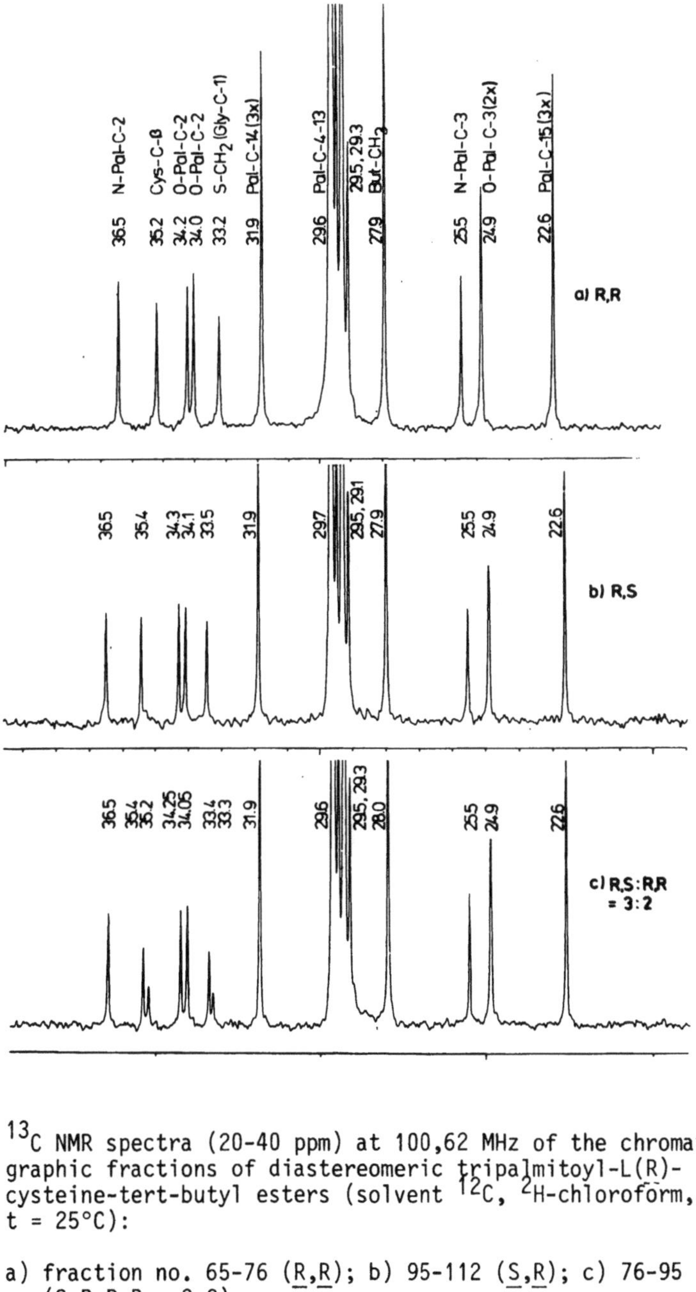

Fig. 2. ^{13}C NMR spectra (20-40 ppm) at 100,62 MHz of the chromatographic fractions of diastereomeric tripalmitoyl-L(R)-cysteine-tert-butyl esters (solvent ^{12}C, ^{2}H-chloroform, t = 25°C):

a) fraction no. 65-76 (R,R); b) 95-112 (S,R); c) 76-95 (S,R:R,R = 3:2).

The optimal dose for stimulation with the lipid compound was around 20-100 µg/ml. The T-lymphocyte mitogen concanavalin A and the B-lymphocyte mitogen lipopolysaccharide, which were included as controls, also showed increased DNA synthesis. Similar to stimulation of DNA synthesis, the synthetic compounds were also able to induce incorporation of ^{3}H-uridine into RNA and of ^{3}H-leucine into proteins; palmitic acid and cysteine were not active (data not shown). Thus, reported in a short communication (7), our results demonstrate that a N-terminal lipoprotein segment prepared by chemical synthesis is a potent lymphocyte mitogen comparable to the bacterial mitogens lipoprotein and lipopolysaccharide. Additionally, we have provided further evidence that the N-terminal lipopeptide region is responsible for the mitogenic activity of lipoprotein.

In order to determine whether the synthetic compound was able to stimulate B-lymphocytes into immunoglobulin secretion, a hemolytic plaque assay was performed using Balb/c mouse spleen cell cultures (Tab. 2). Tripalmitoyl-pentapeptide induced a marked increase in the number of plaque forming cells at mitogen concentrations ranging from 1 to 100 µg/ml. A pronounced increase in plaque numbers could also be observed in lipoprotein and LPS treated cultures, which is in agreement with our previous results (5). Similar findings were made with C3H/HeJ LPS nonresponder cells (Johnson, R.B., Köhl, S., Wiesmüller, K., Jung, G., Bessler, W.G., submitted for publication) confirming that the synthetic fragment is a B-lymphocyte mitogen . Tripalmitoyl-pentapeptide was superior to tripalmitoyl-cysteine or N-palmitoyl-cysteine methyl ester (8). Obviously in order to obtain full biological activity the presence of a lipophilic part is necessary, which serves as an anchor in the plasma membrane followed by a hydrophilic oligopeptide assuming probably a β-bended conformation.

The new synthetic mitogens, which are well characterized products, will be a valuable tool for continuing the investigation of the molecular mechanism of B-lymphocyte activation. Interestingly, we could also demonstrate that the synthetic lipoprotein fragments induced splenomegaly and <u>in vivo</u> B-lymphocyte activation in mice (R. Johnson, S. Köhl, W. Bessler, unpublished). Thus, the synthetic compounds may be useful as <u>in vivo</u> stimulants of the immune response. The new synthetic compounds are also useful for investigating the molecular mechanism of the mitogen-lymphocyte interaction.

Table 1. In vitro stimulation of ³H-thymidine incorporation into DNA in spleen cells of Balb/c mice of tripalmitoyl-pentapeptide, lipoprotein prepared from S. typhimurium SL761, and concanavalin A (n.d. not determined). Means of triplicate experiments ± standard deviations. Experiments were carried out as described (5) in microtiter plates (FALCON PLASTICS, Los Angeles, California, USA). In short, spleen lymphocyte cultures were performed for 48 h at a cell density of 3.3 x 10^6/ml in 0.18 ml aliquots per well in RPMI-1640 medium supplemented with 3.3 % inactivated fetal calf serum, fresh glutamine (2 mM), penicillin (100 units/ml), streptomycin (100 µg/ml), and 2-mercaptoethanol (5 x 10^{-5} M). Before harvesting, cultures were pulsed for 24 h by the addition to each well of 0.5 µCi ³H-thymidine (AMERSHAM, Braunschweig, Germany) of specific activity 40 Ci/mol.

MITOGEN CONCENTRATION (µg/ml)	RADIOACTIVITY INCORPORATED (cpm)		
	TRIPALMITOYL-PENTAPEPTIDE	LIPOPROTEIN	CONCANAVALIN A
274	3446 ± 524	62 ± 11	n.d.
137	4039 ± 833	2901 ± 1088	64 ± 34
68	2972 ± 842	3982 ± 273	545 ± 411
34	3612 ± 595	3084 ± 394	316 ± 249
17	3443 ± 1458	2444 ± 1142	670 ± 97
8	3473 ± 1189	2030 ± 810	3836 ± 635
4	2270 ± 1059	1226 ± 182	14047 ± 563
2	n.d.	1314 ± 212	18001 ± 1689
1	n.d.	999 ± 232	16939 ± 1760
	541 ± 40	541 ± 110	541 ± 110

In order to determine the binding proteins for lipoprotein in lymphocyte plasma membranes, tripalmitoyl-cysteine was coupled to CPG-aminopropyl pore controlled glass beads, and ^{125}iodine or ³H-labelled membrane solubilisates were separated by affinity chromatography (9). In the presence of the detergent Nonidet P40, lipoprotein-binding proteins were specifically attached to the column and could be eluted by washing the column with sodium dodecylsulphate. The binding proteins were characterized by polyacrylamide gel electrophoresis, radioautography, and by immunoprecipitation. Table 3 shows the molecular masses of the major binding proteins for lipoprotein obtained from lymphocyte plasma membranes by this method. The binding proteins found

Table 2. Stimulation of Balb/c B-lymphocytes to the development of plaque forming cells (PFC) by tripalmitoyl-pentapeptide, lipoprotein, and lipopolysaccharide from E. coli B/r. Plaque forming cells (PFC) per 1 x 10^6 recovered spleen cells. Means of quadruplicate assays ± standard deviations. Lymphocytes were cultured at 5 x 10^6 cells/ml in 1 ml aliquots in supplemented RPMI-1640 medium (5), but with 10 % inactivated fetal calf serum. After four days, 100 µl of the lymphocyte suspension appropriately diluted in HEPES buffered Eagles Minimal Essential Medium containing additional glucose (0.45 % w/v), 20 µl of guinea pig complement (SERVA, Heidelberg, Germany), and 20 µl of a 30 % suspension of trinitrophenol coated sheep red blood cells were added to 300 µl of 0.50 % of agarose in medium at 45°C, mixed and plated on disposable petri dishes for 3 h, and plaques were scored using a plaque viewer (TECNOMARA, Zürich, Switzerland).

MITOGEN APPLIED	MITOGEN CONCENTRATION (µg/ml)	PFC PER 1 x 10^6 RECOVERED SPLEEN CELLS
None	0	1.75 ± 1.5
Tripalmitoyl-pentapeptide	100	52.5 ± 12
	25	50 ± 10.2
	6.25	30.4 ± 3.7
	1.56	12.75 ± 4.4
	0.4	7.1 ± 1.4
	0.1	4.6 ± 1.8
Lipoprotein	50	63.3 ± 16.7
Lipopolysaccharide	50	37.4 ± 3.4

Table 3. Molecular masses of binding proteins as determined by gel electrophoresis in the presence of sodium dodecyl sulphate (cf. 9). Molecular masses are given in 1,000 daltons (KD).

^{125}J-LABELLED MEMBRANE PROTEINS (KD)	^{3}H-LEU-LABELLED MEMBRANE PROTEINS (KD
-	110 - 114
78 - 83	78 - 82
70 - 74	-
62 - 64	62 - 64
32 - 36	34 - 36
26 - 30	28 - 30
-	16 - 18

correspond to the binding proteins for LPS found in C3H/HeJ mouse B lymphocytes by Yokoyama et al. (10) and thus might represent Ia-antigen, the δ-chain from cell surface IgD, and the μ-chain from cell surface IgM.

Thus, our results suggest that there exists a great variety of binding components for bacterial lipoprotein in mouse spleen cells; we are presently extending our studies on purified B and T lymphocytes. More investigations to further characterize the proteins by serologic techniques are in progress. Furthermore, we are presently investigating lipoprotein-binding proteins from plasma membrane vesicles obtained by nitrogen cavitation.

Acknowledgements

We wish to thank Prof. Dr. V. Braun for helpful discussions and support. (R),3-propanediol-(1,2) was kindly provided by Dr. Koppenhöfer. The project was funded by Deutsche Forschungsgemeinschaft (SFB 76).

References

1. Melchers, F., Braun, V. & Galanons, C., J. Exp. Med. 142, 473-482 (1975).
2. Bessler, W., & Braun, V., Z. Immun.-Forsch. 150, 193 (1975).
3. Bessler, W., & Ottenbreit, B.P., Biochem. Biophys. Res. Comm. 76, 239-246 (1977).
4. Braun, V., Biochim. Biophys. Acta 415, 335-377 (1975).
5. Bessler, W., Resch, K., Hancock, E. & Hantke, K., Z. Immun.-Forsch. 153, 11-22 (1977).
6. Jung, G., Carrera, C., Brückner, H. & Bessler, W.G., submitted for publication (1982).
7. Bessler, W.G., Johnson, R.B., Wiesmüller, K. & Jung, G., Hoppe-Seyler's Z. Physiol. Chem. 363, 767-770 (1982).
8. Cybulla, J., Brückner, H., Jung, G., Wipperfürth, Th. & Bessler, W.G., Biochem. Biophys. Res. Comm. 92, 1389-1396 (1980).

9. Bessler, W.G., Kustermann-Kuhn, B. & Schneuer, V.W., in: K. Resch, H. Kirchner (eds.): Mechanisms of Lymphocyte Activation, Elsevier/North-Holland, Biomedical Press, Amsterdam, New York, Oxford, 476-479 (1981).

10. Yokoyama, K., Mashimo, J., Kasai, N., Terao, T. & Osawa, T., Hoppe-Seyler's Z. Physiol. Chem. 360, 587-595 (1979).

SYNTHESIS OF DEFINED PEPTIDE DERIVATIVES BY AMINOLYSIS OF 3-[N$^{\alpha}$-BENZYLOXYCARBONYL PEPTIDYLOXY]-2-HYDROXY-N-METHYL-BENZAMIDES AT ELEVATED TEMPERATURES II[1)]

Synthesis of the Peptide Derivatives Z-Ala-X-Gly-N(Et)$_2$,
X = Phe, Leu, Val, Ser(Bu^t), Glu(OBu^t)

Heinz Berndt

Lehrstuhl für Textilchemie und Makromolekulare Chemie der RWTH Aachen, D-5100 Aachen, FRG

Christoph Turck[2)]

Deutsches Wollforschungsinstitut, D-5100 Aachen, FRG

Introduction

Segment condensation has been successfully applied to the synthesis of many peptides and proteins up to about 120 amino acid residues. However, with larger peptides, segment condensation is often impeded by poor solubility of larger protected peptides. The need for methods which avoid the solubility problem during the coupling of big peptide segments in solution prompted us to look for coupling methods at elevated temperatures which should allow to circumvent the solubility problem.

Results

Continuing our previous experiments[2-4)] to synthesize defined peptide derivatives via aminolysis of activated peptide ester derivatives at elevated temperatures we investigated the aminolysis of 3-[N$^{\alpha}$- cylpeptidyloxy]-2-hydroxy-N-methylbenzamides

Chemistry of Peptides and Proteins, Vol. 2

with glycinediethylamide at temperatures within 60-100°C utilizing dimethylsulfoxide,dimethylformamide or ethylenecarbonate as solvents.
The aim of our experiments with model peptide derivatives is to examine whether the most common protecting groups(Z,Boc,OBu^t, Bu^t)used in peptide chemistry are stable under the reaction conditions(60-100°C) and to determine the degree of racemization occuring during peptide synthesis at elevated temperatures.

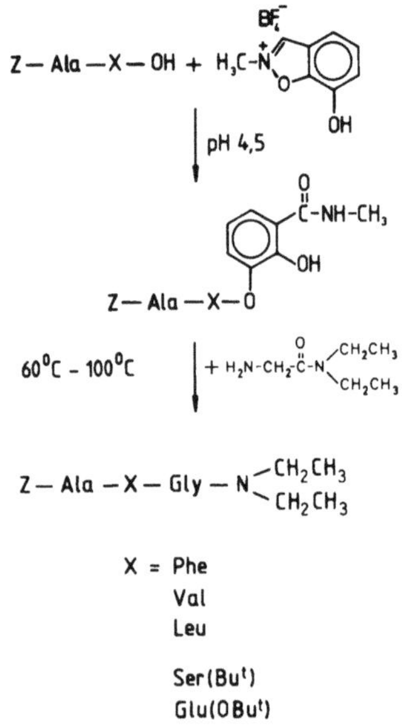

Fig.1. Model reactions for peptide synthesis at elevated temperatures.

As epimeric tripeptide derivatives can be separated into their optical isomers by HPLC on silica gel[2-4],we choose the model reaction given in figure 1 for our condensation experiments.
To establish the degree of racemization during the condensation steps at elevated temperatures we synthesized the tripeptide derivatives Z-Ala-X-Gly-$N(Et)_2$ as LL derivatives as well as LD derivatives via the dicyclohexylcarbodiimide/ 1-hydroxybenzotriazole method[5] as references.This permitted us to compare the optical purity of the synthesized tripeptide derivatives via separation of racemic peptide mixtures into their optical isomers by HPLC as well as to compare the yields obtained by a conventional coupling method and the yields obtained by the catechol ester method[6] at elevated temperatures[2].

Benzyloxycarbonyldipeptide derivatives, Z-Ala-X-OH, X= Phe, Leu,Val, Ser(Bu^t), Glu(OBu^t) were reacted at pH 4,5 with the modified Kemp[6] reagent, 7-hydroxy-2methylbenzisoxazoliumtetrafluoroborate[2], to yield the corresponding 3-[N^α- benzyloxycarbo-

nyldipeptidyloxy }-2-hydroxy-N-methylbenzamides in 90% yield, which were used without further purification.
Activated dipeptide derivatives, Z-Ala-X-ODhmb, were reacted with 1,1 equivalents of H-Gly-N(Et)$_2$, 1.0 equivalent of tetramethylguanidine as base in dimethylformamide,dimethylsulfoxide or ethylenecarbonate as solvents. Reaction temperatures varied within 20-100°C. Condensation experiments in ethylenecarbonate were done without adding tetramethylguanidine.
Coupling reactions were stopped by adding three equivalents of acidic acid,diluted with ethylacetate and filtered.After extraction with buffer pH 2[7] and pH 9 [2] the organic supernatants were dried and evaporated to dryness. The epimer content of each batch was determined from the reaction products after evaporation by HPLC prior to any further purification steps.* Reference peptide derivatives as well as peptide derivatives obtained by the catechol ester method at elevated temperatures underwent further purification steps, recristallisation from ethylacetate/n-hexane, until no more epimeric derivatives were detectable by HPLC. Reference peptide derivatives -LL forms - and corresponding peptide derivatives obtained from coupling experiments at elevated temperatures showed congruent analytical data(full analytical and experimental data will be published elsewhere) after the final purification steps.
Model peptide derivatives obtained via the catechol ester method carried out at 20°C show nearly the same epimer content as those derivatives obtained via the DCC/HOBT procedure in tetrahydrofuran - s.tab. 1-. At 60°C, a temperature at which even peptide derivatives with poor solubility become soluble in dimethylformamide or dimethylsulfoxide, the epimer content is surprisingly low, 0,55% to 2,42% in dimethylformamide, 0,61% to 2,30% in dimethylsulfoxide. Utilizing ethylenecarbonate as solvent the epimer content is reduced to one third of the values obtained

---

* Therefore falsifications of the epimer content due to the working up procedures are most unlikely.

Tab. 1. Effect of coupling methods and reaction conditions upon yields and racemization during the synthesis of model peptide derivatives following fig. 1: Z-Ala-X-ODhmb + H-Gly-N$(Et)_2$ → Z-Ala-X-Gly-N$(Et)_2$ + HO-Dhmb

Method	Solvent	Temp. °C	Yield* %	% D-Epimer of Z-Ala-X-Gly-N$(Et)_2$** ± 0,1% X				
				Phe	Leu	Val	Ser(Bu^t)	Glu(OBu^t)
-ODhmb	DMF	20	52-80	0,0	0,21	0,1	1,52	0,92
		60	52-80	0,55	0,82	2,42	1,84	1,90
		80	52-80	1,14	1,65	6,02	2,06	3,49
		100	52-80	2,49	3,58	10,33	4,56	5,90
	DMSO	20	50-85	0,0	0,1	0,0	1,26	1,22
		60	50-85	0,61	0,59	2,13	2,30	1,37
		80	50-85	0,74	2,10	4,90	3,00	3,72
		100	50-85	2,27	2,45	7,52	6,45	7,53
	Etc	60	60-65	0,18	0,32	0,86		
		80	60-65	0,30	0,33	2,32		
		100	60-65	1,96	3,20	9,50		
DCC/HOBT	THF	-15	51-91	0,0	0,05	0,25	1,40	1,20

-ODhmb= 2,3-dihydroxy-n-methylbenzamide

Etc = ethylenecarbonate

*epimer free products, ** prior to purification procedure

in dimethylsulfoxide or dimethylformamide. Experiments in dimethylformamide as well as in dimethylsulfoxide without base showed nearly the same racemization or even higher racemization rates than with base. The low racemization rates obtained by the catechol ester method in ethylene carbonate therefore must be regarded as a solvent effect.

As the obeyed racemization rates during peptide synthesis at elevated temperatures are surprisingly low and optically pure peptides are obtained from the crude reaction product by simple purification procedures in good yields, we hope that aminolysis of 3- N^{α}-acylpeptidyloxy -2-hydroxy-N-methylbenzamides at elevated temperatures is the first step forward to circumvent the solubility problem during the coupling of big segments in solution.

References

1. Turck, Ch. and Berndt, H.: Hoppe-Seyler's Z. Physiol. Chem. 362, 821-828 (1981).
2. Turck, Ch.: part of the thesis, RWTH Aachen 1983
3. Matoni, G. and Berndt, H. :Tetrahedron Lett.21, 37-40(1980).
4. Berndt, H.,Matoni,G. and Lahm,H.W.:Peptides 1980(K.Brunfeldt, Edt.) 159-164,Scriptor, Kopenhagen 1981
5. König,W. and Geiger,R.: Chem. Ber. 103, 788-798 (1970).
6. Kemp,D.S. ,Wang, S.-W., Mollan,R.C.,Banquer,C. and Subramanyam: Tetrahedron 30, 3677-3688 (1974).

RECENTLY DEVELOPED AMINO PROTECTING GROUPS

Wolfgang Voelter, Hubert Kalbacher, Charles Beni, Wolfgang Heinzel, Jürgen Müller

Abteilung für Physikalische Biochemie des Physiologisch-chemischen Instituts der Universität Tübingen, Hoppe-Seyler-Str.1, D-7400 Tübingen, FRG

Introduction

Considerable progress has been made in recent years in finding protecting groups for peptide synthesis which can be removed under mild acidolytic conditions compared to the tert -butyl and benzyloxycarbonyl group (1).Although the Bpoc (2-p-biphenylylisopropyloxycarbonyl)(2), Ddz(3,5-dimethoxy(α,α-dimethyl)benzyloxycarbonyl)(3), Ppoc(α,α-dimethylbenzyloxycarbonyl) (4) and Azoc (α,α-dimethyl-4-phenylazobenzyloxycarbonyl)(5) groups cleave under the desired conditions, only the first two ones have found wider application, because their amino acid derivatives are commercially available. However, Bpoc and Ddz amino acids decompose autocatalytically resp. are photolabile. Moreover, under the conditions of catalytic hydrogenolysis - a widely used procedure in peptide synthesis - these protecting groups are split off.

Recently, we described several new amino protecting groups, which can be cleaved under various acidic conditions (6-9, Table 1).
Several peptide hormones, such as somatostatin (10), TRH (11), LH- RH (12), and the thymopoietin pentapeptide (13-15) have been synthesized using Adpoc protection; the t-Bumeoc group was succesfully applied for the synthesis of molluscan neuropeptide (8), and thymosin ß9 (16).

Chemistry of Peptides and Proteins, Vol. 2

Further studies have been made to investigate the influence of α,α-dimethyl groups in protecting groups of the urethane type leading to the development of the 1-benzyl-1-methylethoxycarbonyl (Bzpoc) group(7). The structures of the above mentioned protecting groups are given in Figure 1, their cleavate constants and stability properties are collected in Table 1.

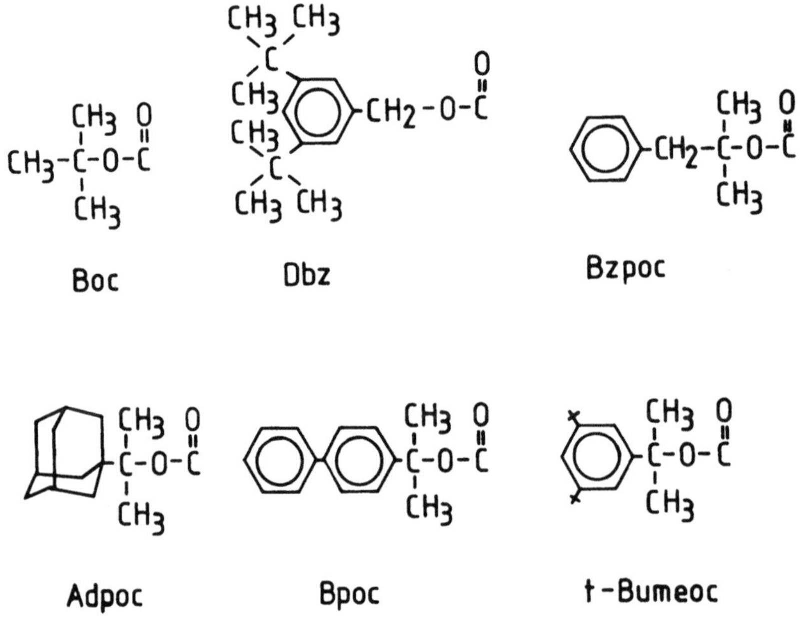

Fig.1. Structures of protecting groups (Boc: tert-butyloxycarbonyl, Dbz: 3,5-di-tert-butylbenzyloxycarbonyl, Bzpoc: 2-benzylpropyl(2)-oxycarbonyl, Adpoc: 1-(1-adamantyl)-1-methylethoxycarbonyl, Bpoc: 2-p-biphenylylisopropyloxycarbonyl, t-Bumeoc: 2-(3,5-di-tert-butylphenyl)-propyloxycarbonyl.

Table I. Comparison of cleavage constants and stabilities to catalytic hydrogenolysis of different protecting groups.

Protecting group	k_1 in 80 % acetic acid	k_{rel}	Stability to catalytic hydrolysis	References
Boc	5.10^{-4}	1	+	1
Dbz	$2.1\ 10^{-4}$	0.43	-	8,9
Bzpoc	1.10^{-2} *	0.5 *		7
Adpoc	1.2	2400	+	6,7
Bpoc	1.4	2800	-	2
t-Bumeoc	2.0	4000	+	8,9

*) determined in 85 % formic acid

Results

The t-Bumeoc and the Dbz groups

A convenient synthesis for the starting alcohols was recently described(8). For the incorporation into amino acids suitable reagents are the fluoroformate for the t-Bumeoc and the chloroformate for the Dbz group. Both reagents are not stable for a longer period of time at room temperature. Successful incorporation of the protecting groups is possible in organic solvents using benzyltrimethylammonium salts as well as in a mixture of water and dioxan. Better yields are obtained with the latter method. Both protecting groups are stable under basic conditions. The Dbz amino acids are stable at room temperature,however, the t-Bumeoc amino acids show autocatalytic decomposition and are therefore better stored as DCHA salts. The physical data,yields and ppm values of ^{13}C NMR spectra of the t-Bumeoc and Dbz derivatives were also published recently (8). The t-Bumeoc group is removable with 3 % trifluoroacetic acid in methylene chloride, acetic acid/

(83 % (v/v) formic acid/water (7:1:2, v/v) or 80 % acetic acid. The cleavage in these three solvents can easily be followed by thin-layer chromatography and high performance liquid chromatography (HPLC). Table 2 and Figures 2 and 3 show the cleavage rates of t-Bumeoc amino acids in different deprotecting solvents. According to our data the cleavage is a reaction of first order and as cleavage products the corresponding alcohols and olefins could be identified. It t-Bumeoc-Lys(Boc)-OH or t-Bumeoc-Ser(Bu^t)-OH are subjected to the solvents mentioned before, no free serine or lysine is formed. The applicability of the t-Bumeoc group is demonstrated by the synthesis of the molluscan neuropeptide (L-Phe-D-Leu-L-Arg-L-Phe-NH_2, Figure 3).

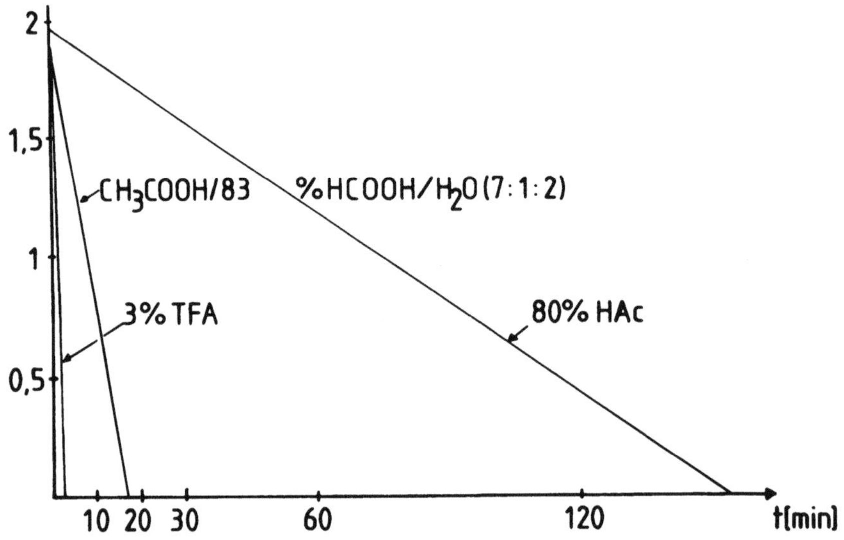

Fig.2. lg c/t diagram of t-Bumeoc-Phe-OH, determined by HPLC (3 % TFA: 3 % trifluoroacetic acid in methylene chloride, 80 % HAc: 80 % acetic acid and acetic acid, formic acid (83 %), water (7:1:2).
Experimental HPLC conditions; column: LiChrosorb RP-18

(5 µm, 250 x 4 mm I.D.), eluent: methanol/0.01 M ammonium acetate (pH 5.1; 85:15), detection:absorption at 254 nm.

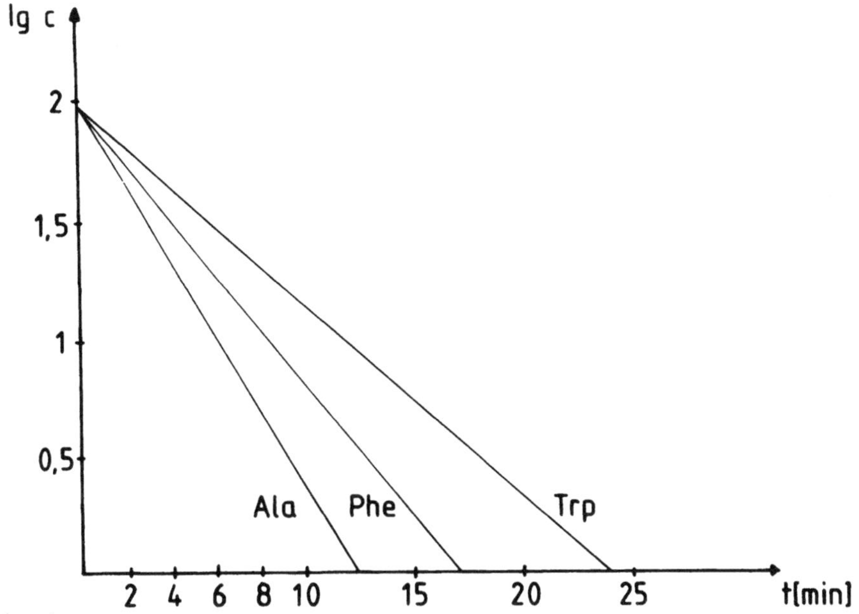

Fig.3. lg c/t diagram of different t-Bumeoc amino acids in acetic acid, formic acid (83 %), water (7:1:2), determined by HPLC.
Experimental HPLC conditions: column: LiChrosorb RP-18 (5 µm, 250 x 4 mm I.D.), eluent: methanol/0.01 M ammonium acetate (pH 5.1, 85:15), detection:absorption at 254 nm.

Table 2. Cleavage rates of t-Bumeoc-Phe-OH(upper values) and Dbz-Phe-OH (lower values) in various media.

Reagent	50 % cleavage (min)	99.9 % cleavage (min)	k_1 (min^{-1})
a	0.07	0.6	9.0
c	2.1	18.8	0.26

Table 2.(Continued).

Reagent	50 % cleavage (min)	99.9 % cleavage (min)	k_1 (min^{-1})
b	22.0	167.0	0.03
a	$1.9\cdot10^4$	$2.0\cdot10^{-5}$	
c	$4.5\cdot10^4$	$5.0\cdot10^{-6}$	
b	$8.5\cdot10^4$	$3.5\cdot10^{-6}$	

a) CF_3COOH (3 %) in CH_2Cl_2

b) CH_3COOH (80 %)

c) $CH_3COOH/HCOOH$ (83 %)/water (7:1:2)

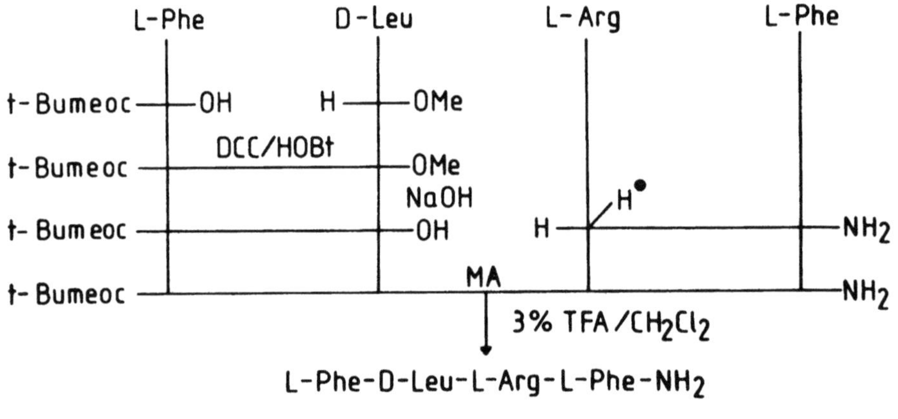

Fig.4. Synthesis of molluscan neuropeptide using t-Bumeoc protection.

The Adpoc group

For the synthesis and properties of the Adpoc group see references 6 and 7. Several methods for introducing the Adpoc group were tested, however, Adpoc fluoride now has been found to be the reagent of choice, as it can be synthesized in high yields as a crystalline (mp 52°C) and stable (at 0°C for months) product.

The Adpoc group can be removed in mild acidic media, preferably 3 % trifluoracetic acid in dichloromethane, in only few minutes or in a mixture of acetic acid/83 % formic acid/water (7:1:2) in 2-5 hours (Table 3).

Table 3. Cleavage rates of different Adpoc amino acids in 3 % TFA/CH_2Cl_2 (a) and CH_3COOH/83 % HCOOH (w/w)/H_2O (7:1:2, v/v, b).

Protected amino acid	Reagent	Temperature	k (min^{-1})	$t_{1/2}$ (min)
Adpoc-Pro-OH	a)	RT	0.257	2.69
	b)		-	-
Adpoc-Leu-OH	a)	RT	0.55	1.26
	b)		$3.3\ 10^{-3}$	210
Adpoc-Lys(Z)-OH	a)	RT	0.58	1.19
	b)		$2.31\ 10^{-3}$	300
Adpoc-Lys(Boc)-OH	a)	RT	0.507	1.36
	b)		$3.65\ 10^{-3}$	189.86
Adpoc-Leu-Trp-OH	a)	RT	0.054	12.91
	b)		$1.47\ 10^{-3}$	472.4

The acid-labile Adpoc group finds increasing application for the synthesis of peptides due to its advantages: it is easy to incorporate in high yields, Adpoc-protected peptides and amino acids are crystalline and stable compounds and the Adpoc group can be removed 10^3 times faster compared to the Boc group. To find a suitable cleavage reagent for the selective cleavage of the Adpoc group in presence of other protecting groups like Boc, Z, etc., cleavage rates in different acidic media were followed by HPLC. The experimental data show (Table 4) that the most suitable reagent is 50 % HCOOH in trifluoroethanol for selective cleavage.

Table 4. Rates of deprotection of Adpoc amino acids and Adpoc dipeptides using 50 % HCOOH/TFEt (1:10), determined by HPLC. Experimental HPLC conditions: column: LiChrosorb RP-18 (7 µm, 250 x 4 mm I.D.), eluent: methanol/0.1 % sodium dodecyl sulfate (pH 4.0, 85:15), detection: absorption at 220 or 278 nm, flow rate: 1.0 ml/min.

Compound	Cleavage reagent	k(min^{-1})	50 % cleavage (min)	99,9 % cleavage (min)
Adpoc-Trp-OH	50 % HCOOH/TFEt (1:10)	0.026	26.35	140
Adpoc-Leu-Trp-OH	50 % HCOOH/TFEt (1:10)	0.024	28.9	200
Adpoc-Leu-Trp-OMe	50 % HCOOH/TFEt (1:9)	0.026	26.7	215
Adpoc-Trp-Lys-(Boc)-OH	50 % HCOOH/TFEt/ $CHCH_3$ (1:9:1)	0.013	53.3	340
	50 % HCOOH/TFEt (1:2)	0.028	25	180
Adpoc-Ala-Lys-(Boc)-Leu-OMe	50 % HCOOH/TFEt (1:10)	0.036	19.25	160

From table 4 it becomes obvious that the rate of cleavage of the Adpoc group depends on the chemical structure of the amino acid, the size of the peptide and the reagent. Examining the acidolytic cleavage of the Adpoc group from Adpoc-Trp-Lys(Boc)-OH in different acid concentrations one observes the following: The Adpoc group can be selectively cleaved from Adpoc-Trp-Lys-

(Boc)-OH without the formation of side products in 50 % HCOOH/ TFEt/$CHCl_3$ (1:9:1). If the concentration in acid is increased (50 % HCOOH/TFEt, 1:2) about 4 % of the Boc group are split off.

In connection with the synthesis of the natural sequence of thymopoietin II (residues 32-36), attempts have been made to acylate the N^{α}, N^{δ}, and N^{ω} functions of the N-terminal arginine. Z-Arg(Z_2)-OH was prepared in 30 % yield (17) and according to reference (18) arginine can be acylated with Adoc-OSu in nearly 100 %. This result, however, could not be reproduced by several authors (19). Adpoc fluoride, in a more than threefold excess, yields with arginine only a mixture of Adpoc-Arg(Adpoc)$_2$-OH and Adpoc-Arg(Adpoc)-OH, probably owing to steric hindrance of the voluminous Adpoc group. Therefore, further studies are necessary to examine the reaction conditions of Adpoc-F with arginine and to investigate the suitability of Adpoc-Arg(Adpoc)-OH for the application in peptide synthesis.

The Bzpoc group

It is well-known that the solvolysis of alkylhalogens R-X proceeds the faster the more the corresponding carbeniumion R^+ is stabilized. Acidic hydrolysis of urethans of the type R-O-CO-NH is influenced by the constitution of the residue R. Taking into account that the acidic cleavage is a monomolecular splitting of the alkyl-oxygen bond, the solvolysis is influenced by the constitution of the residue R, similar to that in alkylhalogens R-X. Based on this considerations we inserted stabilizing α,α-dimethyl groups in protecting groups.

Insertion of an isopropyl between the oxycarbonyl and benzyl group of the residue decreases the acidolytic stability to 1/1000 of the original value. The Bzpoc group can be removed under acidic conditions similar to the Boc group and is also cleaved by catalytic hydrogenation. These considerations are of interest for the development of new protecting groups (e.g. Adpoc, t-Bumeoc).

The starting material for the synthesis of the Bzpoc group is 2-benzyl-2-propanol which can be obtained e.g.by the following two routes:

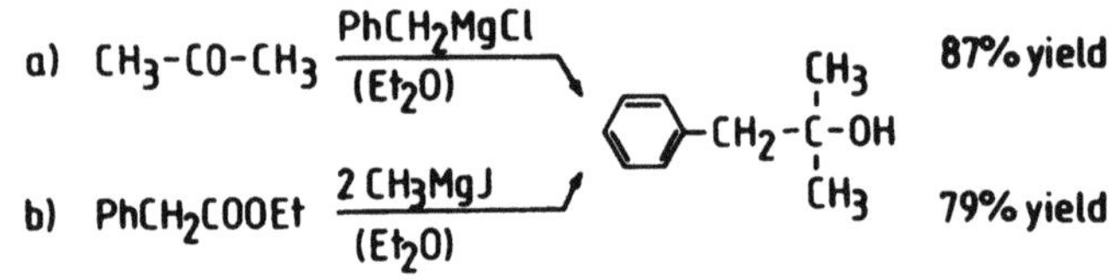

Purification of the tertiary alcohol is achieved by distillation (b.p.$_{10}$ 104-105°C, n_D: 1.5170). The colourless oil cristallizes at 0°C in long needles (mp. 24°C). Incorporation of the Bzpoc group can easily be performed by benzylisopropyl phenylcarbonate which is a cristalline stable acylation reagent. The benzyltrimethylammonium (Triton B) salts of the amino acids are reacted in absolute dimethylformamide with the phenylcarbonate according to the following scheme:

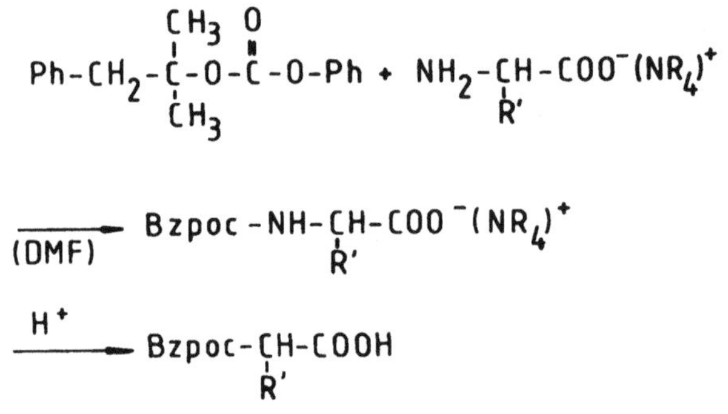

In Table 5 the physical data of some Bzpoc derivates and the yields of pure isolated material are collected.

Table 5. Synthetic yields and physical data of selected Bzpoc derivatives.

Bzpoc amino acid	Yield	mp (°C)	$[\alpha]_D^{23}$
Bzpoc-Gly-OH	72	110	-
Bzpoc-Ala.DCHA	82	129-130	-1.5° (c=1.6, MeOH)
Bzpoc-Phe	55	126	+18.5° (c=0.8, MeOH)

The Bzpoc group can completely be removed in 50 % TFA/CH_2Cl_2 within 4-5 minutes or in 85 % (w/w) formic acid within 4-7 hours (Table 6). Contrary to the Boc group, the Bzpoc residue can also be removed by catalytic hydrogenation. In tetrahydrofuran/water (80:20), deprotection is complete at pH 4-5 in about 60 minutes.

Table 6. Cleavage rates of Bzpoc and Boc amino acids in 85 % (g/g) formic acid .

Amino acid derivate	50 % cleavage	99 % cleavage	k_1 (min^{-1})	k_{rel} (Boc=1)
Bzpoc-Phe-OH	64	426	$1.08 . 10^{-2}$	0.5
Bzpoc-Gly-OH	52	346	$1.33 . 10^{-2}$	0.6
Boc-Phe-OH	31.5	209	$2.2 . 10^{-2}$	1

References

1. Wünsch,E.: Synthese von Peptiden, in: Houben-Weyl-Müller: Methoden der Organischen Chemie, Vol.XV/1, Thieme, Stuttgart 1974
2. Sieber,P., Iselin,B.: Helv.Chim.Acta 51, 622 (1968).
3. Birr,C.: Liebigs Ann.Chem.763, 162 (1972).

4. Sieber,P., Iselin,B.: Helv.Chim.Acta 51, 614 (1968).

5. Kyi, A.T., Schwyzer,R.: Helv.Chim.Acta 54, 1642 (1976).

6. Kalbacher,H., Voelter,W.: Angew.Chem.90, 998 (1978); Angew. Chem.Int.Ed.Engl.17, 944 (1978).

7. Kalbacher,H.: Thesis, Tübingen University, 1980

8. Voelter,W., Müller,J.: Liebigs Ann.Chem.248 (1983).

9. Müller,J., Voelter,W.: 2-(3,5-Di-tert-butylphenyl)-propyloxycarbonyl (t-Bumeoc) and 3,5-Di-tert-butylbenzyloxycarbonyl(DBZ) Amino Acids and their Use for Syntheses of Peptide Hormones, in: Biologically Active Principles of Natural Products, p. 176, Voelter,W., Daves,D.G. (eds), Georg Thieme Verlag, Stuttgart, New York, 1983

10. Voelter,W., Kalbacher,H.: 1-(1-Adamantyl)-1-methylethoxycarbonyl (Adpoc): A New Group for Amino Protection in Peptide Synthesis with Advantageous Properties, in: Peptides, Structure and Biological Function, p.325, Gross,E., Meienhofer,J.(eds), Pierce Chemical Company, Rockford, Illinois, 1979

11. Kalbacher,H., Voelter,W.: 1-(1-Adamantyl)-1-methylethoxycarbonyl (Adpoc): A new Useful Amino Protecting Group, in: Structure and Activity of Natural Peptides, p. 383, Voelter,W., Weitzel, G.(eds), Walter de Gruyter, Berlin, New York, 1981

12. Wollmann,H.: Thesis, Tübingen University, 1983

13. Rampold,G.,Lundanes,E., Folkers,K., Voelter, W., Kalbacher, H., Bliznakow,E.: Z.Naturforsch.35b, 1476 (1980).

14. Folkers,K., Rampold,G., Sakur,N., Leban,J., Dahmen,J., Lundanes,E., Ohta,M., Chang,D., Wan,Y.-P., Voelter,W., Kalbacher,H.: Research on Thymic Peptides and Immunocompetence in: Structure and Activity of Natural Peptides, p. 59, Voelter,W., Weitzel,G. (eds), Walter de Gruyter, Berlin, New York, 1981

15. Heinzel,W., Kronbach,T., Voelter, W.: Z.Naturforsch. 37b, 1652 (1982).

16. Heinzel,W., Voelter,W.: Liebigs Annalen, in press.

17. Zervas,L., Winitz,M., Greenstein,J.P.: J.Org.Chem.22, 1515, (1957).

18. Jäger,G., Geiger,R.: Chem.Ber.103, 1727 (1970).

19. Wünsch, E., private communication.

SYNTHESES AND REACTIONS OF PEPTIDE CYCLOLS

M. Fähnle and M. Rothe
Lehrstuhl Organische Chemie II, University of Ulm
D-7900 Ulm, Germany

Cyclols (acidic mono- or bicyclic ortho acid amides) are of interest as natural products (e.g. oxacyclols as part of the ergot alkaloids) as well as transient intermediates in peptide syntheses.

Special interest for tripeptide cyclols arose from the cyclotripeptide equilibrium suggested by Rothe et al. (1) (Equ. 1). Accordingly, the complete failure of isolation of cyclotripeptides containing secondary amide bonds (III) up to now is a consequence of the formation of tripeptide cyclols (I) and subsequent decomposition reactions. Conversely, the synthesis of tripeptide cyclols via aminoacyl diketopiperazines (II) represents a suitable approach to secondary cyclotripeptides containing one or more CO-NH groups under different conditions. Moreover, tripeptide cyclols seem to be important to a similar degree as tetrahedral adducts, the stereoelectronic properties of which have decisive influence on reaction paths and resulting products in enzymatic and non-enzymatic reactions of carboxylic acid derivatives (2,3).

Synthesis of tripeptide cyclols

In this work the preparation of tripeptide cyclols was performed for the first time both by direct cyclization of

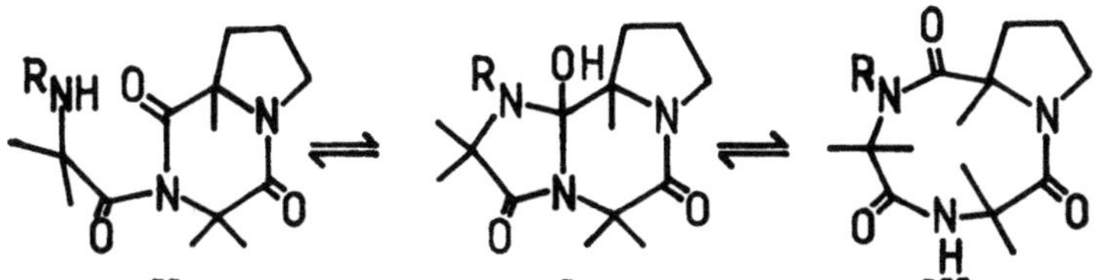

II I III (1)

Chemistry of Peptides and Proteins, Vol. 2

activated tripeptide sequences under usual conditions and by independent syntheses via N-protected aminoacyl diketopiperazines (II, R=Z) and N-protected tripeptide cyclols (I, R=Z) and subsequent removal of the protecting groups.
Cyclization reactions were carried out with sequence isomeric tripeptides containing N-terminal, C-terminal or internal proline (tripeptide active ester hydrobromides, c = 10^{-2} molar, DMF/TEA, room temperature). Together with oligomeric cyclopeptides containing up to ten tripeptide units, monomeric cyclic products were obtained only from tripeptide sequences with C-terminal proline (internal: any amino acid; N-terminal: proline or any amino acid except glycine; for examples compare table 1). This points to a formation mechanism of the cyclols via the isomeric aminoacyl diketopiperazines even under the conditions of peptide cyclization. The separation of the monomeric cyclic products from the cyclopeptide oligomers was achieved by GPC on Sephadex LH 20 in methanol.

Tripeptide cyclols Pro-L-X-Pro

In this way, stable cyclols or the isomeric cyclotripeptides with one CO-NH group were obtained from sequences Pro-L-X-Pro (Equ. 2), partially in good yields (X = Ala: ≈75%; Leu: ≈30%; Phe: ≈85%; Val: ≈25%). In all cases, the initially formed tripeptide cyclols (IV) proved to be in equilibrium with the corresponding cyclotripeptides (V) in protic solvents (methanol, water) or in solvents acting as hydrogen bond acceptors (DMF, DMSO). In agreement with the theory of stereoelectronic control of the cleavage of tetrahedral intermediates (2) there is no equilibrium with the corresponding aminoacyl diketopiperazines. Moreover, all-L-cyclotripeptides containing one secondary amide bond prove to exist in solution in two forms in equilibrium, the well-known crown conformation and a so far unknown and unexpected trans-cis-cis conformation. In crystalline state only the product of the sequence Pro-L-Phe-Pro is present as the cyclol (mp: 100 - 101°C), stabilized by one m.

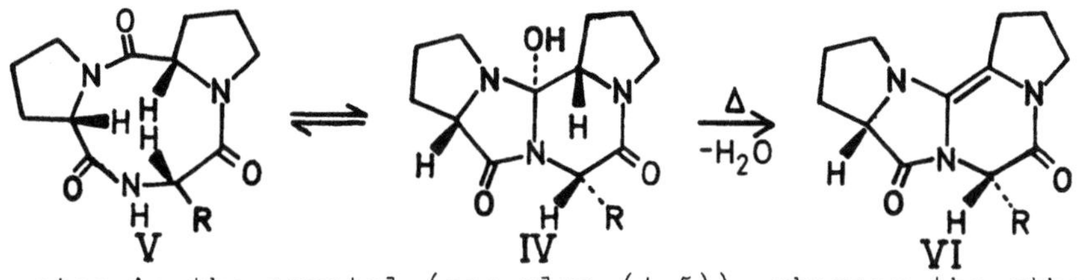

(2)

of water in the crystal (see also (4,5)), whereas the other representatives crystallize as secondary cyclotripeptides, c-(Pro-L-Leu-Pro) in the all-cis crown form, c-(Pro-L-Ala-Pro) and c-(Pro-L-Val-Pro) in the trans-cis-cis form (for more details see (6)). At room temperature all these compounds are stable in the solid state as well as in solution. At higher temperatures or in the presence of acids, however, they split off water to form ketene aminals (VI), which are further transformed to various oxidation products.

Tripeptide cyclols X-Y-Pro

Under corresponding conditions at high dilution the analogous tripeptide cyclols (VII), the isomers of secondary cyclotripeptides containing two CO-NH groups (VIII), were not stable. Instead of the cyclols, formed only as intermediates, complex mixtures of compounds were obtained (Equ. 3): a) peptide imidazolones (acylamidines, IX) by splitting off water, b) the corresponding peptide dehydroacylamidines (X) by air oxidation of (IX) and c) peptide dehydrocyclols (XI) by subsequent readdition of water to (X). These products could be separated by extractive procedures and reversed-phase chromatography. In addition, at every stage of the decomposition process further reactions occur. Oxidation reactions, addition of water or alcohols (during GPC and HPLC) and ring opening by nucleophilic attack at the carbonyl group of the imidazolinone ring could be detected. The composition of the monomeric fractions strongly corresponds with the oxidation sensitivity and the reactivity of the primarily formed acylamidines of different peptide sequences (cf. table 1). The identification of all compounds was performed by spectroscopical means (MS, IR, NMR) and by

comparison with the corresponding derivatives which were obtained by independent syntheses.
The starting compounds for these syntheses were available either by acylation of diketopiperazines with N-protected amino acid chlorides (1) or treatment of N-protected tripeptide p-nitrophenyl esters in carbonate-buffer systems (7). Dependent on size and configuration of the amino acid residues N-protected aminoacyl diketopiperazines (II, R=Z) or N-protected tripeptide cyclols (I, R=Z) were obtained.
By cleaving the protecting groups both from N-protected aminoacyl diketopiperazines and N-protected tripeptide cyclols under very mild conditions (H_2/Pd, -20°C, TEA) the free tripeptide cyclols (VII), the isomers of cyclic tripeptides with two CO-NH groups, could be prepared in pure, crystalline state.

Reactions of tripeptide cyclols X-Y-Pro

These tripeptide cyclols with a free NH group and an adjacent free OH group drastically differ in stability and reactivity from the N-protected derivatives and the N-alkylated tripeptide cyclols (IV) as well. At room temperature, they proved to be unstable in the solid state and particularly in solution but could be stored at -20°C for some time. Under chromatographic conditions (even during HPLC and TLC) total decomposition

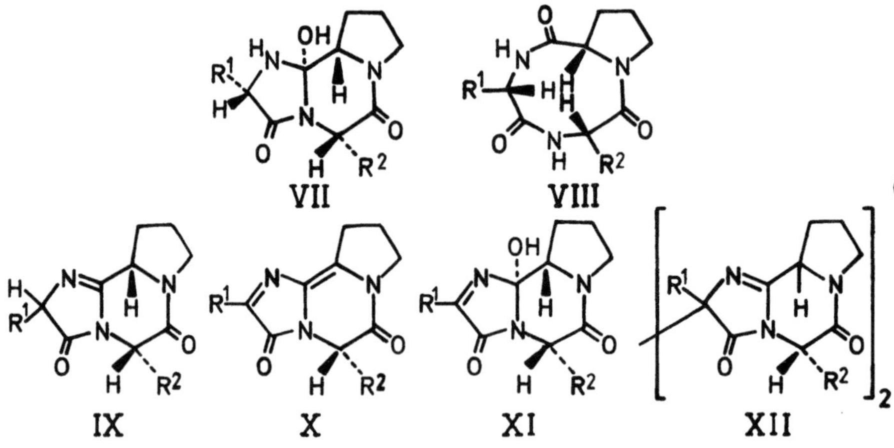

(3)

occurs. In the first step, it always leads to acylamidines (IX) by splitting off water. In pure state, these compounds were prepared directly from the Z-protected tripeptide cyclols by transfer hydrogenation (cyclohexene/Pd) or from the free tripeptide cyclols by thermal decomposition in inert solvents. Peptide acylamidines racemize due to a keto-enol analogous imidazolinone-hydroxyimidazole tautomerism at the chiral centre of the imidazolinone ring. Furthermore, they turned out to be extremly sensitive against oxidation. Thus, they are oxidized to dehydroacylamidines (X) by atmospheric oxygen both in solid state and in solution. From aqueous silver nitrate solutions elementary silver is precipitated immediately. Through oxidation by activated manganese dioxide on carbon an oxidative dimerization of the acylamidines takes place. The product (XII), formally consisting of two molecules of acylamidine symmetrically bridged by a C-C bond, thermally disproportionates to the starting acylamidine and the dehydroacylamidine, resp.

Table 1. Products of cyclization reactions (DMF/TEA, 0.01 molar, room temperature).

sequences	final products		
	monomeric[1]		oligomeric c-(tripeptide)$_n$
L-Ala-L-Ala-L-Pro	IX, X, XI, XII,	VII[2]	n=2-10,(11)
L-Ala-L-Pro-L-Ala	---		n=2-9,(10)
L-Pro-L-Ala-L-Ala	---		n=2,3,6
L-Ala-L-Phe-L-Pro	IX, X, XI, XII,	VII[2]	n=2-10,(11,12)
L-Ala-D-Phe-L-Pro	IX, X, (XI),	VII[2]	n=2-5
L-Ala- Gly-L-Pro	IX, X, XI		n=2-5
Gly-L-Ala-L-Pro	---		n=2,3,(4)
L-Pro-L-Phe-L-Pro	IV, (V[3]), VI		n=2
L-Pro-L-Ala-L-Pro	V, (IV[3]), VI		n=2-5
L-Pro-L-Leu-L-Pro	V, (IV[3]), VI		n=2-5
L-Pro-L-Val-L-Pro	V, (IV[3]), VI		n=2-7

[1] main products are underlined
[2] prepared by independent syntheses only
[3] proved to be present in equilibrium in polar solvents

The very reactive and unstable dehydroacylamidines (X) could be isolated in pure form only in small amounts. By addition of water under neutral or slightly acidic conditions they react to dehydrocyclols (XI), which were isolated as relatively stable end products of the decomposition process. Parallel to this main path and analogously as in the cyclization reactions, further addition of water or alcohols (during chromatography) to double bonds, further oxidation and ring opening occur. Even the dehydrocyclols are destroyed during standing in methanolic solutions for some time.

Acknowledgements

We gratefully acknowledge financial support by the Fonds der Chemischen Industrie. We also thank Dr. G. Schmidtberg for determining the mass spectra.

References

1. Rothe, M., Schindler, W., Pudill, R., Kostrzewa, M., Theyson, R., Steinberger, R.: Peptides 1971, Nesvadba, H., ed., North-Holland Publ. Co., Amsterdam, p. 388 (1973).
2. Deslongchamps, P.: Heterocycles 7, 1271 (1977) and references cited therein.
3. a) Bizzozero, S. A., Zweifel, B. O.: FEBS Letters 59, 105 (1975).
   b) Mock, W. L.: Bioorg. Chem. 4, 270 (1975).
4. The anhydrous product fails to crystallize. The obtained amorphous substance melts at 67 - 72°C (dec.); cf. Rothe, M., Fähnle, M., Mästle, W.: Proc. 7th American Peptide Symposium, Rich, D. H. (ed.), Madison-Wisconsin, (1981).
5. The same product was obtained by Lucente et al. by a different method in a yield of 8.5%; cf. Lucente, G., Pinnen, F., Zanotti, G., Cerrini, S., Fedeli, W., Gavuzzo, E.: Tetrahedron Lett. 22, 3671 (1981).
6. Rothe, M., Fähnle, M., Mästle, W.: these Proceedings.
7. Lucente, G., Romeo, A.: J. Chem. Soc., Chem. Commun. 1605 (1971).

SYNTHESIS OF CYCLIC TRIPEPTIDES CONTAINING CONH GROUPS

Manfred Rothe, Manfred Fähnle, Werner Mästle
Lehrstuhl Organische Chemie II, University of Ulm
D-7900 Ulm, Germany

Cyclic peptides have received increasing interest for stereochemical reasons in recent years. In particular, medium-sized rings with their inherent rigidity may serve as models for chain folding in proteins and for helical arrangements of peptide chains. Special importance is attributed to the 9-membered cyclic tripeptides which are supposed to show uncommon properties: 1) a small tendency of formation from open-chain peptides, 2) special physical properties due to cis- or even non-planar peptide bonds, and 3) transannular reactions between two amide groups across the ring.

For that reason, the synthesis of cyclotripeptides has been the goal of several groups since 1952 (1) - albeit without success. Instead, the corresponding cyclohexapeptides have been obtained by cyclodimerization in every case (2).

This has been attributed to a particularly high ring strain due to the occurrence of three rigid peptide bonds in a non-planar cis-configuration and to the mutual van der Waals repulsions of the inner α-hydrogen atoms. Moreover, an antiparallel association of two tripeptide molecules by hydrogen bonding will favour the formation of cyclohexapeptides because it resembles the transition state of cyclodimerization.

Although there are plausible objections against all these arguments such an antiparallel ß-type conformation was shown for several cyclohexapeptides by NMR spectroscopy (2).

From that the question arose whether or not cyclotripeptides could be obtained if an association by hydrogen bonding is excluded, e.g. by incorporation of N-alkylated amino acids.

Chemistry of Peptides and Proteins, Vol. 2

Following these ideas we succeeded in the synthesis of the first cyclotripeptide, c-tri-L-prolyl, in 1965 (3). In the meantime, several analogs containing only tert. peptide bonds of three N-alkylamino acids have been obtained. In all these cases the formation of cis-peptide bonds required for ring closure is facilitated because the cis and the trans conformations are energetically equally likely. Tripeptides with only two CON<u>R</u> and one CON<u>H</u> groups, e.g. Pro-Pro-Gly, also cannot associate by hydrogen bonding in the antiparallel fashion mentioned above. Nevertheless, cyclodimerization and even cyclooligomerization occurred yielding ring peptides with 2 - 8 tripeptide units. The same phenomenon was observed with nearly all tripeptide sequences containing CONH as well as CONR groups, e.g. with D-Pro-L-Pro-L-Pro, Sar-Pro-Pro, Ala-Ala-Pro, Ala-D-Phe-Pro, and Gly-Gly-Gly. The formation of large ring peptides with up to 90 ring atoms must have been overlooked by previous authors who had not applied GPC separation of the cyclization products.

From the high yields up to 90% in which tert. cyclotripeptides were obtained, it can be concluded that sec. cyclotripeptides should also be formed but had only failed to be isolated from the cyclization products so far. This could be due to conversion to secondary products by transannular reactions between the peptide groups across the ring. As an explanation (4) we proposed a tautomeric equilibrium between cyclotripeptides, peptide cyclols, and aminoacyl diketopiperazines, resp. (Equ. 1).

According to this suggestion azacyclols (II) were synthesized by an independent route (5). They proved to be extremely

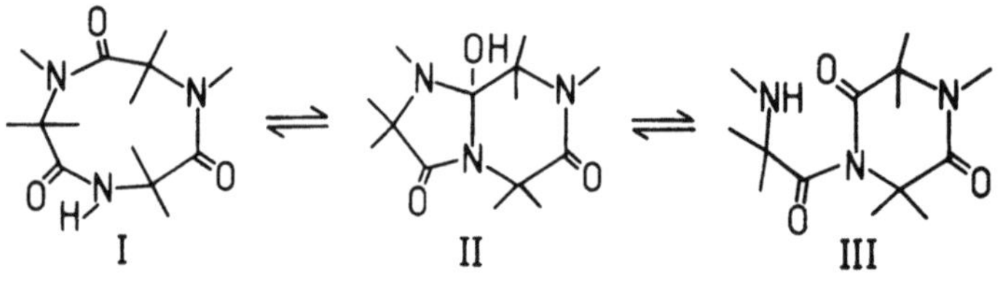

(1)

unstable and are easily decomposed with dehydration and oxidation to give dark oils containing several compounds of acylamidine and dehydroacylamidine structure.
Indeed, brown oils of similar composition are obtained as raw products from ring closure of tripeptides with CONH groups at extreme dilution which had been discarded up to now. In the case of the cyclization of active esters of Ala-Phe-Pro and Ala-D-Phe-Pro under the usual conditions (c=0.01 m, DMF/NEt_3, r.t.) a complex mixture of tripeptide ring oligomers and a monomeric fraction was formed which could be separated by GPC and HPLC. A peptide acylamidine was isolated and identified by spectroscopy and by comparison with an authentic sample synthesized independently. The detection of this compound shows the initial formation and subsequent dehydration of a cyclol during ring closure. Similar results have been obtained with other sequences containing two CONH groups. If, however, the acylamidines are more sensitive to air, e.g. those from Ala-Ala-Pro or Ala-Gly-Pro, the corresponding dehydrocyclols are formed as the more stable products.
Most interesting are tripeptides containing two proline residues at the N-terminal and the C-terminal position because dehydration to acylamidines is not possible in this case owing to the N-substituent (Equ. 2). These sequences can cyclize with particular ease under the normal ring-closure conditions to give aminoacyl diketopiperazines by reaction of the CONH group with the activated carboxylic group because an unstrained 6-membered ring is formed in place of a 9-membered one. These compounds spontaneously react intramolecularly by addition of the terminal amino group to the diacylimide group of the diketopiperazine to give the cyclol containing a new 5-membered ring.
A peptide cyclol, however, could only be isolated from the sequence Pro-Phe-Pro. In the other cases studied so far, cyclotripeptides containing a CONH group are obtained in crystalline form for the first time. c-(Pro-Ala-Pro) was isolated in 75% yield (m.p. 165-6°C), c-(Pro-Val-Pro) in 25% yield

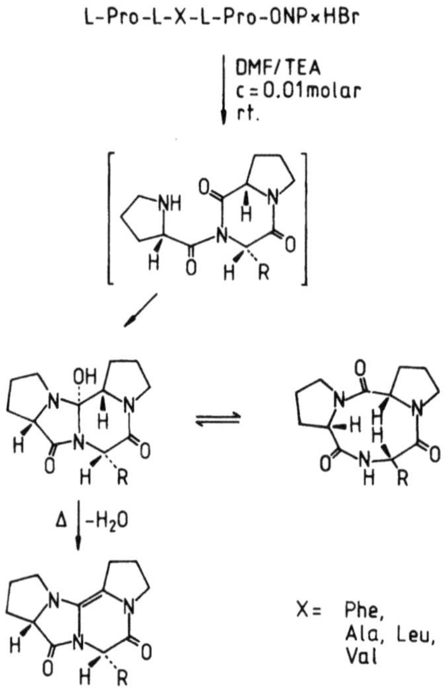

(2)

(m.p. 220-2°C), and c-(Pro-Leu-Pro) in 30% yield (m.p. 184-5°C). These compounds are more stable than the cyclols but are dehydrated at higher temperatures to give ketene aminals which in turn undergo further decomposition reactions.

The cyclotripeptide structures have been unequivocally determined by NMR, IR, and high resolution MS. In this respect, the detection of 3 carbonyl C-atoms in the ^{13}C-NMR spectrum at 170 -180 ppm is of decisive importance. On the other hand, the tetrahedral C-atom of the cyclols is found at about 95 ppm; the 2 carbonyl-C signals of the remaining 2 peptide bonds appear in the normal region. Cyclols and cyclotripeptides also differ in their mass spectra. In the former case a $(M-18)^+$ peak arises from dehydration to the ketene aminal, in the latter case a characteristic additional $(M-43)^+$ ion due to loss of HNCO is observed.

According to the NMR studies, an equilibrium between cyclotripeptides and cyclols is established in solution as has been suggested before (4). In polar solvents (water, methanol, DMF, DMSO) the cyclols are favoured. This may readily lead to decomposition of cyclotripeptides in solution and makes their preparation considerably difficult. Cyclotripeptides, however, crystallize from unpolar solvents, like ether. Furthermore, the cyclotripeptides mentioned above show a different conformational behaviour. In the solid state, only c-(Pro-Leu-Pro) exists in the well-known all-cis form as can be shown from the IR spectra because of the absence of the amide II band at 1550 cm^{-1}. Quite unexpectedly, c-(Pro-Ala-Pro) and c-(Pro-Val-Pro) crystallize in a trans-cis-cis form which is even favoured in solution with all the three cyclotripeptides. This form shows a distinct amide II band at 1515 cm^{-1} which was shifted to 1430 cm^{-1} on deuteration and is regarded as characteristic of the trans-peptide bond. This also follows from the NMR spectra which show an asymmetric distribution of the 3 carbonyl-C signals with the "trans" signal appearing at lower field. According to earlier studies by Kessler (6), the all-cis form corresponds to a crown conformation. Our new trans-cis-cis form, however, - which had been completely excluded for cyclotripeptides on steric reasons so far - is in agreement with a twist conformation. This twist form must also be the intermediate required for rearrangement of all-cis cyclotripeptides to cyclols.

A third conformation of the cyclotripeptides is the boat form which should be formed by incorporation of a D-amino acid (6). We therefore studied the cyclization of D-Phe-Pro-Pro by diphenylphosphoryl azide (7). In this case, ring closure cannot proceed via the cyclol but must occur by reaction of the chain ends with each other. Actually, a new sec. cyclotripeptide, c-(D-Phe-Pro-Pro), m.p. 176-8°C, was obtained in crystalline form in 23% yield. Again the structure determination was performed by high resolution MS, IR (before and after deuteration), and ^{13}C-NMR spectroscopy (showing 3 typical carbonyl-C signals).

This cyclotripeptide is more stable than the all-L analogs because the boat form can be transformed to the cyclol only with more difficulty on steric reasons.
The results mentioned above seem to be only valid for cyclotripeptides containing one CONH group. Proceeding to the analogs with two CONH groups we found that both the cyclols and the isomeric cyclotripeptides are much more unstable even at room temperature. Thus, the formation of the cyclotripeptide of Ala-Ala-Pro could only be detected by ^{13}C-NMR spectroscopy in aqueous bicarbonate solution. In water or organic solvents the equilibrium is shifted strongly in favour of the cyclol.

We gratefully acknowledge financial support by the Fonds der Chemischen Industrie. We also thank Dr. G. Schmidtberg for determining the mass spectra.

References

1. Boissonnas, R.A., Schumann, I.: Helv. Chim. Acta 35, 2229 (1952).

2. Schwyzer, R., Sieber, P.: Helv. Chim. Acta 41, 2186 (1958); Schwyzer, R., Ludescher, U.: ibid. 52, 2033 (1969).

3. Rothe, M., Steffen, K., Rothe, I.: Angew. Chem. 77, 347 (1965).

4. Rothe, M., Schindler, W., Pudill, R., Kostrzewa, M., Theysohn, R., Steinberger, R.: Peptides 1971, Nesvadba, H., ed., North-Holland Publ. Co., Amsterdam, p. 388 (1973).

5. Rothe, M., Fähnle, M., Pudill, R., Schindler, W.: Peptides, Gross, E., Meienhofer, J., eds., Pierce Chem. Co., Rockford, Ill., p. 285 (1979).

6. Kessler, H.: Stereodynamics of Molecular Systems, Sarma, R.H., ed., Pergamon Press, N.Y., p. 187 (1979).

7. Brady, S.F., Varga, S.L., Freidinger, R.M., Schwenk, D.A., Mendlowski, M., Holly, F.W., Veber, D.F.: J. Org. Chem. 44, 3101 (1979).

CONFORMATIONAL PREFERENCES OF MODEL PEPTIDES AND THEIR IMPACT IN PEPTIDE SYNTHESIS

F.Maser, K.Bode, B. Klein and M.Mutter
Institut für Organische Chemie, Johannes-Gutenberg-Universität, D-6500 Mainz, FRG

Introduction

Conformational studies on homo-oligopeptides revealed a strong dependence of the preferred conformation on the physico-chemical properties of peptides (1,2). As a general finding, β-structure formation results in a considerable decrease of solubility. From a practical point of view, the ability of predicting the secondary structure reliably would be the key to a proper planning of the synthetic pathways in peptide synthesis. In this context, however, an increased knowledge about conformational preferences of the amino acids and their derivatives under relevant experimental conditions is necessary.

To this end, we started a series of investigations using a newly designed guest-host model: various guest amino acids were inserted into a homo-oligopeptide (host peptide) as depicted in scheme I. L-Ala was chosen as host residue due to

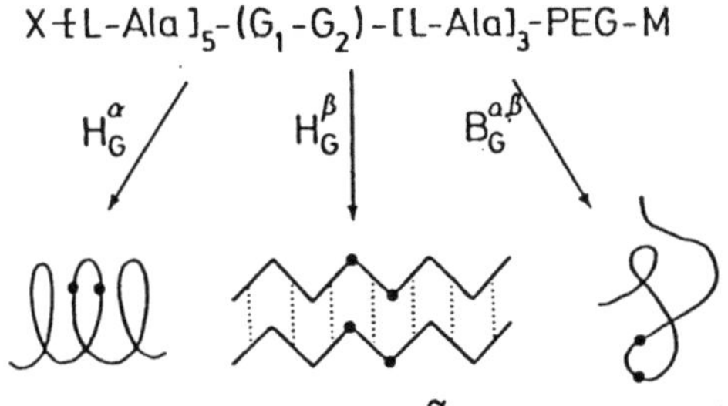

Scheme 1. Strong helix forming (H_G^α), β-structure forming (H_G^β) and ordered structures breaking ($B_G^{\alpha,\beta}$) effect of guest amino acids G.

Chemistry of Peptides and Proteins, Vol. 2

its potential for adopting helical and ß-structures. The insertion of guest amino acids was expected to result in conformational transitions, depending on the conformational preferences of the guest residue. This has been verified earlier for Gly and Pro as guest amino acids (3).

So far, experimental information about the ß-structure-forming tendency of individual amino acid residues is restricted mainly due to the generally low solubility of peptides taking up a pleated sheet structure. This problem has been overcome by attaching the growing peptide chain to a solubilizing C-terminal protecting group, e.g. polyethylene glycol (PEG) (4).

For the present study the following guest-host peptides were synthesized according to the liquid-phase peptide synthesis:

$$X\text{-}(L\text{-}Ala)_5\text{-}[G_1\text{-}G_2]\text{-}(L\text{-}Ala)_3\text{-}PEG\text{-}M$$

X = tert. butyloxycarbonyl (t-Boc); H_2^+

PEG-M = $HN\text{-}(CH_2CH_2O)_m\text{-}CH_2CH_2\text{-}O\text{-}CH_3$

I: $G_1 = G_2$ = L-Glu(OBzl)

II: $G_1 = G_2$ = L-Glu(Ot-Bu)

III: $G_1 = G_2$ = L-Lys(Z)

IV: $G_1 = G_2$ = L-Asp(OBzl)

V: $G_1 = G_2$ = L-Cys(Bzl)

VI: $G_1 = G_2$ = L-Val

VII: $G_1 = G_2$ = L-Ala .

The preferred conformation of I-VII was investigated in a number of solvents and in the solid state applying mainly circular dichroism (CD) and infrared (ir) techniques.

Results and Discussion

The dramatic influence the guest residues exert upon the host peptide is shown in Fig. 1. Whereas the homo-L-Ala_{10}- peptide adopts a pure β-structure, the insertion of two guest residues results in a disruption of this ordered structure.

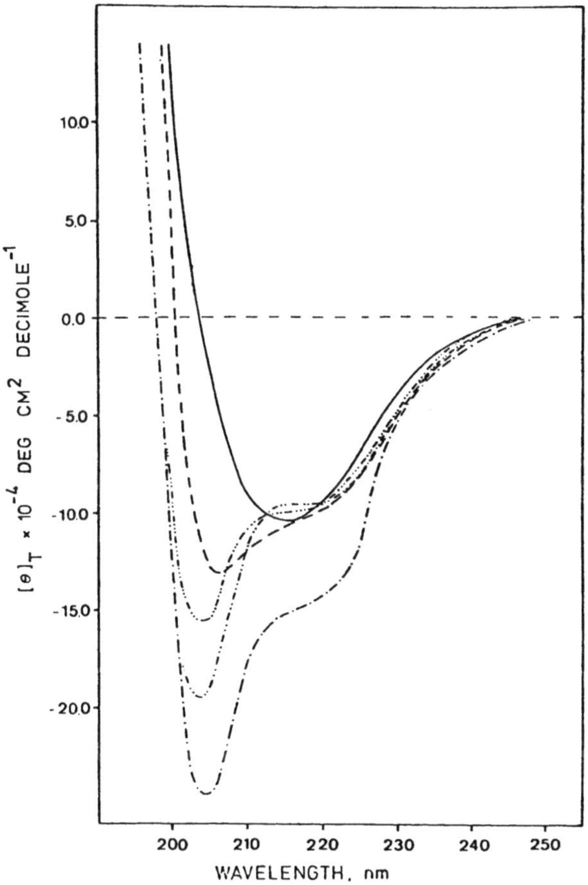

Figure 1. CD spectra of I (-·-), III (-··-), IV (-···-), V (- -), and VII (——) in TFE;(X = t-Boc, c = 1,1 mg of peptide/ml).

Depending on the helix-inducing power of the guest residues, amounts of α-helical structure up to ca. 50% (for compound I) are observed. For the L-Cys(Bzl)-containing peptide, a partial β-structure can be deduced from its CD spectrum.

The experimentally found scale for the helix-inducing power of the residues investigated in TFE would be as follows:

L-Glu(OBzl) > L-Lys(Z) > L-Asp(OBzl) > L-Cys(Bzl)

This scale is in harmony with values taken from the statistical occurrence of the side chain unprotected residues in proteins (prediction codes) (5,6). Most notably, the amount of helical structure of I (X = t-Boc) equals that of the corresponding homo-L-Glu(OBzl)-peptide, an observation pointing to the conformational dominance of the guest amino acids in the host peptide.

From the present result, the influence of the side chain upon the preferred conformation depends on the chemical nature and the distance of the side chain protecting group from the backbone. For example, only minor effects were found for III (X = t-Boc) compared to its N^{α}-deprotected form (not schown), whereas considerable differences in the α-helical nucleation were observed for the side chain protected and deprotected forms of I and IV (Fig. 2).

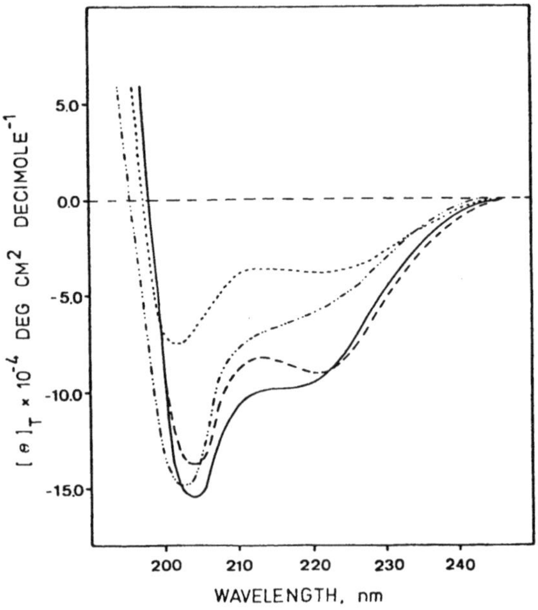

Figure 2.

Figure 2. CD spectra of IV, X = t-Boc, in TFE, (——); IV, X = t-Boc, in TFE/H_2O (1:1), (– –); IV, X = H^+, in TFE, (---); t-Boc-(L-Ala)$_5$-[L-Asp(OH)]$_2$-(L-Ala)$_3$-PEG-M, in TFE, (—··—); (c = 1.1 mg of peptide/ml).

Furthermore, the Bzl-group appears to diminish the tendency for β-structure-formation compared to the t-Bu-side chain protection (Fig. 3).

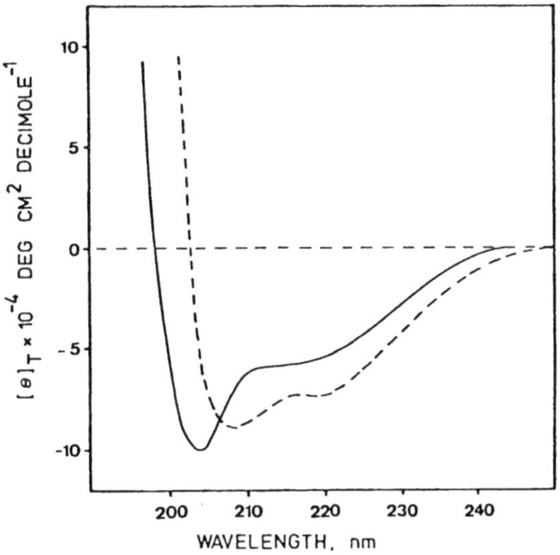

Figure 3. CD spectra of $Cl^-H_2^+$-(L-Ala)$_4$-[X]$_2$-(L-Ala)$_3$-PEG-M in TFE, (c = 1.1 mg of peptide/ml); X = Glu(OBzl) (——); X = Glu(OtBu), (---).

On the other hand, the influence exerted by the N-terminal protecting group of these peptides is generally more pronounced (compare Fig. 1 to Fig. 4). From the CD-curves, a tentative scale for β-structure formation of the guest amino acid residues investigated would be:

L-Cys(Bzl) > L-Glu(Obzl) > L-Lys(Z) > L-Asp(OBzl).

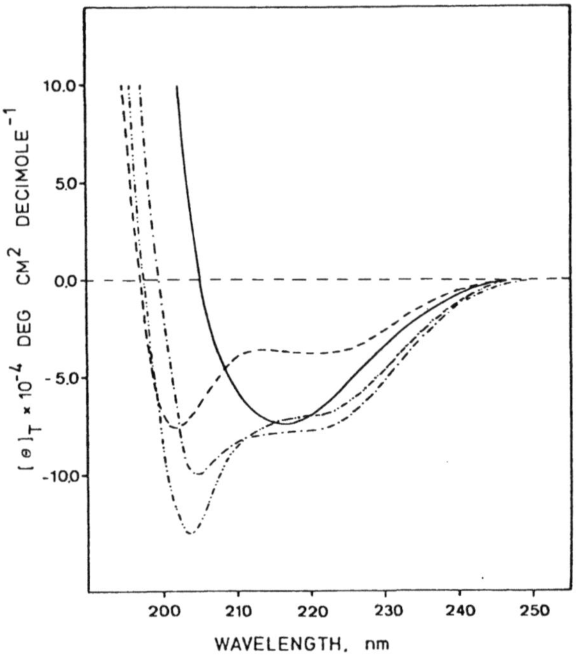

Figure 4. CD spectra of I (–·–), III (–··–), IV (– –), V (——) in TFE; X = H_2^+, (c = 1,1 mg of peptide/ml).

Similar tendencies are reflected by the critical chain lengths for ß-structure formation in the solid state as measured by ir. For example, the L-Cys(Bzl)-containing peptide adopts an antiparallel ß-structure in the N-protected form at a chain length of n = 4, whereas for the L-Asp(OBzl) analogues the critical n was found to be 6.

Most notably, the degree of disruption of a given ß-structure by a guest residue appears to be <u>independent</u> of its intrinsic potential for adopting a ß-structure. This fact is in contradiction to prediction rules, which are based on the assumption of the dominance of short range interactions resulting in neighbour -independent additive prediction schemes. For further clarification of this important finding, a guest residue with high β-forming potential (L-Val) was inserted into the host-L-Ala-peptide.

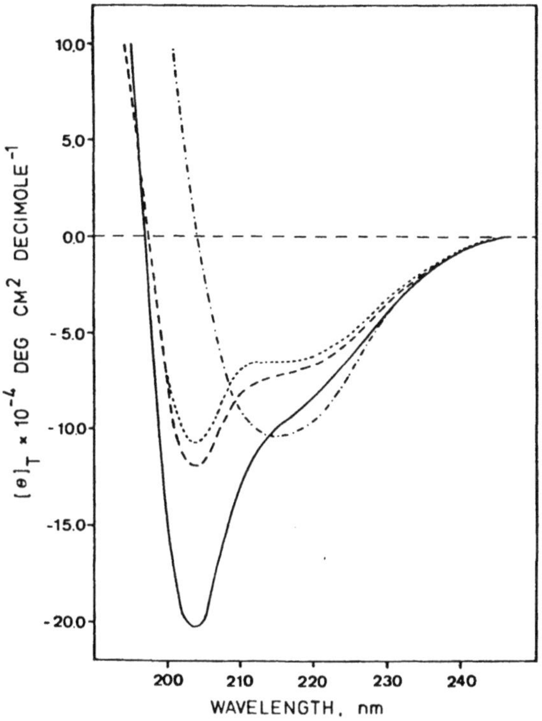

Figure 5. CD spectra of VI (——) and VII (-·-) in TFE, X = t-Boc, VI in TFE/45%H_2O, X = H_2^+ (----); VI in TFE, X = H_2^+ (— — —); (c = 1,1mg of peptide/ml).

Again, in contrast to prediction rules, the guest-host peptide shows considerably lower tendency to form a β-structure compared to the corresponding host-peptide, as shown in Fig. 5: Even the addition of 45% of H_2O to a TFE-solution of the N-deprotected form results in only a marginal increase of the content of β-structure.

The present finding might provide an explanation for the discrepancies between predicted and experimentally observed β-structure domains in proteins. Overpredicted regions may originate in the assembly of potentially β-structure-forming residues with very different chemical and spatial character of the side chains resulting in a loss of symmetry and, consequently, in a destabilization or even disruption of ß-structures. Therefore, theoretical approaches for predicting secondary structures which take into account medium- and long-range interactions (i.e. domain theories) should be more promising(7).

Another important conformational feature of proteins is the ß-turn. Model hexapeptides of the type Boc$(L\text{-}Ala)_2$-X-Y$(L\text{-}Ala)_2$-PEG, were investigated by means of ^{1}H NMR temperature- and concentration-dependent measurements of the chemical shift and by CD spectroscopy.

X and Y represent amino acid pairs with different probabilities to occupy the corner positions i+1 and i+2 of a chain reversal (8). The L-Pro-Gly pair proved to form a bend stabilized by a 14-membered ring system between Ala^2-N_H_ and Ala^5-C_O_. Interestingly, a similar conformational behavior was found for the repetitive hexapeptide of elastin, Boc-L-Val-L-Ala-L-Pro-Gly-L-Val-Gly-PEG. However, the stability of the turn developed by the elastin sequence proved to be considerably lower compared to the model peptide pointing to the stabilizing effect of the adjacent L-Ala blocks. The insertion of X = Gly and Y = L-Ile resulted in the formation of an associated conformation. The results on potentially ß-turn forming sequences obtained so far are in good qualitative agreement with prediction codes indicating that short range interactions play a dominant role in the folding process of proteins.

It was reported earlier (9) that conformational transitions occurring during chain elongation are often paralleled by dramatic changes in the physico-chemical properties of the peptide. The present study clearly supports this finding: As a general result, destabilizing effects on ß-structure formation increase the solubility and reactivity considerably. For example, the solubility of peptide VI turned out to be significantly higher compared to the corresponding homo-oligopeptide VII. Also, the formation of ß-turns is paralleled by an increase in solubility. Obviously, the delineation of the close relationship between conformational preferences and physicochemical properties of peptides opens up new perspectives for peptide synthesis.

This work was supported by the Deutsche Forschungsgemeinschaft through the Sonderforschungsbereich 41, "Chemie und Physik der Makromoleküle".

References

1) Toniolo, C., Bonora, G.M., Salardi, S. & Mutter, M.: Macromolecules 12, 620-625 (1979).

2) Toniolo, C., Bonora, G.M. & Mutter, M.: J. Am. Chem. Soc. 101, 450-454 (1979).

3) Toniolo, C., Bonora, G.M., Mutter, M. & Pillai, V.N.R.: Makromol. Chem. 182, 2007-2014 (1981).

4) Mutter, M. & Bayer, E. in: The Peptides: Analysis, Synthesis, Biology, Vol. 2, Gross, E. & Meienhofer, J., Eds , Academic Press, New York, p p. 285-332 (1980).

5) Chou, P.Y. & Fasman, G.D.: Adv. Enzymol. 47, 45-148 (1978).

6) Tanaka, S. & Scheraga, H.A.: Macromolecules 9, 168-182 (1976).

7) Argos, P. & Palan, J.: Int. J. Pept. Protein Res. 19, 380-393 (1982); Lim, V.I.: J. Mol. Biol. 88, 857-872 (1974).

8) Kolaskar, A.S., Ramabrahman, V. & Soman, K.V.: Int. J. Pept. Protein Res. 16, 1-11 (1980).

9) Mutter, M., Pillai, V.N.R., Anzinger, H., Bayer, E. & Toniolo, C. in: Peptides 1980, Proceedings of the Sixteenth European Peptide Symposium, Brunfeldt, K., Ed., Scriptor, Copenhagen, p p. 660-665 (1981).

ANALYTICAL STUDIES ON SIDE PRODUCTS IN THE SYNTHESIS OF SECRETIN

Walter Göhring, Regina Scharf and Erich Wünsch
Max-Planck-Institut für Biochemie, Abt. Peptidchemie,
8033 Martinsried, Germany

Christian Gespach
Centre de Recherches INSERM, Paris Saint-Antoine,
F-75571 Paris Cedex 12, France

Introduction

The hormone secretin has been studied for several years in our institute for its possible application as a promising drug in the case of ulceras or as a diagnostic of the pancreas in combination with pancreozymin related peptides. These studies resulted in a reproducible, standardized synthesis, followed by a simple purification via ion exchange chromatography on SP-Sephadex C-25 (1).

For the last fragment condensation step, i.e. coupling of Boc-[1-6]-OH to H-[7-27]-NH_2 both N-hydroxysuccinimide (HONSu) as proposed by us (2) and 1-hydroxybenzotriazole (HOBt) as proposed by König and Geiger (3) were used as additives for the dicyclohexylcarbodiimide method. The resulting secretin preparations (HONSu- and HOBt-secretin) exhibited a high degree of purity as judged from all the analytical tests performed and the biological activity was found to be slightly superior to that of the best natural secretin of Mutt (4).

Chemistry of Peptides and Proteins, Vol. 2

Results

New developments in hplc induced us to analyze natural and synthetic secretin materials on a Rad Pak C18 column. The elution profiles (fig. 1) of the crude HOBt- and HONSu-secretins obtained upon trifluoroacetic acid treatment of the fully protected heptacosapeptide amides indicate in the HOBt material a side peak with higher t_R, practically absent in the HONSu-preparation. Gas chromatographic analyses on a Chirasil-Val glass capillary column (5) of the two derivatized acid hydrolysates of the crude HONSu- and HOBt-secretins indicated a strong difference in the content of D-phenylalanine; whereas HOBt-secretin had 15 % D-Phe, HONSu-secretin was practically free of racemized phenylalanine. For the other amino acids the obtained values were within the limits of error of the method.

For comparative analysis D-Phe^6-secretin was prepared according to scheme 1. The crude product obtained after trifluoroacetic acid treatment exhibited at above mentioned hplc conditions a main peak with the identical retention time as the side product in the HOBt-material which was suspected to correspond to the D-Phe^6-diastereoisomer.

The HOBt-secretin material resulting from SP-Sephadex C-25 chromatography (for hplc see fig. 2) was further purified by partition chromatography on Sephadex G 25 using 2-butanol/0.15M ammonium acetate/ethyl acetate 5:5:0.68 (v:v) as eluent. The resulting material behaved homogeneously in hplc (fig. 3) and in hptlc (fig. 4). In the racemization test of this purified secretin no remarkable amounts of D-amino acids could be observed, whereas for the side product with higher t_R (enriched to ca. 70 % according to the integrated areas of the hplc-peaks) a content of 66 % D-Phe was determined by gas chromatography on a Chirasil-Val-column. This result ultimately proofs the nature of the HOBt-secretin side product.

Both the HOBt- and the HONSu-secretin revealed an additional impurity with a shorter t_R on hplc as the natural material.

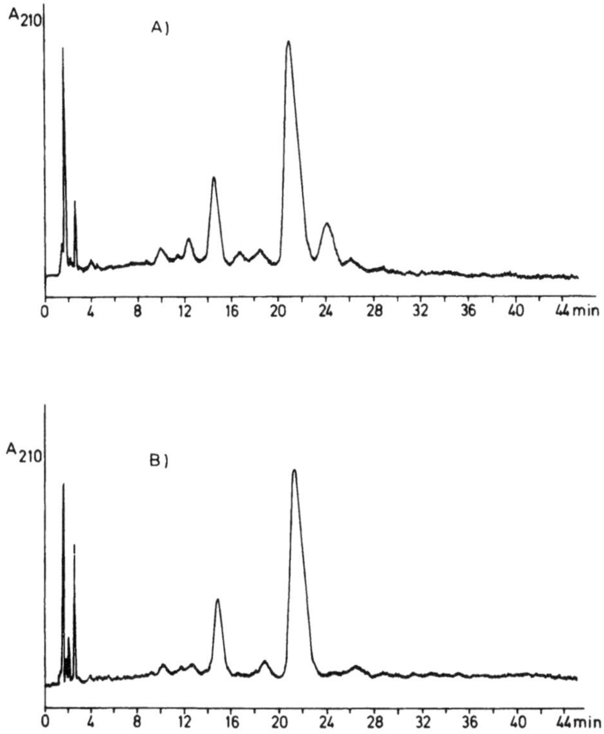

Fig. 1. Hplc of crude secretin preparations: A) HOBt-secretin, B) HONSu-secretin; column: Rad Pak C18; eluent: acetonitrile/ 0.13 M dibutylammonium trifluoroacetate (pH 2.6); 31 : 69 (v:v); flow rate: 2ml/min.

The amino acid analysis of the acid hydrolysate of this impurity (again enriched to ca. 70 % by above mentioned partition chromatography) gave no indication about its chemical structure. In the AP-M-digest however, the data for Ser and Arg differed slightly from the theoretically expected values. To elucidate the structure of this side product tryptic digests of the contaminant, the purified secretin and the natural se-

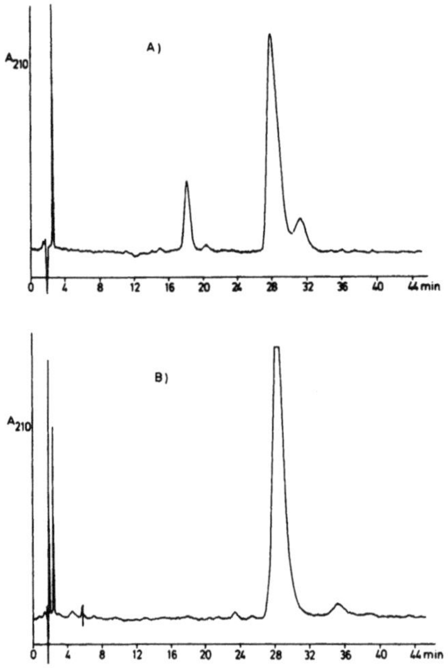

Fig. 2. Hplc of secretin: A) HOBt-secretin, B) natural secretin; column: Rad Pak C18; eluent: acetonitrile/0.13M dibutylammonium trifluoroacetate (pH 2.6); 31:69 (v:v); flow rate: 2 ml/min.

Adoc-His(Adoc)-Ser(Bu^t)-Asp(OBu^t)-Gly-Thr(Bu^t)-D-Phe-OH [1-6 a]

GC-test: 99 % D-Phe

[1-6 a] + H-Thr(Bu^t)-Ser(Bu^t)-Glu(OBu^t)-Leu-Ser(Bu^t)-Arg(HBr)-Leu-Arg(HBr)-Asp(OBu^t)-Ser(Bu^t)-Ala-Arg(HBr)-Leu-Gln-Arg(HBr)-Leu-Leu-Gln-Gly-Leu-Val-NH_2

↓ DCC/HONSu

Adoc-His(Adoc)-Ser(Bu^t)-Asp(OBu^t)-Gly-Thr(Bu^t)-D-Phe-Thr(Bu^t)-Ser(Bu^t)-Glu(OBu^t)-Leu-Ser(Bu^t)-Arg(HBr)-Leu-Arg(HBr)-Asp(OBu^t)-Ser(Bu^t)-Ala-Arg(HBr)-Leu-Gln-Arg(HBr)-Leu-Leu-Gln-Gly-Leu-Val-NH_2

↓ trifluoroacetic acid/ethanedithiol

crude D-Phe^6-secretin

GC-test: 91.3 % D-Phe

Scheme 1. Synthetic route to D-Phe^6-secretin.

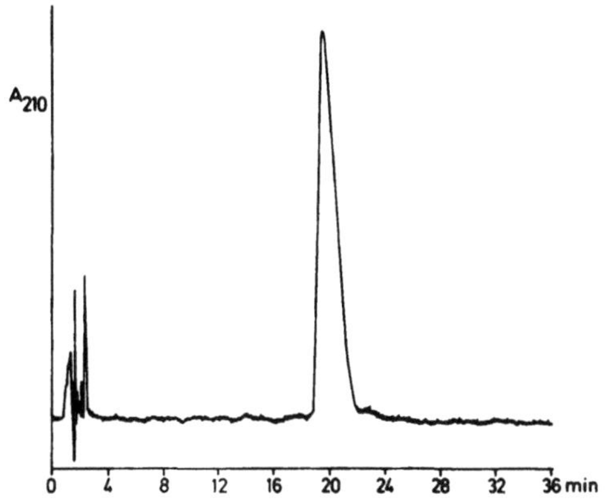

Fig. 3. Hplc of purified secretin:
column: Rad Pak C18; eluent: acetonitrile/0.13 M dibutylammonium triflouroacetate (pH 2.6); 31:69 (v:v); flow rate: 2 ml/min.

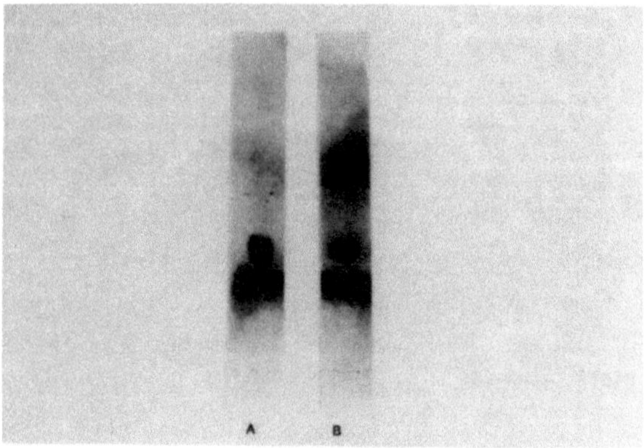

Fig. 4. hptlc of A) synthetic secretin (HOBt-secretin) and B) purified secretin; solvent: n-butanol/acetic acid pyridine/water, 45:10:15:20; chlorine/tolidine; time: 15 h.

cretin were performed. The resulting tryptic fragments were examined by hplc and tlc (fig. 5) whereby the only significant difference was the disappearance of a band (with the lowest R_f-value) in the chromatogram of the digested side product. By comparative tlc of the fragments 22-27 and 19-21 as well as of the tryptic digests of secretin and its partial sequence 5-27 with additional micropreparative tlc and subsequent amino acid analysis of the isolated bands, it has been possible to locate the modification in the sequence 15-18, i.e. aspartyl-seryl-alanyl-arginin.

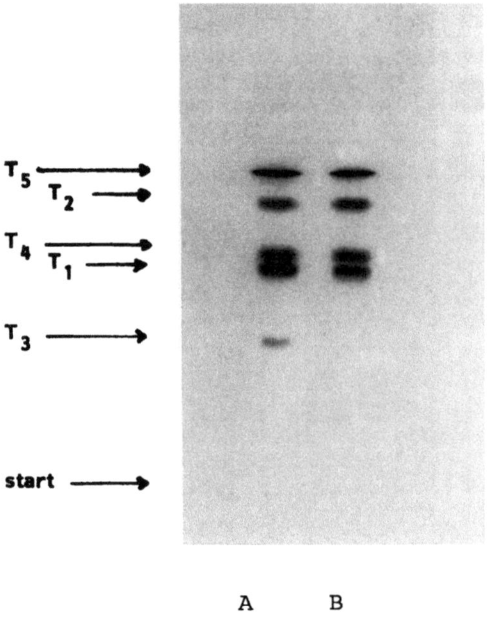

Fig. 5. hptlc of the tryptic fragments of A) secretin and B) secretin side product; (solvent: n-butanol/acetic acid/pyridine/water, 45:10:15:20)

T_1: H-His-Ser-Asp-Gly-Thr-Phe-Thr-Ser-Glu-Leu-Ser-Arg-OH [1-12]
T_2: H-Leu-Arg-OH [13-14]
T_3: H-Asp-Ser-Ala-Arg-OH [15-18]
T_4: H-Leu-Gln-Arg-OH [19-21]
T_5: H-Leu-Leu-Gln-Gly-Leu-Val-NH_2 [22-27].

Previous studies on the stability of secretin have clearly confirmed the high tendency of the aspartyl-glycine sequence in position 3 and 4 to cyclize in slightly acidic conditions to the corresponding succinimide derivative. This unstable intermediate reopens with concomitant α, ß-transpeptidation in a ratio of about 8 to 2 (6,7). These findings strongly suggest that a similar transpeptidation may occur even at the level of the Asp-Ser sequence (positions 15-16).
Adenylate cyclase activity in rat fundic glands according to the procedure of Gespach et al. (8) indicates a very low potency for both contaminants. Surprisingly the HOBt-secretin preparations prior and after the additional purification by partition chromatography were equipotent in this in vitro assay (fig. 6).

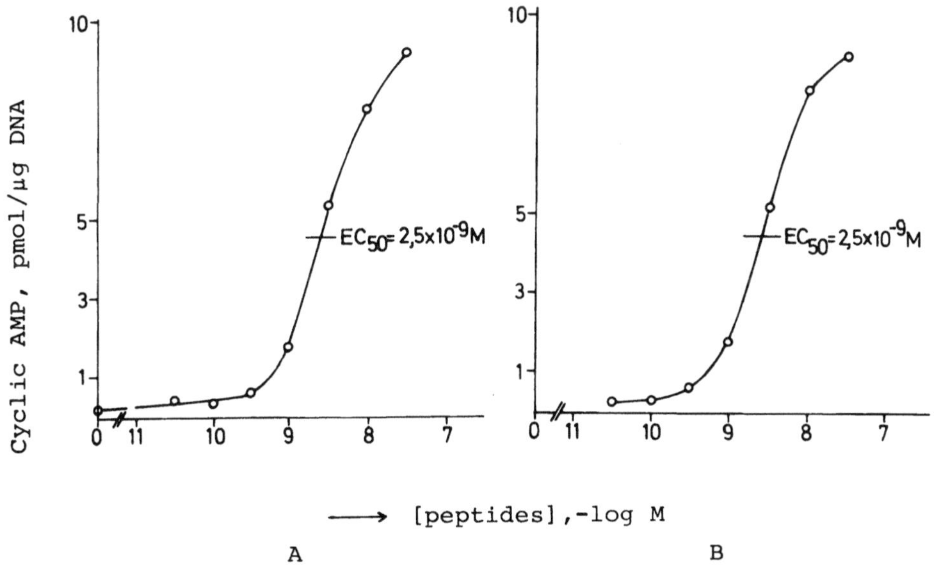

Fig. 6. Adenylate cyclase activity in rat fundic glands: A) HOBt-secretin, B) pure secretin.

References

1. Wünsch, E., Jaeger, E., Deffner, M., Scharf, R., Lehnert, P.: Chem. Ber. 105, 2515-2522 (1972).
2. Wünsch, E., Drees, F.: Chem. Ber. 100, 816-819 (1967).
3. König, W., Geiger, R.: Chem. Ber. 103, 788-798 (1970).
4. Wünsch, E.: Naturwissenschaften 59, 239-246 (1972).
5. Frank, H., Nicholson, G.J., Bayer, E.: J. Chromatogr. Sci. 15 174-176 (1977).
6. Jaeger, E., Knof, S., Scharf, R., Lehnert, P., Schulz, I., Wünsch, E.: Scand. Gastroenterol. 13 (Suppl. 49), 93 (1978).
7. Beyermann, H.C., Grossman, M.I., Solomon, T.E., Voskamp, D.: Life Sciences 29, 885-894 (1981).
8. Gespach, C., Bataille, D., Dupont, C., Rosselin, G., Wünsch, E., Jaeger, E.: Biochim. Biophys. Acta 630, 433-441 (1980).

For this study the financial support of Fonds der Chemischen Industrie is gratefully noted by one of the authors (E.W.).

HIGH PERFORMANCE LIQUID CHROMATOGRAPHY OF PROTECTED PEPTIDES

V.V.Ylyashin

Shemyakin Institute of Bioorganic Chemistry, USSR Academy of Sciences, Moscow, USSR

V.I.Deigin

Institute of Protein Research, USSR Academy of Sciences, Puschino, Moscow Region, USSR

During the synthesis of the 61-membered polypeptide, neurotoxin II from the Central Asian *Naja naja oxiana* cobra venom, (see also this volume) we encountered some difficulties in controlling the reaction kinetics and testing the purity of the compounds obtained as well as in their preparative purification. Such common methods of peptide chemistry, as thin-layer and open column chromatography, were of little use here, because of their low efficiency. Therefore we looked for more effective methods and of course "high performance liqiud chromatography" (HPLC) was selected as a method of choice.

For peptides highly soluble in nonpolar organic solvents adsorption chromatography on nonmodified silica gel was successfully applied. Peptides with unprotected carboxyl groups give asymetric peaks under these conditions (Fig. 1). Besides the loading strongly influences retention times. This is apparently connected with nonlinearity of the adsorption isoterm. In some cases this effect can be suppressed by using complex systems of solvents or gradient elution.

Applicabily of the reverse-phase chromatography is limited by poor solubility of peptides in aqueous solvents, normally used as eluents. Some better soluble peptides could be eluted with aqueous methanol or acetonitrile systems (Fig. 2). Peptides, soluble only in dimetnylformamide or in dimethylsulfoxide, have

Chemistry of Peptides and Proteins, Vol. 2

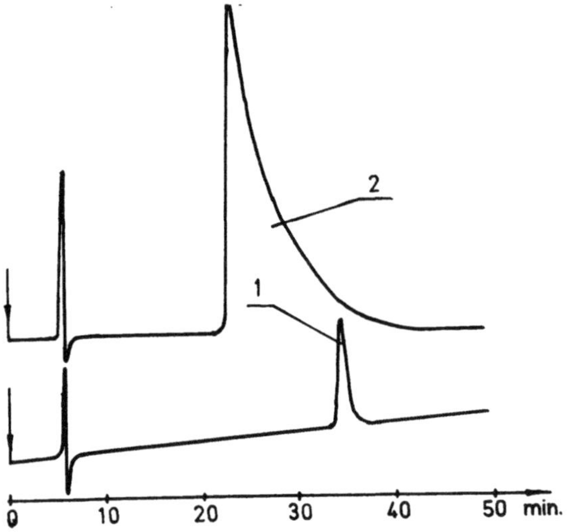

Fig. 1. Analytical (1) and preparative (2) chromatography of Boc-Cys(Acm)-Gly-Cys(Acm)-Pro-OH on a silica gel L column (330x25mm I.D.). Flow-rate 20 ml/min. Eluent $CHCl_3$:CH_3OH:CH_3COOH=9:1:0,5, Detector UV 280 nm.

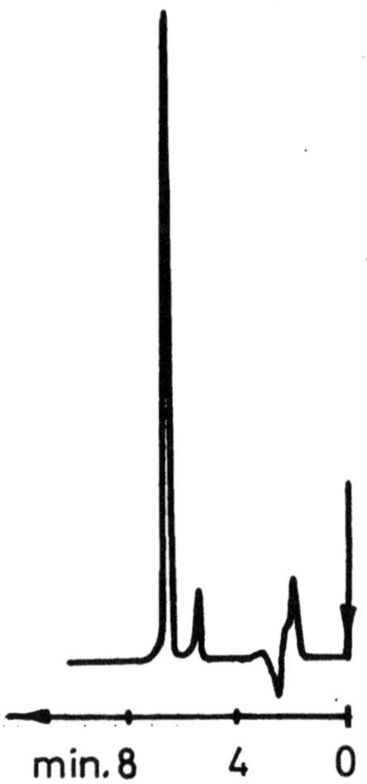

Fig. 2. Chromatography of Boc-Gly-Val-Asn(Bzh)-Leu-OH on a Zorbax ODS column (250x4,6mm I.D.). Flow-rate 1 ml/min. Eluent 60% CH_3CN in H_2O, Detector UV 254 nm.

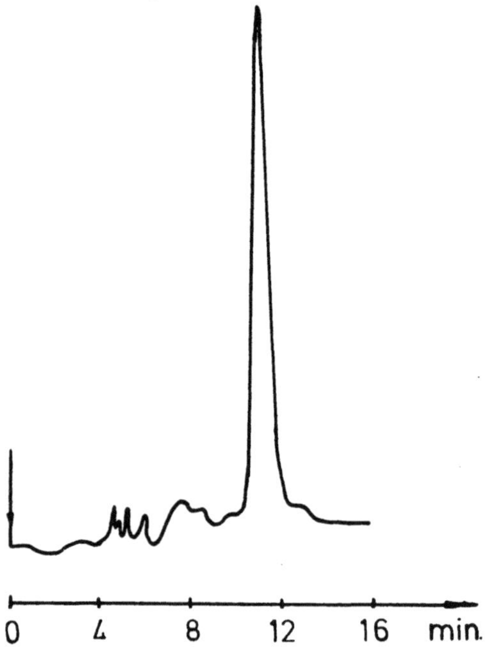

Fig. 3. Chromatography of fragment 52-61 on a KCK C-8 column (250x7 mm I.D.). Flow-rate 1,5 ml/min. Eluent 18% H_2O in DMF, Detector UV 280 nm.

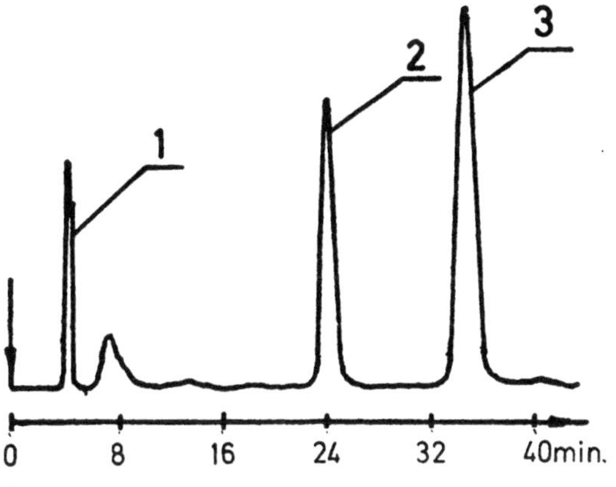

Fig. 4. Separation of 55-61 fragment with unprotected N-terminal amino group (2) and with Boc-protection of N-termini (3). Peak 1 belongs to solvent. Zorbax ODS column (250x4,6mm I.D.). Flow-rate 0,5 ml/min 18% H_2O in DMF. Detection UV 280 nm.

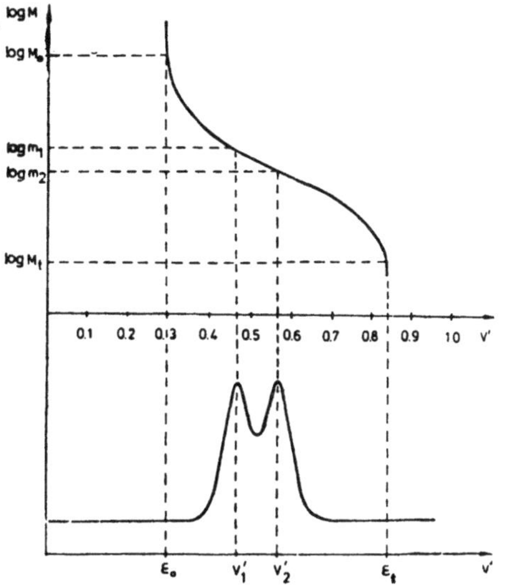

Fig. 5. The calibration plot and its parameters.

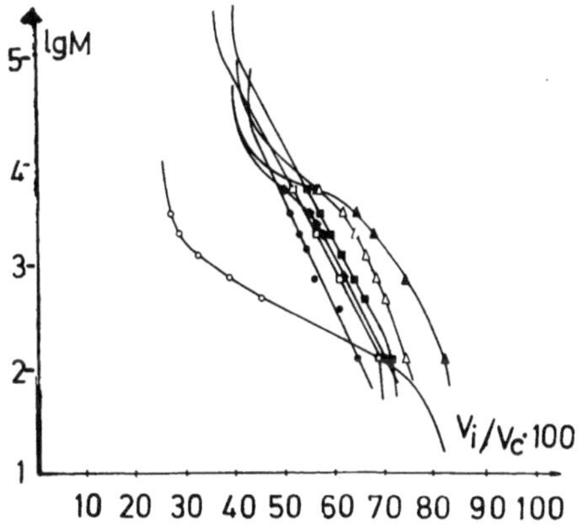

Fig. 6. Calibration plots for synthetic protected peptides in DMF. -o- Sephadex LH-20; -●- Si-40; -□- PSM-60S; ■-Zorbax Sil; -◆-silica gel L; -△- Silasorb 600; -▲- spherical silica gel.

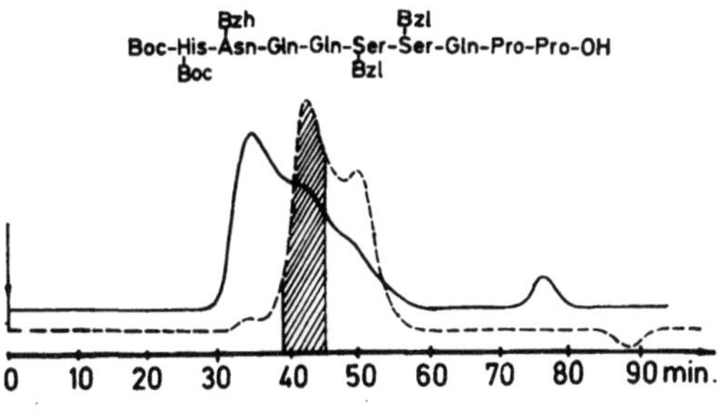

Fig. 7. Preparative chromatography of Boc-4-12-OBu^t fragment on Sephadex LH-20 column (1200x16mm I.D.). Flow rate 2 ml/min. Solid lines - UV 280 nm, broken lines - refractive index, detector DMF.

to be eluted with on aqueous dimethylformamide (Fig. 3). Peptides with unprotected amino and carboxyl group also can be chromatographed under reverse-phase conditions. In this case retention time is shorter than for fully protected peptides (Fig. 4). High selectivity and reproducibility allows to follow the coupling kinetics and elimination of protective groups. We succeeded in monitoring the purity of the majority of neurotoxin II fragments under reverse-phase conditions.

The major part of this paper will be devoted to novel developments in size exclusion high perfomance liquid chromatography (SE HPLC). SE chromatography differs in a number of aspects from the absorption chromatography. Separation can be carried out in any solvent which would dissolve a substance, and that is the main advantage of this method. Besides, information on molecular masses of the separated substances can be obtained directly from the recording chart. The loading value is especially important for preparative procedures. In size exclusion conditions loading capacity is about 10 times higher than in adsorption chromatography, without loss of its efficiency. Special columns and packings are needed for operating at the SE HPLC conditions that is governed by pore structure of the packing. To analyze the factors influencing separation by SE chromatography, we propose to use the term Rs (resolution) determined by formula(1):

$$R_s = 0{,}25 \cdot \log\,(m_2/m_1) \cdot S \cdot \frac{\varepsilon_t - \varepsilon_o}{\varepsilon_t} \cdot \sqrt{N} \qquad (1)$$

m_1 and m_2 are molecular masses of the separated substances, ε_o and ε_t are the free and total volums of the column (Fig. 5). N is the column efficiency, S=packing selectivity calculated according to equation:

$$S = \frac{1}{(\varepsilon_t - \varepsilon_o) \cdot \tan\alpha} \qquad (2)$$

where α is the slope of the calibration plot.

According to equation of the resolution (1) the separating ability depends not only on the packing selectivity, but also on porosity parameters ε_o and ε_t. Therefore in order to compare various packing, we introduce a novel term, the characteristic efficiency of packing, N_e.

$$N_e = 180 \cdot \left[\frac{\varepsilon_t}{(\varepsilon_t - \varepsilon_o) \cdot S} \right]^2 \quad (3).$$

This value depends upon the parameters S, and ε_o and ε_t and is characteristic of packing properties. Numerial value of N_e is determined as the efficiency value necessary for separation of two substances with a two-fold difference in molecular masses.

Using equations (3) and (1), one can obtain the following equation for resolution value:

$$R_s = 3.32 \cdot \log\,(m_2/m_1) \cdot \sqrt{\frac{N}{N_e}} \quad (4).$$

According to equations (1) and (4), in order to obtain the maximal resolution of two peaks, one needs not only high-efficiency columns (N), but also a packing, of possibly higher selectivity (S) and with possibly wider range from ε_o to ε_t. In other words, one should choose packings with minimal characteristic efficiency (N_e).

With these considerations in mind, we studied the applicability of various gel packings for SE HPLC of protected peptides. Soft packing were presented by Sephadex LH-20, LH-60 and Enzacryl K-2. The best properties were found for Sephadex LH-20. This gel is easily packed into the column with efficiency 8.000 t.p. per meter. Calibration plot of Sephadex LH-20 for peptides in DMF is nonlinear (Fig. 6). Therefore, characteristic efficiency for various peptides is different. For peptides with 300 d molecular masses N_e is 620 t.p., and for peptides with 1000 d. N_e is somewhat higher and equal to 1000 t.p.. The upper exclusion limit for Sephadex LH-20 is 25000 d. Mechanical properties of this packing allow separation at relatively high velocity of solvent flow (Fig. 7).

Partial adsorption of the material is observed upon chromatography of peptides with unprotected amino group or tyrosin-and tryptophan-containing peptides .
Sephadex LH-60 is not suitable for SE HPLC, because of its poor mechanical properties. With this packing one has to work with very low velocity and with large particles of about 200 microns. Enzakryl K-2 is not suitable for chromatography of peptides, too, because peptides containing benzyl aspartate undergo transpeptidation, which was obviously catalyzed by basic moieties of the gel template.
Of semi-rigid gels we studied Spherons P-40, P-300 and P-1000. These gels possess high mechanical stability, permitting high velocity of eluent. High-efficiency columns can be packed with these gels. However, Spheron P-40 has a low coefficient of mass transfer (Fig. 8). If the size of the peptide is comparable with that for gel pores, it shows 3 times broader peaks than the peak of toluene freely penetrating the gel pores and than the peak of polystyrene unpenetrating gel pores. It means that there is no resistance to mass transfer by the gel template for these molecules. In addition Spheron P-300 and P-1000 have very low selectivity. N_e of Spheron P-300 is about 15.000 t.p.. Due to these reasons Spherons did not find a broad application in this study.
Of rigid gels we investigated various types of silica gels (Fig. 6). Silica gel Si-40 from "Merck", Zorbax and PSM-60S from "Du Pont" have linear calibration plots which facilitates calculation of the efficiency needed for separation and determination of molecular masses of the compounds analyzed (equation 4). The parameters of PSM-60S and Zorbax Sil are very similar; the N_e for these packing is 6500 t.p. Silica gel Si-40 is less selective and its operating area is narrower. The N_e of this silica gel is 10.000 t.p.. The orther silica gels, the home-made spherical silica gel, Silasorb 600 and silica gel "L" from "Lachema" showed non-linear calibration plots. The best selectivity was found for the spherical silica gel obtained

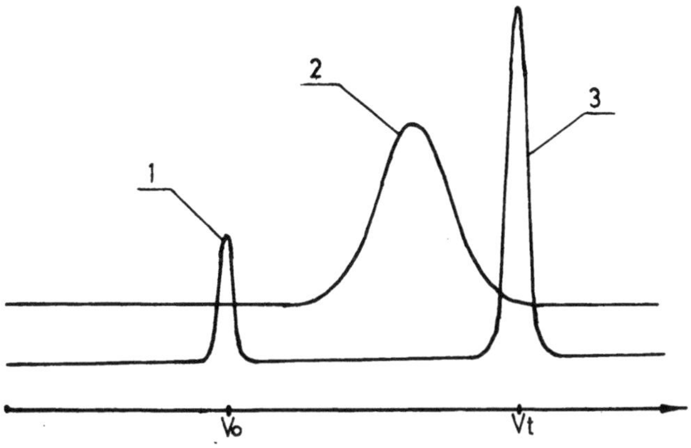

Fig. 8. Chromatograms of a mixture of polystyrene, m.m.=$2 \cdot 10^6$ d (1), toluene, m.m.=93 d (3) and of Aoc-Arg(Toc)-Cys(Acm)-Asn(Bzh)$_2$-Asn(Bzh)--OBzl, m.m. 1265 (2) on a Spheron P-40 column (600x16mm ID) in DMF. Flow-rate 3 ml/min. Detection UV 280 nm.

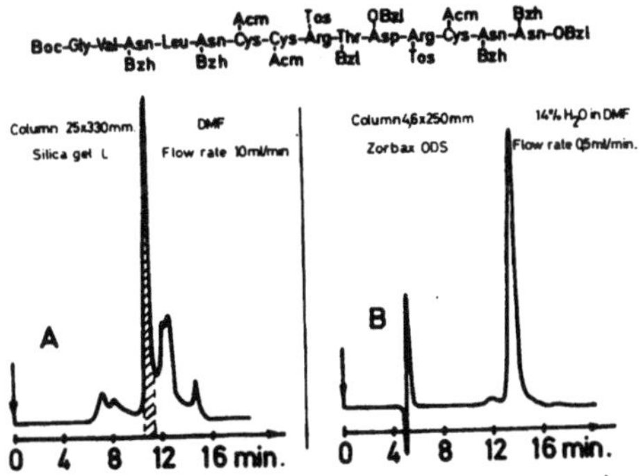

Fig. 9. Preparative size-exclusion separation of 48-61 fragment (A) and reverse-phase analitical chromatography (B) of isolated fraction from A. (Cross-hatched area).

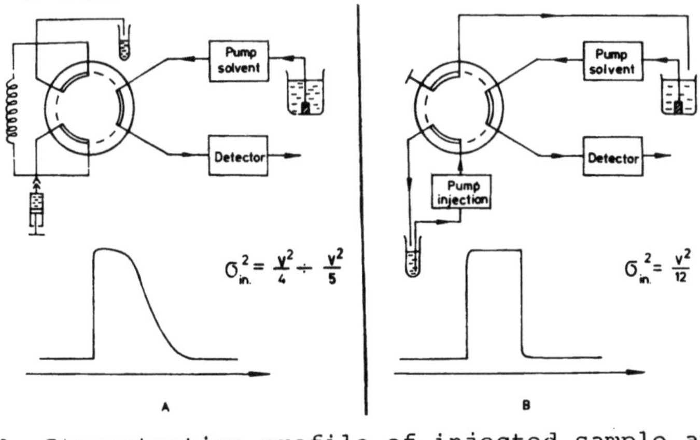

Fig. 10. Concentration profile of injected sample and diagram of pump connection, sample injection and detector sites for loop (A) and pump (B), and its contribution to the dispersion.

by the Unger's method. N_e for peptides with nearly 4000 d molecular masses is about 1100 t.p.. The fully protected pe-peptides in DMF on unmodified silica gel were eluted according to the pure size exclusion mechanism. Peptides with unprotected His or with unprotected amino groups are partially adsorbed and are eluted later then the fully protected peptides of the same molecular mass.

Generally reverse-phase chromatography allows to reach higher selectivity than SE chromatography. Nonetheless, peptides purified under size-exclusion condition were usually quite pure under the reverse-phase conditions (Fig. 9).

The injection technique of a substance into the column is very important for preparative purification of peptides. We investigated two ways of the sample injection into the column (Fig. 10), the widely used "loop injection" (scheme A) and the "pump injection" (scheme b). Fig. 10 shows that only at the pump injection the rectangular concentration profile is obtained. Contribution to the dispersion in case of the pump injection is 3 times less than for the loop injection i.e. in this way one can inject a sample with 1,7-fold larger volume than with loop injection with the same broadining of the peak. Column loading can be increased as well as concentration of the injected sample. However, at higher concentrations the resolution considerably decreases due to increasing viscosity. This effect can be suppressed by diluting of the sample with a non-viscous solvent such as ether or acetonitrile With the 34-61 fragment the resolution was similar at the injection of 50 mg of the compound in pure DMF and 75 mg in the DMF-acetonitrile mixture (Fig. 11). We believe that the data obtained provide convincing evidence for the potential importance of HPLC in the peptide synthesis. These are all reasons to predict further rapid development of this method in the near future.

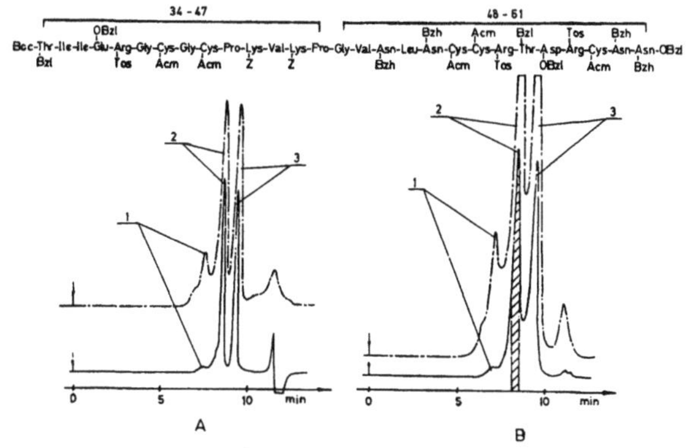

Fig. 11. Analytical (A) and preparative (B) chromatography of 34-61 fragment on a silica gel L column (330x25 mm I.D.) in DMF. Flow-rate 10 ml/min 1.High molecular impurities, 2. Boc-34-61-BzL, 3. Initial fragments. Broken lines: UV 280 nm, solid lines: refractometer. Loading under preparative conditions 50 mg per 2 ml in DMF and 75 mg per 2 ml in DMF-CH_3CN.

References

1. V.V.Ulyashin, V.I.Deigin, V.T.Ivanov and Yu.A.Ovchinnikov, J. Chromatogr. 215 (1981) 263.

FIELD DESORPTION MASS SPECTRA AND FAST ATOM BOMBARDMENT SPECTRA OF PARTIALLY PROTECTED PEPTIDES

Hanspaul Hagenmaier, Gerhard Vossler

Institut für Organische Chemie der Universität Tübingen
Auf der Morgenstelle 18, D-7400 Tübingen

Lutz Grotjahn

Gesellschaft für Biotechnologische Forschung mbH
Mascheroderweg 1, D-3000 Braunschweig-Stöckheim

We recently have synthesized somatostatin by alternating liquid solid phase peptide synthesis (1 - 6). In alternating liquid solid phase peptide synthesis the transfer of an elongated, resin-bound peptide to the liquid phase occurs during the deprotection step, which yields a C- and side chain protected peptide with free terminal amino group. Depending on the deblocking reagent the peptides are obtained either as hydrochlorides or as triflouroacetates and, in contrast to conventional peptide synthesis, can be purified and characterized only in this form. Field desorption mass spectrometry (FD-MS) and fast atom bombardment mass spectrometry (FAB-MS) (7-12) have been found extremely valuable in characterizing the peptide intermediates obtained during this somatostatin synthesis. We want to report some of the characteristics observed in FD and FAB mass spectra of partially protected peptides.

In Figures 1 and 2 the FD and FAB mass spectra of the partially protected somatostatin segment 13-14 H_2N-Ser(Bzl)-Cys(MOB)-ODCB x HCl are shown. In the FD spectrum the base peak is found at $(C + H)^+$ with m/z 578. A number of peaks are found at higher mass. In the FAB spectrum the base peak is registered at m/z 577, the mass of the peptide cation C^+, and no peaks at higher mass are found. Similar results were obtained for all partially protected peptide segments carrying a free terminal amino group.

Chemistry of Peptides and Proteins, Vol. 2

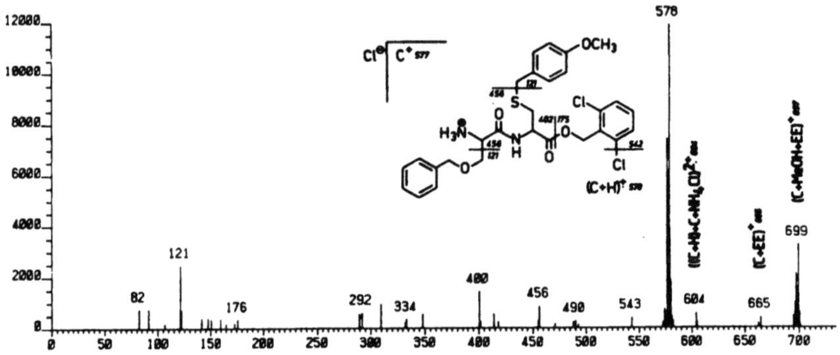

Figure 1. FD-MS of H_2N-Ser(Bzl)-Cys(MOB)-ODCB x HCl, 25 mA, solvent: methanol/ethyl acetate.

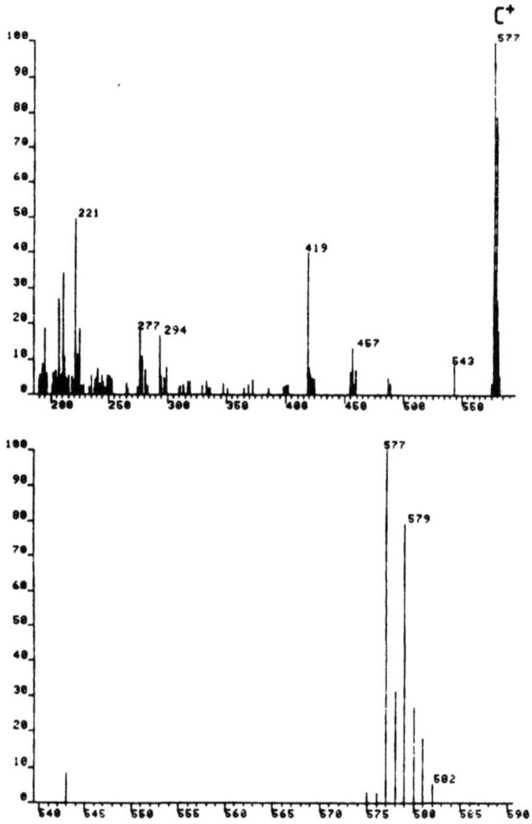

Figure 2. FAB-MS of H_2N-Ser(Bzl)-Cys(MOB)-ODCB x HCl, solvent: methanol/ethyl acetate; matrix: glycerol.

The mechanism of ion formation is rather different for the FD and the FAB technique. In FD spectra even electron ions as well as odd electron ions are found with high intensity. During the field-induced cationization reaction ions as well as radical ions are produced. Often protonated ion clusters are found which are formed by addition or subtraction of one or more hydrogen atoms. Their composition cannot be predicted. Depending on their composition the life times of these clusters differ considerably which makes it difficult to obtain reproducable FD spectra. In analogy to FD spectra of inorganic salts (13) we find in the FD spectra of the peptide salts additional ions at higher mass than C^+, containing the anion and carrying a double charge. The peak at m/z 604 is an example, which we interpret as $((C+H)+C+NH_4))^{2+}$. For the same sample, but with a different heating rate of the emitter, FD spectra can be obtained which only show the quasi molecular ion $(C + H)^+$. Complex ions formed from peptides which are present in form of their salts, is certainly a problem in the interpretation of FD spectra. Ion clusters with solvent molecules which we also find in our FD spectra (Figure 1, m/z 665 and m/z 649) have been reported previously by many authors (14,15). The manifold field-induced reactions at the anode by the FD technique prohibit the registration of the characteristic isotope pattern of the molecular ion. In all our FD spectra of chlorine-containing peptides the relative intensity of the registered quasi molecular ions is far higher than the corresponding isotope abundance.

Ion formation in fast atom bombardment mass spectrometry occurs through accelerated neutral atoms which hit the sample matrix with a kinetic energy of 2 to 19 keV. Charged molecules are formed in an interaction of solvent molecules and sample molecules with a concomitant exchange of electrons (11,16). Glycerol has been used most often as a solvent to obtain FAB spectra due to its high tendency for hydrogen bond formation. According to Baker and others (8,11,17) protonated quasi molecular ions are formed in this solvent by a process similar to chemical ionization. The life time of these protonated quasi molecular ions is in contrast to those obtained by field-induced cationization very high which results in excellent reproducibility of FAB spectra. In FAB spectra the ions with high intensity always possess even electron configuration (8). In salts the ions are preformed and according to our results direct transfer of the cation into the vapor phase occurs. The mechanism of ion formation in the FAB technique allows the registration of the characteristic isotope cluster of the stable cation C^+ (Figure 2). Table 1 shows the very good agreement of calculated and found isotope abundance for the isotpe cluster of the cation C^+. Figure 2 also shows that additional ion clusters of higher mass than the molecular ion are absent, which makes the interpretation of the spectra easier and with respect to sample characterization much more reliable.

Table 1. Isotope ratios for C^+ in Figure 2.

m/z	calculated (%)	found (%)
577	100	100
578	33.5	31.3
579	74.8	79.5
580	23.8	26.9
581	17.6	18.2
582	5.0	5.4

Comparison of the spectra in Figures 1 and 2 indicates that the fragment ions formed are more or less the same by both techniques. Differences are e.g. explained by proton transfer to form even electron ions in the FAB spectra. The 4-methoxy-benzyl(MOB) group gives rise to a characteristic fragment ion with m/z 121. The complementary ion in the FD spectrum is found at mz/456, in the FAB spectrum, however, at m/z 457. Elimination of protecting groups leads to charcteristic fragment ions. No sequence specific fragmentation is found in the partially protected peptides.

The solvent used for dissolving the peptides can be rather critical in FAB mass spectrometry. No problems were encountered using chloroform, methanol or ethyl acetate as solvents. Addition of dimethylsufoxide to dissolve large peptide segments resulted in less reproducible spectra and undesirable "adduct" formation. Figure 3 shows the FAB spectrum of the partially protected somatostatin segment 10-14 (H_2N-Thr(Bzl)-Phe-Thr(Bzl)-Ser(Bzl)-Cys(MOB)-ODCB x TFA), and Figures 4 and 5 show FD

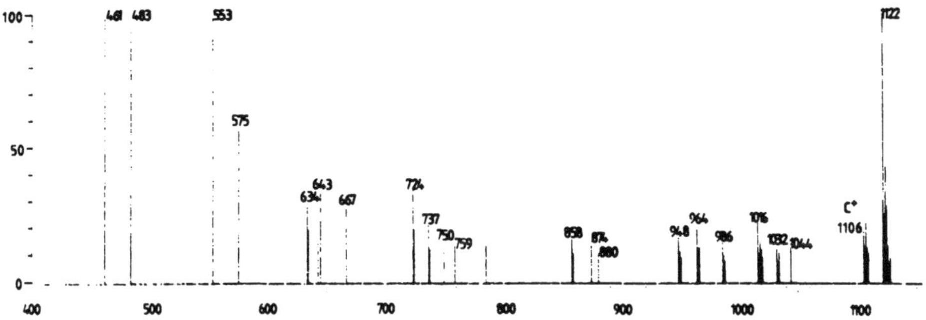

Figure 3. FAB-MS of H_2N-Thr(Bzl)-Phe-Thr(Bzl)-Ser(Bzl)-Cys(MOB)-ODCB x TFA;solvent: DMSO; Matrix:glycerol.

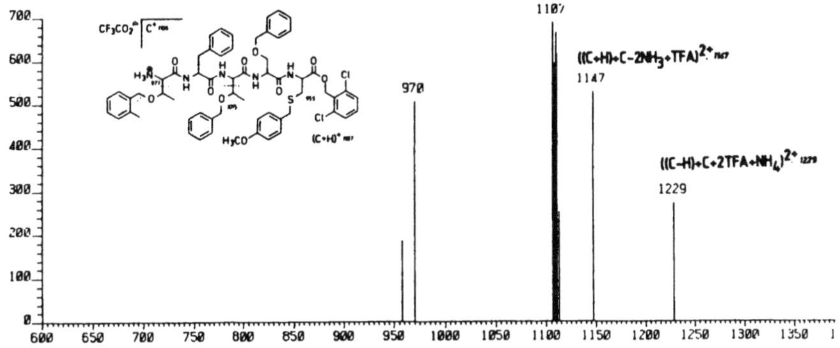

Figure 4. FD-MS of H_2N-Thr(Bzl)-Phe-Thr(Bzl)-Ser(Bzl)-Cys(MOB)-ODCB x TFA; 20 mA; solvent:methanol/ethyl acetate.

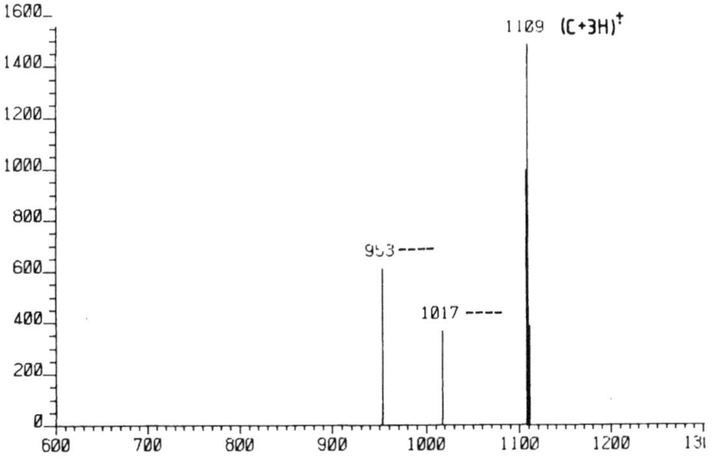

Figure 5. same as Figure 4, but 16 mA .

spectra of the same peptide obtained at 16 mA and 20 mA, respectively. In the upper mass range of the FAB spectrum two isotope clusters with a mass difference of 16 units appear. M/z 1106 represents the peptide cation C^+. A corresponding mass difference is found for lower mass fragments. We first interpreted this as the result of a partial oxidation of our sample at the cysteine residue. Further examination of the sample by various methods did not yield any evidence for a partial oxidation. The FD spectra in Figures 4 and 5 also do not show any peak for the sulfoxide. We therefore assume that this addition of oxygen occurs in the presence of dimethylsulfoxide under the conditions of ion formation. While the FAB spectrum in figure 3 shows a peak for the cation C^+ at m/z 1106, the FD spectra show base peaks at m/z 1107 (Figure 4) and m/z 1109 (Figure 5). The FD spectrum obtained at 20 mA shows in addition complex ions containing the anion of the peptide (triflouroacetate). The interpretation of these ions is of course speculative.

Both, FD and FAB mass spectra of partially protected peptides in form of their ammonium salts have been obtained. In FAB spectra the ammonium ion is found as such and in the presence of e.g. chlorine atoms the correct isotope abundance for the ion is found. This makes FAB spectra especially useful in characterizing intermediate products in peptide synthesis.

References

1. Frank, H., Hagenmaier, H.:Experientia 31,131 (1975).
2. Frank, H., Meyer,H., Hagenmaier,H.: in: Walter, R., Meienhofer, J. (ed.): Peptides: Chemistry, Structure, Biology, p. 439, Ann Arbor Sci. Publ., Ann Arbor, 1975
3. Frank, H., Hagenmaier, H.: Chemiker-Ztg. 101, 188 (1977)
4. Frank, H., Hagenmaier, H., Bayer, E., Desiderio, D.M.: in: Goodman, M.,Meienhofer, J. (eds): Peptides, Proceedings of the Fifth American Peptide Symposium, p. 514, John Wiley and Sons, New York, 1977
5. Hagenmaier, H., Frank, H., Vossler, G.: in Peptides, 1978, p. 227, Wrocłow University Press, Poland, 1979
6. Vossler, G.: Ph.D. Thesis, Universität Tübingen, 1982

7. Beckey, H.D.: Principles of Field Ionization and Field Desorption Mass Spectrometry, Pergamon Press,Oxford, 1977

8. Barber, M., Bordoli, R.S., Sedgwick, R.D., Tyler, A.N.: J. Chem. Soc., Chem. Commun. 1981, 325; Nature 293, 270 (1981).

9. Grigsby, R.D., Scheppele, S.E., Grindstoff, Q.G., Sturm jr., G.P., Taylor, L.C.E., Tudge, H., Wakefield, C., Evans, S.: Anal. Chem. 54, 1108 (1982).

10.Barber, M., Bordoli, R.S., Sedgwick, R.D., Tetler, L.W.: Org. Mass Spectrometry 16,256 (1981).

11.Williams, D.H.: Biochem. J. 201, 105(1982).

12.Barber, M.: Biochem. J. 197, 401, (1981).

13.Schulten, H.R., Lehmann, W.D.: Anal. Chim. Acta 87, 103 (1976).

14.Asanta-Poka, S., Wood, G.W., Schmidt, D.E.: Biomed.Mass Spectrom.1975, 121.

15.Desiderio, D.M., Sabbatini, J.Z., Stein, J.L.:Adv.Mass Spectrom. 1980,8b,1298.

16.Williams, D.H.: J. Am. Chem. Soc. 103, 5700 (1981).

17.Rinehart, K.L., Am. Soc. for Mass Spectrom., 29th Amer.Conf. Mass Spectrometry and Allied Topics, Minneapolis, 1981 p. 794.

Structural Features of Peptides and Proteins

ISOLATION AND STRUCTURAL ANALYSIS OF A NEW NEUROPEPTIDE, THE HEAD ACTIVATOR

H. Bodenmüller
Max-Planck-Institut für medizinische Forschung, Jahnstr. 29 D-6900 Heidelberg, FRG

1. Introduction

Head activator - neuropeptide with conserved structure from coelenterates to human - high-pressure liquid chromatography - sequence analysis with micromethods.

From the freshwater coelenterate hydra we recently isolated and sequenced a peptide which acts as a head-inducing morphogen (1). This peptide - called head activator - is responsible in hydra for head specific growth and differentiation processes. It affects growth by triggering cells to divide (2) and cellular differentiation by determining the multipotent interstitial cells to become nerve cells (3). In normal animals the head activator is produced by nerve cells and is stored there in neurosecretory granules (4). We have previously shown, that the same peptide is also present in mammalian brain and intestine (5). It is the first example of a neuropeptide which conserved its structure from coelenterates to humans.

2. Isolation and purification

The sources, from which the head activator was isolated and sequenced, are listed in Table 1. The tissues were first extracted with methanol. After centrifugation the supernatants were concentrated and extracted with petrol ether and chloroform to remove the lipids. The extracts from large amounts of tissue (sea anemones, intestines) were either applied on a 6 liter Sephadex G-10 column or absorbed on DEAE Sephadex A-25 and eluted with a pH gradient. After the further purification steps (Fig. 1) the yield of the head activator was 10 % with a 10^9-fold enrichment. To obtain a homogeneous material and to compare the elution times of the peptides from the different origins we used reverse-phase HPLC as final purification step. The extracts were first separated isocratically with 50 % methanol in 5 mM ammonium bi-

Chemistry of Peptides and Proteins, Vol. 2

Animal	Starting material	Yield of head activator
Coelenterates:		
hydra attenuata	3 kg	500 pmol
anthopleura elegantissima (sea anemone)	200 kg	20 nmol
Mammals:		
rat intestine	3000 pcs	20 nmol
human hypothalamus	2 pcs	2 nmol
bovine hypothalamus	unknown	10 nmol

Table 1. Sources of the head activator.

carbonate, pH 7.8, and/or with 30 % acetonitrile 0.1 % trifluoracetic acid. As final steps the peptides were subjected to gradient elutions using methanol or acetonitrile in the mobile phase (Fig. 2a, b).

The head activator was located in the various purification steps and column fractions by assaying the biological activity as stimulation of bud outgrowth in hydra (6). In later steps, when the head-activator peptide was free from other protein or peptide contaminants, we usually located it first by hydrolysing a small sample of all the fractions and looked for the characteristic head activator pattern of the dansylated amino acids on thin layer plates. 10 picomoles were sufficient for that purpose. In a second step we then determined the biological activity.

3. Amino acid composition and sequence

For a qualitative amino acid analysis the head activator was hydrolysed in 6 M HCl for 18 h. After dansylation the amino acids were separated by two-dimensional thin-layer chromatography on 3 x 3 cm polyamide plates (7). The following nine amino acids could be identified: Gly, Glu, Ile, Leu, Lys, Phe, Pro, Ser, Val. Since a quantitative evaluation is difficult on such plates, we tried to separate the dansylated amino

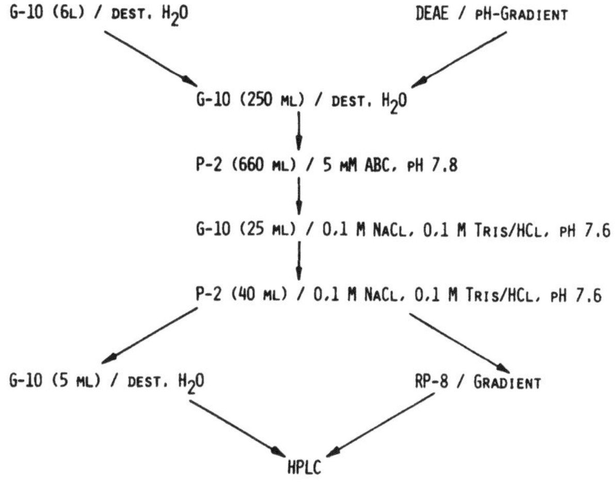

Fig. 1. Purification procedure for the head activator.

acids on two serially connected LiChrosorb RP-8 and RP-18 HPLC columns using both the absorption at 254 nm and the fluorescence with excitation at 365 nm and emission at 451 nm for detection. Fig. 4a shows the separation of an equimolar mixture of the nine amino acids (50 pm each) present in the head activator and Fig. 4b the head-activator hydrolysate. Both from the relative fluorescence and from the absorption at 254 nm can be deduced that only two amino acids occur twice, namely Gly and Pro. Thus, the head activator is composed of the following amino acids: Gly(2), Glu(1), Ile (1), Leu(1), Lys(1), Phe(1), Pro(2), Ser(1) and Val(1).

Amino-terminal sequence (Table 2) - The active head activator was found to lack a free amino terminus: Acid hydrolysis of the dansylated peptide yielded mono-substituted lysine indicating that only the ε-amino group of the lysine was available for dansylation. However, during the last purification steps we involuntarily produced a peptide (Fig. 2c) with identical amino acid composition, but different retention times on HPLC, which had no biological activity and glutamic acid as end group. The opening of the pyroglutamyl ring was probably due to the mild hydrolytic conditions of the HPLC buffers used. This peptide (Glu-peptide) could be converted into the active peptide by treatment with anhydrous trifluoroacetic acid, which results in a cyclization of the amino-terminal glutamic acid to pyroglutamic acid. Partial sequencing

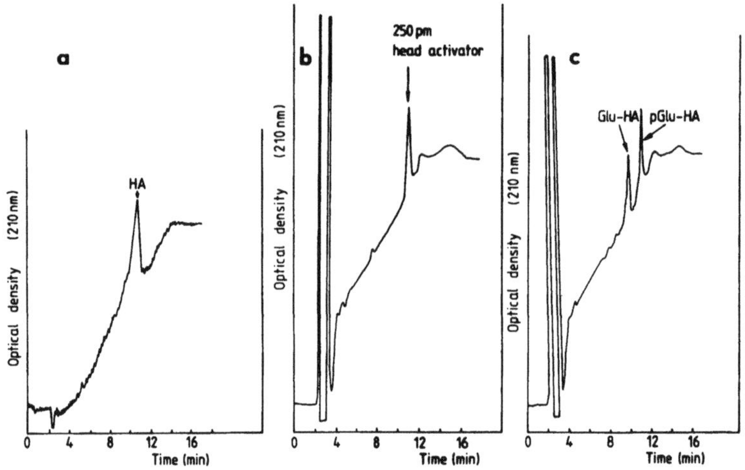

Fig. 2. Chromatography of the head activator on a LiChrosorb RP-8 HPLC column (column size 250 x 4 mm, particle size 7µm, flow-rate 1 ml/min) using a) a gradient from 20 to 40 % acetonitrile in 0.1 % trifluoracetic acid and in b) from 40 to 60 % methanol in 5 mM ammoniumbicarbonate as mobile phase. Gradient time was 10 min, in each case 250 pm of the head activator were applied. c) Separation of the Glu from the pGlu head activator under conditions as in 2a.

of the Glu peptide by the consecutive dansyl Edman procedure (8) showed the NH_2-terminal sequence to be Glu-Pro-Pro-Gly-Gly-. The biologically active peptide was not affected by pyroglutamate aminopeptidase confirming pGlu-Pro-as NH_2-terminal sequence.

Carboxy-terminal sequence - This end of the head activator was sequenced by kinetic studies with carboxypeptidase Y (CpY) (see Table 2). After incubations of 2, 5 and 15 min, the whole incubation mixture was dansylated and applied to polyamide plates. By this method, the carboxy-terminal end was found to be -Lys-Val-Ile-Leu-Phe. For these studies it was important to use the Glu peptide, since the biologically active pGlu peptide was only degraded up to Ile. This rather astonishing finding suggests that the Glu peptide and the pGlu peptide have very different conformations.

Sequence of fragments - The head activator was cleaved by trypsin and by an endopep-

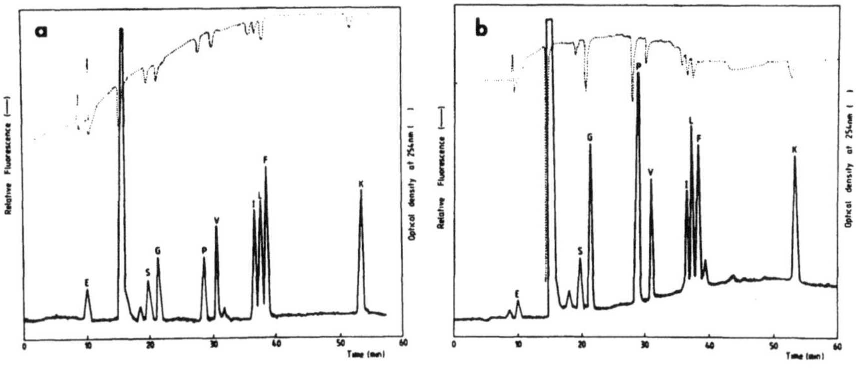

Fig. 3. Separation of dansylated amino acids on two serially connected LiChrosorb RP-8 and RP-18 columns (column size 250 x 4 mm, particle size 7µm, flow rate 1 ml/min, mobile phase 25 min gradient from 29-57 % methanol in 0.01 M Na_2HPO_4, pH 7.1). a) 50 pm of the nine amino acids contained in the head activator were dansylated and applied to the column, in b) the dansylated hydrolysate of 50 pm of the head activator. The amino acid residues are indicated by the one-letter code.

tidase isolated from Astacus fluviatilis (9) called "low molecular mass protease" (LMP). Incubations of the pGlu peptide with LMP yielded two fragments and Gly as free amino acid. A subsequent digestion with aminopeptidase M revealed -Ser-Lys-Val-as neighbouring sequence to Gly. From the carboxy- and amino-terminal sequences and from the overlap of the fragment after low molecular mass protease digestion, we deduced the following sequence for the biologically active peptide: pGlu-Pro-Pro-Gly-Gly-Ser-Lys-Val-Ile-Leu-Phe. In the meantime we have synthesised the head activator and could show that there is both chemically and biologically no difference between the synthetic and the native peptides (10).

Among the features of the head activator are (1) the lipophilic character, which results in a low solubility in aquous solvents, (2) the problem common to many biologically active peptides - the tendency to adsorb to surfaces presented by plastic or glass tubes, (3) the formation of high molecular weight aggregates, which can be separated on gel columns, (4) the weak immunogenicity, which is a problem for raising antibodies.

Method	Amino acid sequence
Edman degradation	Glu - Pro - Pro - Gly - Gly -
Carboxypeptidase Y	- Lys - Val - Ile - Phe
LMP*	- Gly -
LMP and end group	- Gly - Ser -
LMP and Edman	Glu - Pro - Pro - Gly -
LMP and aminopeptidase M	- Gly - Ser - Lys - Val - ILe - Leu - Phe

Table 2. Sequence analysis of the head activator.

The sequence of the head activator differs from those of all other biologically active peptides described so far. The only structural homology we have observed is with bradykinins, in the sequence -Pro-Pro-Gly-X-Ser-. Bradikinins are known as smooth muscle stimulants, i.g. stimulation of uterus contraction. We also found stimulation of uterus contraction with the head activator at a concentration of 10^{-5}M, which may imply some correlation between structural and functional homology.

References

1. Schaller, H.C., Bodenmüller, H., Proc. Natl. Acad. Sci 78, 7000-7004 (1981).
2. Schaller, H.C., Cell Differ. 5, 1-11 (1976).
3. Schaller, H.C., Cell Differ. 5, 13-20 (1976).
4. Schaller, H.C., Gierer, A., J. Embryol. Exp. Morphol. 29, 39-52 (1973).
5. Bodenmüller, H., Schaller, H.C., Nature 293, 579-580 (1981).
6. Schaller, H.C., Schmidt, T., Grimmelikhuijzen, C.J.P., Wilhelm Roux's Arch. dev. Biol. 186, 139-149 (1979).
7. Gray, W.R., Meth. Enzym. 25, 121-138 (1972).
8. Gray, W.R., Smith, J.F., Analyt. Biochem. 33, 36-42 (1970).
9. Zwilling, R., Dörsam, H., Torff, H.J., Ródl, J., FEBS Lett. 127, 75-78 (1981).
10. Birr, C., Zachmann, B., Bodenmüller, H., Schaller, H.C., FEBS Lett. 131, 317-321 (1981).

BIOCHEMICAL, BIOLOGICAL and IMMUNOLOGICAL CHARACTERIZATION of BOVINE NERVE GROWTH FACTOR

Gregory P. Harper and Hans Thoenen
Dept. Neurochemistry, Max-Planck-Institute for Psychiatry, 8033 Martinsried/Munich, FRG

The biological actions of the protein Nerve Growth Factor (NGF) have been investigated almost exclusively using the NGF isolated from the submandibular salivary glands of mice. The recent purification of NGF from bovine seminal plasma (1) provides an opportunity to study the relationship between the biochemical features of NGF molecules and their biological and immunological properties.

Biological Activities

NGF has, in several neural cell types, a wide variety of biological actions, some of which are summarized in Table I. Though these effects may be considered to be "pharmacological" actions of NGF, the corresponding physiological significance of the protein is clearly demonstrated *in vivo* by the ability of antibodies against NGF to destroy or functionally impair NGF's target neural cells by inhibiting the supportive actions of the NGF endogenous to the animal. There appears to be, for all the neural cells responding to NGF, a critical developmental stage before which the inhibition of endogenous NGF causes neural cell death, because these cells are absolutely dependent on NGF for survival *in vivo*. These conditions can be mimicked *in vitro* (see Table I), simply by withdrawing NGF from the culture medium. Once past this critical developmental stage, the various types of neural cells can remain

Chemistry of Peptides and Proteins, Vol. 2

Table I. Biological activities of bovine and mouse NGFs.

cell type	species source	biological effect
sensory neurons (*in vitro*)	chick embryo (10-day old)	maintains survival *in vitro*; stimulates nerve fibre outgrowth
	chick embryo (16-day old)	cannot maintain survival *in vitro*
sympathetic neurons (*in vitro*)	chick embryo (13-day old)	maintains survival *in vitro*; stimulates nerve fibre outgrowth
	neonatal rat	maintains survival *in vitro*; stimulates nerve fibre outgrowth; selectively induces adrenergic neurotransmitter-synthesizing enzymes (tyrosine hydroxylase)
parasympathetic neurons (*in vitro*)	chick embryo (8-day old)	unresponsive due to absence of NGF-receptors
CNS cholinergic neurons (*in vivo*)	neonatal rat	temporary selective induction of cholinergic neurotransmitter-synthesizing enzymes (choline acetyltransferase)
retinal explants (CNS) (*in vitro*)	goldfish	maintains neuronal survival *in vitro*; stimulates nerve fibre outgrowth
pheochromocytoma PC12 cells (*in vitro*)	rat adrenal medullary tumour	stimulates nerve fibre outgrowth; selectively induces choline acetyltransferase; stimulates cellular hypertrophy
adrenal medullary cells (*in vitro*)	bovine (calf)	selectively induces tyrosine hydroxylase

responsive to, though no longer dependent on, continued NGF availability (for review, see 2).

In all the systems listed in Table I, bovine and mouse NGFs are indistinguishable, both, with respect to the magnitude of the biological effects elicited and in the dose-dependencies of the responses (3). It is known that the initial and, indeed, essential step in the mechanism of action of NGF is its combination with a cell-membrane receptor (4). For bovine and mouse NGFs to have identical biological efficacies and potencies in all these systems strongly argues that the "biologically active site" of these two NGFs, *i.e.* the segment of the NGF protein that combines with the receptor molecule, must be highly conserved during evolution.

Immunological Properties

The simplest method of examining the immunological properties of NGF proteins is to check the ability of antibodies produced against one NGF to inhibit the biological activity of another. A number of antisera produced against purified mouse NGF have been found to be as potent at inhibiting bovine NGF in a tissue culture assay system as they are against their homologous mouse NGF (3). In contrast, however, antisera produced against bovine NGF are very much less able to inhibit mouse NGF *in vitro* than they are effective against *their* homologous bovine NGF (3). This can also be studied *in vivo*: thus the administration to neonatal rats of antisera against mouse NGF destroys virtually all the sympathetic neurons, an effect known as immunosympathectomy (2,3), which is, as described above, the *in vivo* correlate of withdrawing NGF from these neurons *in vitro* (see Table I). This result indicates that the NGF endogenous to neonatal rats is, at least immunologically, very similar to mouse NGF. In contrast, comparable doses of an antiserum against bovine NGF are only poorly effective at inhibiting the endogenous rat NGF, and so cause the death of relatively few sympathetic neurons (3). The "nonsymmetrical" pattern of this immunological comparison between bovine and mouse NGFs suggests that, while most of the antigenic determinants on the mouse NGF molecule are also present in some form on the bovine NGF protein, there is at least one major antigenic determinant on the bovine NGF molecule that is absent from the mouse NGF protein. Work is in progress in our laboratories to support this hypothesis using monoclonal antibodies against the two NGFs. Four monoclonal antibodies raised against mouse NGF have all been found to recognize bovine NGF, albeit with lower affinities. In principle, however, it should be possible to produce monoclonal antibodies against bovine NGF that do <u>not</u> recognize mouse NGF at all, and in this way the major antigenic determinant(s) restricted to the bovine NGF molecule may be identified.

More highly resolving immunochemical techniques indicate that, even excluding the major antigenic determinants restricted to bovine NGF (see above), the immunological properties of mouse and bovine NGFs differ substantially. Thus it is possible, using a two-site radioimmunoassay that can quantify immunological crossreactivity (5), to demonstrate that most of the antibodies in polyclonal antisera produced against mouse NGF either do not recognize bovine NGF at all, or have a sufficiently lower affinity for bovine NGF that their recognition is not evident under the conditions of the assay. Presumably for similar reasons, it also proved impossible to immunoprecipitate (in an Ouchterlony double immunodiffusion system) bovine NGF by anti-mouse NGF antibodies, or, conversely, mouse NGF by anti-bovine NGF antibodies (3).

Clearly, therefore, the close similarity between bovine and mouse NGFs with regard to their <u>biological</u> activities is not evident in their <u>immunological</u> properties. This suggests that the parts of the NGF molecules not directly involved in receptor binding, but able to act as antigenic determinants, have diverged quite extensively during evolution. It should, therefore, be possible in principle to identify by comparative biochemical analysis (i) the conserved biologically active site of the NGF proteins and (ii) the major antigenic determinants which reside outside this active site.

Biochemical Characterization

The biochemical properties of mouse NGF have been extensively characterized during the last 15 years (for review, see 6). The mouse NGF protein is a homodimer of non-covalently linked polypeptide chains, each of which contains 118 amino-acid residues, and has a molecular weight of 13,256 (from sequence data). Mouse NGF is a very basic protein (pI 9.3). In the submandibular gland this form of mouse NGF (called the ß subunit) exists as part of a high molecular weight complex in association with a (α) subunit of unknown, possibly regulatory, function and with a serine protease (γ subunit), which probably plays a role in cleaving (at the COOH-terminus of the ß subunit) a larger biosynthetic precursor of mouse ß NGF.

The gross biochemical properties of bovine NGF (1) are broadly similar to those of mouse NGF: thus in untreated seminal plasma bovine NGF also exists as part of a high molecular weight complex. The bovine NGF subunit is, however, more basic (pI 9.5-10.0) and larger (containing two non-covalently linked peptide chains of molecular weight ~15000, from SDS-polyacrylamide gels) than mouse ß NGF. Though the amino acid compositions (Table II) indicate family resemblances between the two NGFs, considerable sequence differences might be expected in order to accomodate the differences between the compositions. It is therefore interesting that, at the NH_2-terminus, the sequences of bovine and mouse NGFs are relatively highly homologous (Fig. 1), though some differences are evident. It remains to be seen whether the rest of the bovine NGF sequence diverges further from the mouse sequence which might indicate that the "conserved" NH_2-terminus of the molecule is involved in the biologically active site of NGF. However, as it is known (6) that the NH_2-terminal octapeptide is not required for biological activity, it is more likely that this particular region is conserved in accordance with the biosynthetic pathway leading to the final NGF molecules, e.g. to provide a recognition site for a possible proteolytic enzyme cleaving a larger precursor at the NH_2-terminus of the eventual NGF protein (cf. γ subunit, above).

Table II. Amino-acid compositions of bovine and mouse NGFs.

Amino-acid	Residues per peptide chain	
	Bovine NGF (integer)	Mouse NGF (sequence)
Aspartic acid	15	11
Threonine	10	14
Serine	11	11
Glutamic acid	13	8
Proline	7	2
Glycine	9	5
Alanine	7	8
Half-cystine	4	6
Valine	12	13
Methionine	2	1
Isoleucine	7	5
Leucine	8	3
Tyrosine	5	2
Phenylalanine	7	7
Histidine	4	4
Lysine	10	8
Arginine	7	7
Tryptophan	3	3
TOTAL	141	118

Fig. 1. NH_2-terminal sequences of bovine and mouse NGFs.

bovine NH_2-ser-ser-ser-his-pro-val-phe-his-
1 5
mouse NH_2-ser-ser-thr-his-pro-val-phe-his-

bovine -arg-gly-glu-phe-ser-val- ? -asp-ser-
10 15
mouse -met-gly-glu-phe-ser-val-cys-asp-ser-

bovine -ile-ser-val- ? -val- ? -asp- ? -thr-..
20 25
mouse -val-ser-val-trp-val-gly-asp-lys-thr-..

References

1. Harper, G.P., Glanville, R.W., Thoenen, H.: J. Biol. Chem. (in press).
2. Harper, G.P., Thoenen, H.: Ann. Rev. Pharmacol. Toxicol. 21, 205-229 (1981).
3. Harper, G.P., Barde, Y.A., Edgar, D., Ganten, D., Hefti, F., Heumann, R., Naujoks, K.W., Rohrer, H., Turner, J.E., Thoenen, H.: (submitted for publication).
4. Heumann, R., Schwab, M., Thoenen, H.: Nature 292, 838-840 (1981).
5. Harper, G.P., Thoenen, H.: J. Neurochem. 34, 893-903 (1980).
6. Server, A.C., Shooter, E.M.: Adv. Protein Chem. 31, 339-409 (1977).

EMBRYONIC HAEMOGLOBINS IN MAMMALS

Haemoglobin Components in Human and Pig Embryos; Primary Structures of the Human and Pig ζ-Chains and their Phylogenetic Relationship to the α-Chains

Heinz Aschauer, Franz Bieber, Sirri Bektaş,
Traute Kleinschmidt, Gerhard Braunitzer
Max-Planck-Institut für Biochemie
D-8033 Martinsried bei München, Germany

Torpong Sanguansermsri
Department of Pediatrics and Human Genetics Unit, Chiang Mai University, Chiang Mai, Thailand

Introduction

Mammals are known to synthesize various globin chains sequentially during their different stage of development, forming the embryonic-, fetal- and adult haemoglobins. The possibility of a human embryonic haemoglobin was first indicated by Drescher and Künzer (1) and designated Gower 1 ($\zeta_2\varepsilon_2$) and Gower 2 ($\alpha_2\varepsilon_2$) by Huehns (2). These studies showed the existence of an unique ß-like embryonic globin chain , called ε-chain. It has been assumed that the difference of embryonic to fetal or adult haemoglobin is solely due to a ß-like chain. Gower 1 was originally thought to be a tetramer of ε-chains (ε_4) only. However, in haemoglobin Portland ($\zeta_2\gamma_2$) - a third normal embryonic haemoglobin - which was described first as a minor abnormal haemoglobin component, found for instance in infants with homozygous α-thalassaemia (hydrops fetalis syndrome), a different globin chain - designated ζ-chain - was identified (3-6).

Chemistry of Peptides and Proteins, Vol. 2

Another evidence for an unique kind of an α-type globin chain in early mammalian embryos was given by Melderis et al. (7). They found an α-like chain - designated x-chain - in mice and rabbit embryos. These chains are very similar to the ζ-chains of human haemoglobin Portland and they suggested that the ζ-chain is a human "x-type" chain. Later ,partial structures indicated that the ζ-chain is indeed of the α-type (8). This suggested that the Hb Gower 1 is not an ε-chain tetramer but consists of two ζ- plus two α-chains (9).
Even if the ontogeny of the human haemoglobins became clear in the last years, very little is known about the structural and functional properties. To get additional information about the ontogeny, the structural and functional properties of mammalian embryonic haemoglobins, we started investigations on pig embryos. The study of embryonic haemoglobins in pig embryos is much easier than in human embryos, because they do not synthesize a fetal haemoglobin. The embryonic haemoglobins are immediately succeeded by the adult haemoglobin (10). Another advantage is the ease to obtain enough suitable material.
Here, the separation and characterization of human and pig embryonic haemoglobins on ion exchange chromatography is shown. The amino acid sequence of both - human (11, 12) and pig - ζ-chains are represented and the phylogenetic and structural features are discussed.

Results

All data so far known about the haemoglobin composition of human embryos of different ages were obtained by electrophoretical methods. We have carried out chromatographic separations of blood from human embryos with 20 and 40 mm crown-rump length on a DEAE-Sephacel column (0.5x10 cm) using Tris-HCl buffers, pH 8.4, with NaCl developers (Fig. 1). In both cases it was possible to identify the different globin chains of the haemo-

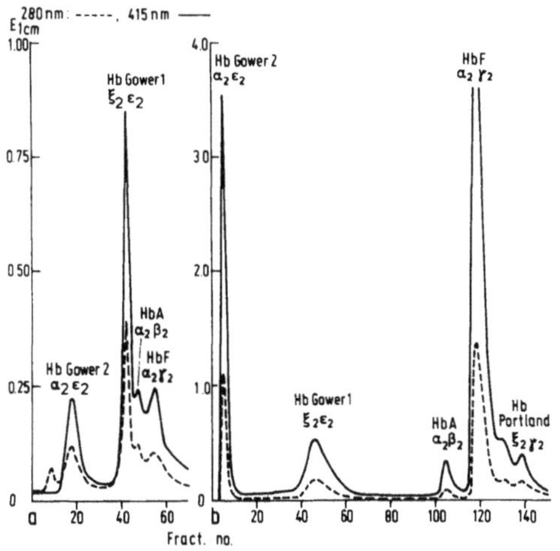

Fig. 1. Chromatography of blood obtained from human embryos of a) 20 mm and b) 40 mm crown-rump length on DEAE Sephacel.

globin fractions by electrophoresis in polyacrylamide gels, containing 6 M urea, 2% Triton X-100 and 5% acetic acid.
The presence of adult Hb A in our haemolysate might be due to contamination with maternal blood, because no Hb A was found earlier in small embryos (6). If we take in account that no Hb A is present during the considered period, the total amount of the embryonic haemoglobins Gower 1 and Gower 2 decreases from 70% to 33% while the fetal haemoglobin increases from 30% to 61%. The most prominent component in the early embryonic development is Hb Gower 1 (46%), but in older embryos the amount of Gower Hb 2 (18%) is greater than the amount of Hb Gower 1 (13%). The third embryonic haemoglobin - Hb Portland - could only be identified in the larger embryos, but not in the smaller ones.
For the structural work on the human ζ-chains we needed larger amounts of suitable material than one could obtain from embryos. So we used blood from asian hydrops foetalis due to homozygous α-thalassaemia. The hydrops foetalis blood consists of the

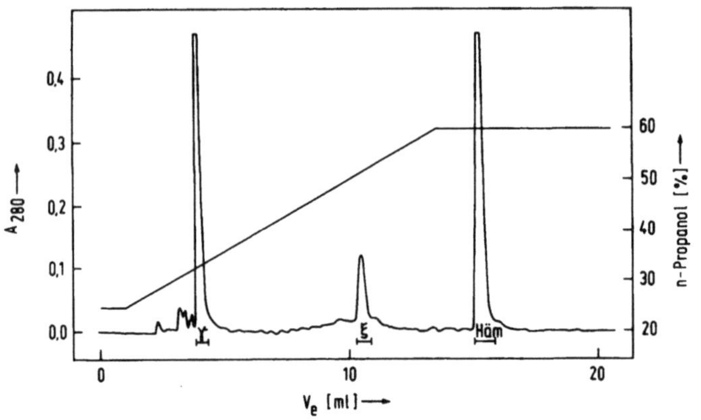

Fig. 2. HPLC separation of hydrops foetalis haemoglobin on reversed phase (RP8, LiChrosorb, Merck).

components Hb Barts (γ_4) and a minor component - less than 10% - Hb Portland ($\zeta_2\gamma_2$). In our hand all classical chromatographic methods failed to isolate either Hb Portland or directly the ζ-chains in a preparative scale. Only the application of HPLC on reversed-phase (C8 or C2 material: RP8 or RP2, LiChrosorb) was successful for separating the γ- and ζ-chains (fig. 2). Peptides for the sequence work were generated by chemical cleaveages (cyanogen bromide cleavage and acid cleavage at the Asp-Pro bond) and enzymatical cleavages (tryptic cleavage of the unmodified and succinilated ζ-chains). The peptides were separated by HPLC. C8 phases (RP8 LiChrosorb, Merck or MOS Hypersil, Shandon) were used for the smaller peptides and C2 phase (RP2 LiChrosorb, Merck) was used for the larger peptides. The sequences of the peptides were determined by automatic Edman degradation methods. A problem was the elucidation of the N-terminal sequence of the ζ-chain, because it has a blocked N-terminus. The nature of the blocking group (Ac-) and the N-terminal sequence were established by mass spectrometry with the N-terminal tryptic tetrapeptide after permethylation. The N-terminal sequence was prouven by digestion of the limited split tryptical N-terminal heptapeptide with chymotrypsin and sequenc analysis of the whole mixture. The sequence of the human ζ-chain is given in tab. 1 and homologous aligned with all other expressed human globin chains.

In our studies with haemoglobin from pig embryos of 20 mm crown-rump length we have discovered a new globin chain - designated ϑ-chain (13) - which is a second embryonic β-like chain similar to the ε-chains. On DEAE-Sephacel it was possible to isolate four different embryonic haemoglobins (fig. 3). Two of them correspond to the human ones: Gower 1 ($\zeta_2\varepsilon_2$) and Gower 2 ($\alpha_2\varepsilon_2$) and the other two are named Heide 1 ($\zeta_2\vartheta_2$) and Heide 2 ($\alpha_2\vartheta_2$) (13, 14). The different chains were isolated from the purified haemoglobin components by chromatography on CM cellulose in phosphate puffer and 8M urea. The primary structure of the pig ζ-chain is nearly complete and was determined by the tryptic

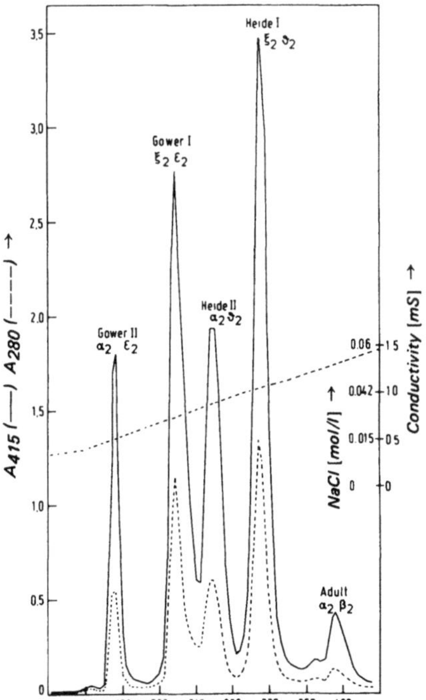

Fig. 3. Chromatography of blood obtained from pig embryos of 20 mm crown-rump length on DEAE Sephacel.

```
                                     10                                                       20                                   30
α-H    Val-    -Leu-Ser-Pro-Ala-Asp-Lys-Thr-Asn-Val-Lys-Ala-Ala-Trp-Gly-Lys-Val-Gly-Ala-His-Ala-Gly-Glu-Tyr-Gly-Ala-Glu-Ala-Leu-Glu-
ζ-H Ac-Ser-    -Leu-Thr-Lys-Thr-Glu-Arg-Thr-Ile-Ile-Val-Ser-Met-Trp-Ala-Lys-Ile-Ser-Thr-Gln-Ala-Asp-Thr-Ile-Gly-Thr-Glu-Thr-Leu-Glu-
β-H    Val-His-Leu-Thr-Pro-Glu-Glu-Lys-Ser-Ala-Val-Thr-Ala-Leu-Trp-Gly-Lys-Val-Asn-         -Val-Asp-Glu-Val-Gly-Gly-Glu-Ala-Leu-Gly-
γ-H    Gly-His-Phe-Thr-Glu-Glu-Asp-Lys-Ala-Thr-Ile-Thr-Ser-Leu-Trp-Gly-Lys-Val-Asn-         -Val-Glu-Asp-Ala-Gly-Gly-Glu-Thr-Leu-Gly-
δ-H    Val-His-Leu-Thr-Pro-Glu-Glu-Lys-Thr-Ala-Val-Asn-Ala-Leu-Trp-Gly-Lys-Val-Asn-         -Val-Asp-Ala-Val-Gly-Gly-Glu-Ala-Leu-Gly-
ε-H    Val-His-Phe-Thr-Ala-Glu-Glu-Lys-Ala-Ala-Val-Thr-Ser-Leu-Trp-Ser-Lys-Met-Asn-         -Val-Glu-Glu-Ala-Gly-Gly-Glu-Ala-Leu-Gly-
                                        10                                                 20

                                        40                                    50
α-H   -Arg-Met-Phe-Leu-Ser-Phe-Pro-Thr-Thr-Lys-Thr-Tyr-Phe-Pro-His-Phe-    -Asp-Leu-Ser-His-                            -Gly-Ser-Ala-Gln-Val-
ζ-H   -Arg-Leu-Phe-Leu-Ser-His-Pro-Gln-Thr-Lys-Thr-Thr-Phe-Pro-His-Phe-    -Asp-Leu-His-Pro-                            -Gly-Ser-Ala-Gln-Leu-
β-H   -Arg-Leu-Leu-Val-Val-Tyr-Pro-Trp-Thr-Gln-Arg-Phe-Phe-Glu-Ser-Phe-Gly-Asp-Leu-Ser-Thr-Pro-Asp-Ala-Val-Met-Gly-Asn-Pro-Lys-Val-
γ-H   -Arg-Leu-Leu-Val-Val-Tyr-Pro-Trp-Thr-Gln-Arg-Phe-Phe-Asp-Ser-Phe-Gly-Asn-Leu-Ser-Ser-Ala-Ser-Ala-Ile-Met-Gly-Asn-Pro-Lys-Val-
δ-H   -Arg-Leu-Leu-Val-Val-Tyr-Pro-Trp-Thr-Gln-Arg-Phe-Phe-Glu-Ser-Phe-Gly-Asp-Leu-Ser-Ser-Pro-Asp-Ala-Val-Met-Gly-Asn-Pro-Lys-Val-
ε-H   -Arg-Leu-Leu-Val-Val-Tyr-Pro-Trp-Thr-Gln-Arg-Phe-Phe-Asp-Ser-Phe-Gly-Asn-Leu-Ser-Ser-Pro-Ser-Ala-Ile-Leu-Gly-Asn-Pro-Lys-Val-
        30                                  40                                  50                                              60

                        60                                      70                                      80
α-H   -Lys-Gly-His-Gly-Lys-Lys-Val-Ala-Asp-Ala-Leu-Thr-Asn-Ala-Val-Ala-His-Val-Asp-Asp-Met-Pro-Asn-Ala-Leu-Ser-Ala-Leu-Ser-Asp-Leu-
ζ-H   -Arg-Ala-His-Gly-Ser-Lys-Val-Val-Ala-Ala-Val-Gly-Asp-Ala-Val-Lys-Ser-Ile-Asp-Asp-Ile-Gly-Gly-Ala-Leu-Ser-Lys-Leu-Ser-Glu-Leu-
β-H   -Lys-Ala-His-Gly-Lys-Lys-Val-Leu-Gly-Ala-Phe-Ser-Asp-Gly-Leu-Ala-His-Leu-Asp-Asn-Leu-Lys-Gly-Thr-Phe-Ala-Thr-Leu-Ser-Glu-Leu-
γ-H   -Lys-Lys-His-Gly-Lys-Lys-Val-Leu-Thr-Ser-Leu-Gly-Asp-Ala-Ile-Lys-His-Leu-Asp-Asp-Leu-Lys-Gly-Thr-Phe-Ala-Gln-Leu-Ser-Glu-Leu-
δ-H   -Lys-Ala-His-Gly-Lys-Lys-Val-Leu-Gly-Ala-Phe-Ser-Asp-Gly-Leu-Ala-His-Leu-Asp-Asn-Leu-Lys-Gly-Thr-Phe-Ser-Gln-Leu-Ser-Glu-Leu-
ε-H   -Lys-Ala-His-Gly-Lys-Lys-Val-Leu-Thr-Ser-Phe-Gly-Asp-Ala-Ile-Lys-Asn-Met-Asp-Asn-Leu-Lys-Pro-Ala-Phe-Ala-Lys-Leu-Ser-Glu-Leu-
                                            70                                      80                                      90

                    90                                     100                                     110
α-H   -His-Ala-His-Lys-Leu-Arg-Val-Asp-Pro-Val-Asn-Phe-Lys-Leu-Leu-Ser-His-Cys-Leu-Leu-Val-Thr-Leu-Ala-Ala-His-Leu-Pro-Ala-Glu-Phe-
ζ-H   -His-Ala-Tyr-Ile-Leu-Arg-Val-Asp-Pro-Val-Asn-Phe-Lys-Leu-Leu-Ser-His-Cys-Leu-Leu-Val-Thr-Leu-Ala-Ala-Arg-Phe-Pro-Ala-Asp-Phe-
β-H   -His-Cys-Asp-Lys-Leu-His-Val-Asp-Pro-Glu-Asn-Phe-Arg-Leu-Leu-Gly-Asn-Val-Leu-Val-Cys-Val-Leu-Ala-His-His-Phe-Gly-Lys-Glu-Phe-
γ-H   -His-Cys-Asp-Lys-Leu-His-Val-Asp-Pro-Glu-Asn-Phe-Lys-Leu-Leu-Gly-Asn-Val-Leu-Val-Thr-Val-Leu-Ala-Ile-His-Phe-Gly-Lys-Glu-Phe-
δ-H   -His-Cys-Asp-Lys-Leu-His-Val-Asp-Pro-Glu-Asn-Phe-Arg-Leu-Leu-Gly-Asn-Val-Leu-Val-Cys-Val-Leu-Ala-Arg-Asn-Phe-Gly-Lys-Glu-Phe-
ε-H   -His-Cys-Asp-Lys-Leu-His-Val-Asp-Pro-Glu-Asn-Phe-Lys-Leu-Leu-Gly-Asn-Val-Met-Val-Ile-Ile-Leu-Ala-Thr-His-Phe-Gly-Lys-Glu-Phe-
                                    100                                     110                                     120

            120                                     130                                     140
α-H   -Thr-Pro-Ala-Val-His-Ala-Ser-Leu-Asp-Lys-Phe-Leu-Ala-Ser-Val-Ser-Thr-Leu-Val-Thr-Ser-Lys-Tyr-Arg
ζ-H   -Thr-Ala-Glu-Ala-His-Ala-Ala-Trp-Asp-Lys-Phe-Leu-Ser-Val-Val-Ser-Ser-Val-Leu-Thr-Glu-Lys-Tyr-Arg
β-H   -Thr-Pro-Pro-Val-Gln-Ala-Ala-Tyr-Gln-Lys-Val-Val-Ala-Gly-Val-Ala-Asn-Ala-Leu-Ala-His-Lys-Tyr-His
γ-H   -Thr-Pro-Glu-Val-Gln-Ala-Ser-Trp-Gln-Lys-Met-Val-Thr-Gly-Val-Ala-Ser-Ala-Leu-Ser-Ser-Arg-Tyr-His
δ-H   -Thr-Pro-Gln-Met-Gln-Ala-Ala-Tyr-Gln-Lys-Val-Val-Ala-Gly-Val-Ala-Asn-Ala-Leu-Ala-His-Lys-Tyr-His
ε-H   -Thr-Pro-Glu-Val-Gln-Ala-Ala-Trp-Gln-Lys-Leu-Val-Ser-Ala-Val-Ala-Ile-Ala-Leu-Ala-His-Lys-Tyr-His
                                130                                     140
```

Tab. 1. Homologous alignment of primary structures of all expressed human globin chains (11, 12, 15-18).

```
      NA              A                              10                                AB      B  20
ζ-H                         Thr                         Val             Ala              Thr
ζ-P   Ac-Ser-    -Leu-Thr-Lys-Ala-Glu-Arg(Thr-Ile-Ile-Gly-Ser-Met-Trp-Thr-Lys )Ile-Ser-Ser-Gln-Ala-Asp-Thr-Ile-Gly-Thr-Glu-Thr-Leu-
ε-P       Val-His-Phe-Thr-Ala-Glu-Glu-Lys-Ala-Val-Ile-Thr-Gly-Leu-Trp-Ser-Arg-Val-Asn-            -Val-Glu-Glu-Thr-Gly-Gly-Glu-Ala-Leu-
θ-P                                   Ser                        Gly-Lys

        30                  C         40                CD                       50       E
ζ-H                     Leu     His         Thr                                     His
ζ-P     Glu-Arg-Leu-Phe-Ala-Ser-Tyr-Pro-Gln-Ala-Lys-Thr-Tyr-Phe-Pro-His-Phe-Asp-Leu-Asn-Pro-Gly-Ser-Ala-Gln-Leu-Arg-Ala-His-Gly-
ε-P     Gly-Arg
θ-P     Gly-Arg

        60                                    70           EF                            F 80                          FG
ζ-H                 Val                 Asp                             Asp-Ile-Gly-Gly             Ser
ζ-P     Ser-Lys-Val-Leu-Ala-Ala-Val-Gly-Glu-Ala-Val-Lys-Ser-Ile-Asp-Asn-Val-Ser-Ala-Ala-Leu-Ala-Lys-Leu-Ser-Glu-Leu-His-Ala-Tyr-

        90          G                           100                                110            GH                  H
ζ-H     Ile                                     Leu                                     Ala-Arg                     Phe
ζ-P     Val-Leu-Arg-Val-Asp-Pro-Val-Asn-Phe-Lys-Phe-Leu-Ser-His-Cys-Leu-Leu-Val-Thr-Leu-Ala-Ser-His-Phe-Pro-Ala-Asp-Leu-Thr-Ala-

        120                                     130                                 HC   140
ζ-H                                             Ser-Val         Ser
ζ-P     Glu-Ala-His-Ala-Ala-Trp-Asp-Lys(Phe-Leu-Thr-Ile-Val-Ser-Gly-Val-Leu-Thr-Glu-Lys)Tyr-Arg
```

Tab. 2. Homologous alignment of the primary structure of the human ζ-, the pig ζ-, the partial sequences of the pig ε- and θ-chains.

peptides - isolated in a similar way as those of the human ζ-chain - in homology to the human ζ-chain (tab. 2). The primary structures of the other embryonic chains (ε and ϑ) are still under investigation. In tab. 2 the pig ζ-chain and the partial sequences of the pig ε- and ϑ-chain are homologous aligned with the human ζ-chain.

Complete or nearly complete sequences of the α-like embryonic chains in mammals are known from human and pig. Partial sequences are given from rabbit and mouse (7). Well investigated are the embryonic chains in chicken. A preliminary phylogenetic analyses shows that the phylogenetic distances between the ζ-chains of the different species are nearly the same as those of the corresponding α-chains and therefore are related to the phylogenetic distances of these species. In addition one can see that the ζ-chains and α-chains mutually are more related than the ζ- and α-chains in the same species. This suggests that the divergence of α- and ζ-chains has taken place quite early - before the separation of these species. On the other hand there is a significant alteration of the functional sites - blocked N-terminus, which is responsible for an altered Bohr effect and exchanges in the $\alpha_1\beta_1$ binding region, responsible for the cooperativity - common to the ζ-chains considered. This suggestion that selection has operated on the α-like chains, producing structural changes in respect to their embryonic and adult environments and an accelerated evoultion, can not be excluded.

References

1. Drescher, H., Künzer, W.: Klin. Wochenschr. 32, 92 (1954).
2. Huehns, E.R., Flynn, F.V., Butler, E.A., Beaven, G.H.: Nature 189, 496-497 (1961).
3. Capp, G.L., Rigas, D.A., Jones, R.T.: Nature 228, 278-290 (1970).

4. Weatherall, D.J., Clegg, J.B., Wong, H.B.: Brit. J. Haematol. 18, 375-367 (1970).

5. Todd, D., Lai, M.C.S., Beaven, G.H., Huehns, E.R.: Brit. J. Haematol. 19, 27-31 (1970).

6. Kaltsoya, A., Fessas, P., Stavropolous, A.: Science 153, 1417-1418 (1966).

7. Melderis, H., Steinheider, G., Ostertag, W.: Nature 250, 774-776 (1974).

8. Kamuzora, H., Lehmann, H.: Nature 256, 511-513 (1975).

9. Huehns, E.R., Farooqui, A.M.: Nature 254, 335-337 (1975).

10. Tautz, C.H., Kleihauer, E.: Res. exp. Med. 159, 44-49 (1972).

11. Aschauer, H., Sanguansermsri, R., Braunitzer, G.: Hoppe-Seyler's Z. Physiol. Chem. 362, 1159-1162 (1981).

12. Aschauer, H., Schäfer, W., Sanguansermsri, T.: Hoppe-Seyler's Z. Physiol. Chem. 362, 1657-1659 (1981).

13. Steer, W., Braunitzer, G.: Hoppe-Seyler's Z. Physiol. Chem. 361, 1165-1169 (1980).

14. Bektas, M.S., Bieber, F.A., Braunitzer, G.: Hoppe-Seyler's Z. Physiol. Chem. 362, 991-995 (1981).

15. Braunitzer, G., Gehring-Müller, R., Hilschmann, N., Hilse, K., Hobom, G., Rudloff, V., Wittmann-Liebold, B.: Hoppe-Seyler's Z. Physiol. Chem. 325, 283-286 (1961).

16. Schroeder, W.A., Shelton, J.R., Shelton, J.B., Cormick, J., Jones, R.T.: Biochemistry 2, 992-1008 (1963).

17. Braunitzer, G., Schrank, B., Stangl, A., Grillemeier, M.: Hoppe-Seyler's Z. Physiol. Chem. 359, 777-783 (1978).

18. Baralle, F.E., Shoulders, C.C., Proudfoot, N.J.: Cell. 21, 621-626 (1980).

THE COMPLETE PRIMARY STRUCTURE OF ELONGATION FACTOR G FROM *E.coli*

Yuly B. Alakhov, Lyudmila P. Motuz, Nataliya V. Dovgas, Leonid M. Vinokurov

Institute of Protein Research, Academy of Sciences of the USSR, 142292 Poustchino, Moscow Region, USSR

Introduction

Elongation factor G (EF-G) is a large protein consisting of one polypeptide chain with $M_r \sim 80\ 000$. Studies of this protein by traditional methods used to determine the primary structure did not seem the most rational. We chose the method of limited proteolysis as the first stage of studies permitting, in many cases, cleavage of the protein molecule into a small number of large fragments relatively stable against further effects of protease. On mild tryptic hydrolysis the G-factor is split into 5 fragments [1, cf. 2,3], 4 of which (T_4-T_7) form the complete protein polypeptide chain (for fragment nomenclature and arrangement in the protein chain see fig. 1). Thus, the studies of the primary structure of the elongation factor G led to determination of the structure of these 4 fragments and a search for the overlapping peptides for reconstitution of the entire polypeptide chain.

Studies of the structure of fragments T_5-T_7 were relatively simple due to their comparatively low molecular mass. The molecular mass of the N-terminal fragment T_6 is 6500, that of the following fragment T_7, 7500, and that of the C-terminal fragment T_5, 25 000. Traditional chemical and enzymatic methods of polypeptide chain cleavage permitted us to determine the complete primary structures of fragments T_6 [4], T_7 [5,6] and T_5 [7] involving a total of > 350 amino acid residues.

To resolve the complete structure of elongation factor G it

Chemistry of Peptides and Proteins, Vol. 2

was necessary to determine the amino acid sequence of the largest tryptic fragment T_4 representing the middle part of the protein molecule and to obtain the peptides joining all the fragments of limited trypsinolysis into a 1 polypeptide chain. To this end, the products of protein cleavage by cyanogen bromide and those of limited acidic hydrolysis at the Asp-Pro bond were studied.

The results of these studies permitted us to join all the fragments of limited trypsinolysis into one polypeptide chain and to present the complete amino acid sequence of the elongation factor G [8].

Results

Studies of the kinetics of tryptic hydrolysis of the G-factor and determination of the N-terminal sequences in the fragments [2] have shown that fragment T_6 is the N-terminal one. Following it along the chain is fragment T_7 containing a functionally important cysteine residue. The GTP-binding site is in the region of this fragment. Fragment T_7 is followed by fragment T_4 (M_r 41 000) and then by fragment T_5 which is the C-terminal part of the protein molecule (fig. 1).

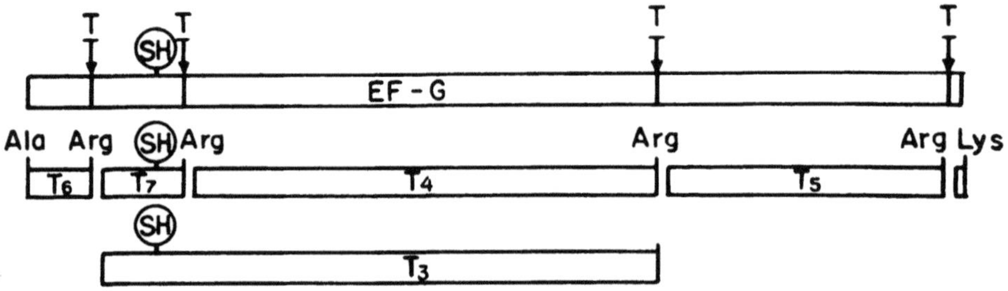

Fig. 1. Arrangement of fragments of limited tryptic hydrolysis along the polypeptide chain of the G-factor.

To elucidate the structure of fragment T_6 we investigated the amino acid sequence of peptides obtained after cleavage of the fragment molecule by cyanogen bromide, BNPS-skatole, trypsin, by trypsin after modification of the fragment at the lysine residues and by trypsin after its modification at the arginine residues. These studies permitted to determine the amino acid sequence of the N-terminal fragment of limited EF-G trypsinolysis. Its peptide chain consists of 58 amino acid residues (see Scheme 1), has one tryptophan residue and does not contain cysteine residues. The precise molecular mass of the fragment is 6559.37.

A structure study of fragment T_7 located along the EF-G polypeptide chain after fragment T_6 boiled down to clarifying the structure of peptides isolated after cleavage of the original T_7 molecule or its fragments by cyanogen bromide, trypsin, thermolysine, by S.aureus protease and limited acidic hydrolysis. The precise molecular mass of this region of the EF-G polypeptide chain is 7596.72. The polypeptide chain consists of 69 amino acid residues (positions 59-127 in Scheme 1), contains 2 tryptophan residues and 1 cysteine residue.

To study the C-terminal fragment of EF-G (fragment T_5) we examined the amino acid sequence of peptides isolated after cleavage of the T_5 molecule or its fragments by cyanogen bromide, trypsin, chymotrypsin, thermolysine and S.aureus protease. The polypeptide chain of the fragment consists of 228 amino acid residues (positions 472-699 in Scheme 1) and does not contain tryptophan and cystein residues.

The products of cyanogen bromide cleavage of the EF-G molecule were studied to establish the structure of fragment T_4, representing the middle part of the molecule, and to obtain overlappings between the fragments of limited proteolysis.

The molecule contains 22 methionine residues and it could be therefore expected that on cyanogen bromide cleavage ≥ 23 peptides would be formed. The scheme of cyanogen bromide hydrolysate separation proposed by us has allowed us to isolate all the 23 peptides. The mixture of peptides was modified by maleic anhydride and then separated on a Sephadex G-25 column. The fractions obtained were subjected to rechromatography on QAE-Sephadex A-25, and the fraction containing low molecular mass peptides was separated on thin-layer cellulose by peptide mapping. After chromatography on QAE-Sephadex A-25 the fractions were subjected once again to re-chromatography on QAE-Sephadex columns after a preliminary removal of the protecting maleic group.

Of the 23 isolated peptides 10 peptides are included in fragment T_4 while 2 peptides also involve some sites of polypeptide chains of fragments T_7 and T_5, and one peptide overlaps fragments T_6 and T_7 (Scheme 1). Thus, the study of the structure of fragment T_4 becomes the determination of the amino acid sequences of the 12 peptides of cyanogen bromide cleavage and the search for the overlaps between them. The structure studies of these peptides have required their additional cleavage by chemical or enzymic methods.

To obtain overlapping peptides we have used tryptic digest of fragment T_4 after its modification with maleic anhydride, digest by S.aureus protease and cleavage of the G-factor at the Asp-Pro bond. There are 3 such bonds in the molecule. Studies of the N-terminal sequences of the formed peptides by the automated Edman method with the use of fluram as a reagent blocking the primary α-amino groups have allowed us to obtain overlaps between the peptides from CBG-13 - CBG-17 (Scheme 1) and thus to reliably incorporate fragments T_4 and T_5 into one polypeptide chain.

From the data of the amino acid analyses and determination of the molecular mass of the fragments of limited trypsinolysis it can be concluded that trypsinolysis does not lead to a nick of any essential regions of the polypeptide chain between these fragments [2]. Separation of the products of cyanogen bromide elongation factor cleavage resulted in the isolation of several peptides, and the determination of their structures allowed us to unite all the fragments of limited trypsinolysis (peptides CBG-3, CBG-6 and CBG-17 in the Scheme) into one polypeptide chain. There are no insertions between the fragments of limited trypsinolysis, i.e., trypsin makes point disruptions in the elongation factor G polypeptide chain [4]. We demonstrated [2] that all the fragments of limited trypsinolysis have arginine as the C-terminal amino acid residue, whereas the C-terminal amino acid residue of EF-G is lysine. A structural study of the peptide CBG-23 indicates an elongation by two amino acid residues Gly-Lys as compared with the corresponding peptide of cyanogen bromide cleavage of fragment T_5 [7]. Thus, we have concluded that the cleavage of the C-terminal dipeptide is the result of limited trypsinolysis of the G-factor.

Thus, the study of fragments of limited trypsinolysis of the G-factor and the products of cyanogen bromide cleavage and hydrolysis at the Asp-Pro bond of the G-factor have allowed us to report the complete amino acid sequence of the protein. Its polypeptide chain consists of 701 amino acid residues, has M_r 77 321.46 and contains 5 cysteine residues, two of which (positions 296 and 392 in the Scheme) form the S-S bond. To find the position of the S-S bond, the G-factor was treated with Elman's reagent (to protect the SH-groups from oxidation) then subjected to cyanogen bromide cleavage. After removal of the protecting group the products of cyanogen bromide cleavage were chromatographed on a thiol-activated Sepharose 4B column. After removal of thiol-containing peptides from Sepharose they were carboxymethylated, then reduced by sodium sulphite and again carboxymethylated by ^{14}C-labelled ICH_2COOH. The cysteine residues containing the radioactive label were found in peptides CBG-12 and CBG-13 (Scheme 1). This technique provided the isolation of all thiol-containing peptides of the elongation factor G.

Elucidation of the primary structure of the G-factor will permit the regularities of the relationship between the structure and biological function to be studied on an entirely new basis. A new possibility appears for the study of the tertiary structure of the G-factor by X-ray structural analysis as large enough crystals have been obtained completely including the protein polypeptide chain.

SCHEME

COMPLETE PRIMARY STRUCTURE OF ELONGATION FACTOR G

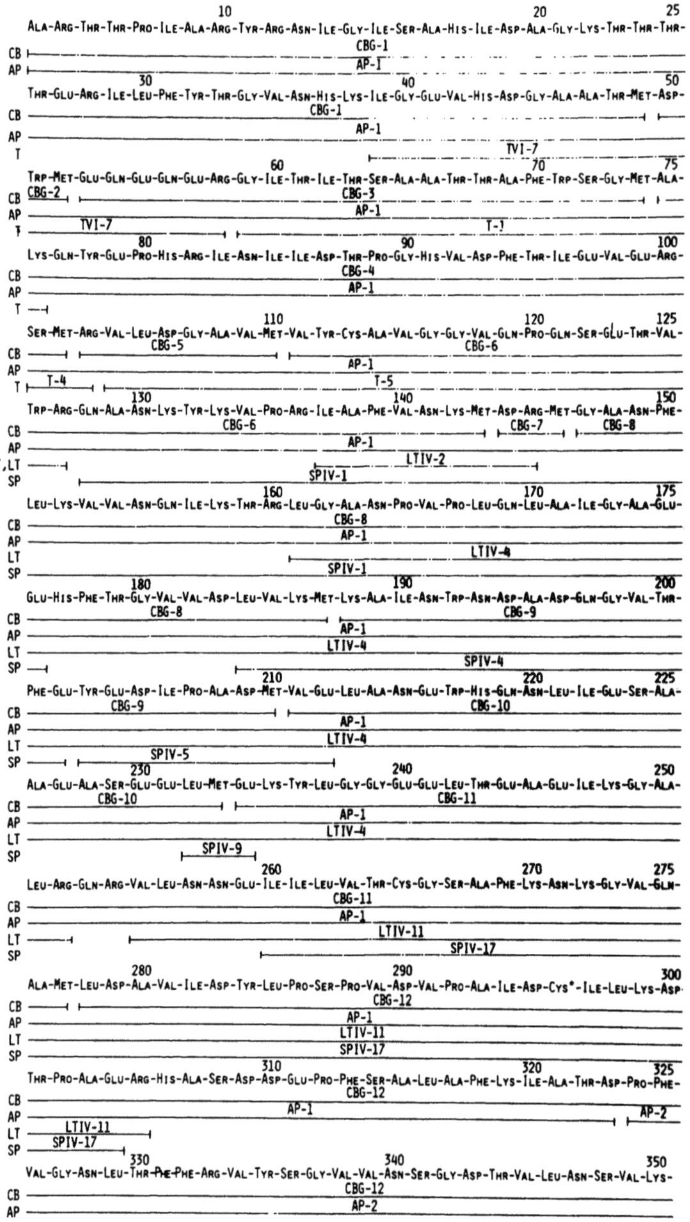

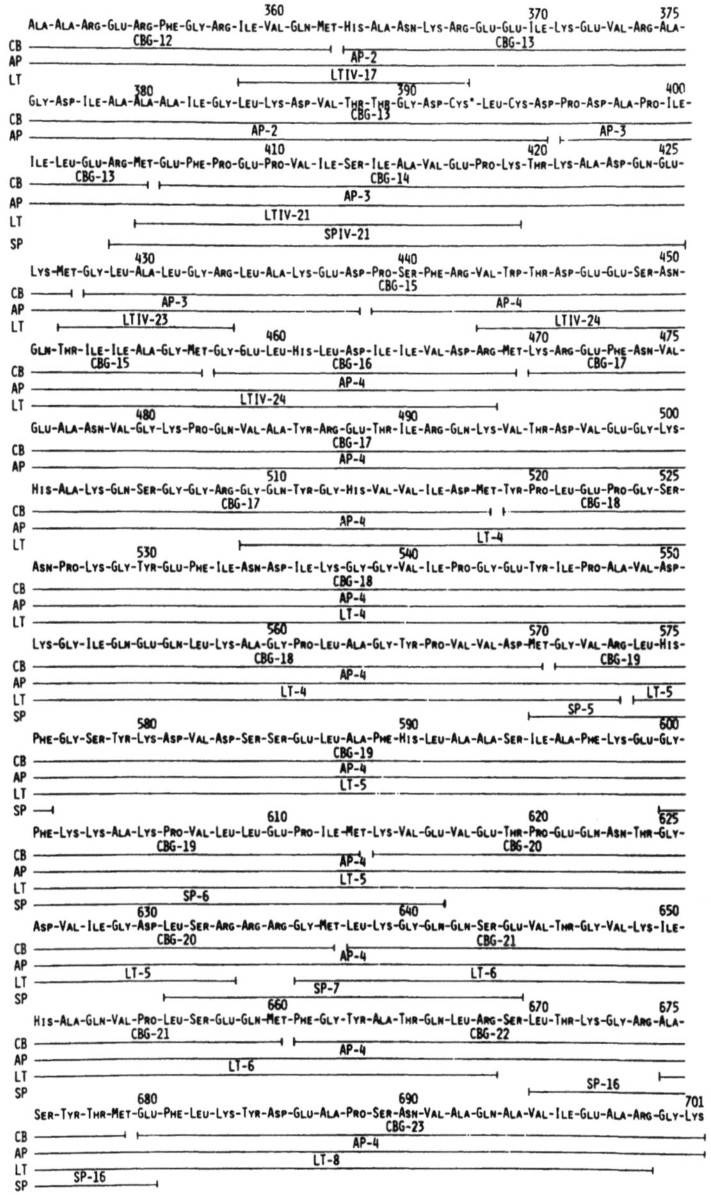

Solid lines indicate only cyanogen bromide peptides of EF-G and the overlapping peptides obtained from other hydrolyses of T_4-T_7 fragments of EF-G.

Cys* - cysteine residues which form the S-S bond.

Fragments of limited trypsinolysis occupy the following positions in the polypeptide chain: T_6 - 1-58, T_7 - 59-127, T_4 - 128-471, T_5 - 472-699.

CB - peptides of cyanogen bromide cleavage of EF-G. AP - peptides of EF-G hydrolysis at the Asp-Pro bond. T - peptides of tryptic hydrolysis of fragments T_6 and T_7. LT - peptides of tryptic hydrolysis of fragments T_4 and T_5 modified at lysine residues. SP - peptides of hydrolysis of fragments T_4 and T_5 by S.aureus protease.

References

1. Skar, D. C., Rohrbach, M. S., Bodley, J. W.: Biochemistry 14, 3922-3926 (1975).

2. Alakhov, Yu. B., Motuz, L. P., Stengrevics, O. A., Vinokurov, L. M., Ovchinnikov, Yu. A.: Bioorg. Khim. 3, 1336-1345 (1977).

3. Alakhov, Yu. B., Stengrevics, O. A., Bundulis, Yu. P., Motuz, L. P., Vinokurov, L. M.: Bioorg. Khim. 5, 330-339 (1979).

4. Motuz, L. P., Bundulis, Yu. P., Alakhov, Yu. B.: Bioorg. Khim. 5, 814-827 (1979).

5. Alakhov, Yu. B., Motuz, L. P., Stengrevics, O. A., Ovchinnikov, Yu. A.: FEBS Lett. 85, 287-290 (1978).

6. Alakhov, Yu. B., Motuz, L. P., Stengrevics, O. A., Vinokurov, L. M.: Bioorg. Khim. 4, 1301-1313 (1978).

7. Alakhov, Yu. B., Dovgas, N. V., Motuz, L. P., Vinokurov, L. M., Ovchinnikov, Yu. A.: FEBS Lett. 126, 183-186 (1981).

8. Ovchinnikov, Yu. A., Alakhov, Yu. B., Bundulis, Yu. P., Bundule, M. A., Dovgas, N. V., Kozlov, V. P., Motuz, L. P., Vinokurov, L. M.: FEBS Lett. 139, 130-135 (1982).

THE PRIMARY STRUCTURE OF E.coli RNA POLYMERASE. NUCLEOTIDE SEQUENCE OF THE rpoC GENE AND AMINO ACID SEQUENCE OF THE β'-SUBUNIT

V.V.Gubanov, G.S.Monastyrskaya, S.O.Guryev, I.S.Salomatina, T.M.Shuvaeva, A.P.Bogachuk, V.M.Lipkin, E.D.Sverdlov.
M.M.Shemyakin Institute of Bioorganic Chemistry, USSR Academy of Sciences, Moscow, USSR.

Introduction

The primary structure determination of E.coli DNA dependent-RNA polymerase is necessary for understanding the mechanism of its activity. Recently we determined complete amino acid sequences of its α-/1/ and β-subunits /2/. The primary structure of the β-subunit containing 1342 amino acid residues was established by using parallel research of the protein amino acid sequence and the nucleotide sequence of its structural gene. Combination of protein and nucleotide chemistry methods greatly enhanced the reliability of the analysis. At present this approach was successfully applied in the sequencing of the β'-subunit of RNA polymerase. The DNA fragments containing rpoC gene fragments and ajacent sequences were cloned in pBR-322. Their sequences were determined from both complementary chains by the modified Maxam-Gilbert procedure /3/. The amino acid sequence of the β'-subunit deduced from the nucleotide sequence was compared and appeared to be in complete accord with structures of the peptides obtained by the cleavage of the protein with cyanogen bromide and trypsin. The β'-subunit of E.coli RNA polymerase comprises 1407 amino acid residues. The β'-subunit sequence determination completes the study of the primary structure of the E.coli RNA polymerase core enzyme. Recently the sequence of the rpoD gene was also determined /4/. Thus, the primary structure of the whole RNA polymerase holoenzyme is now available,

Chemistry of Peptides and Proteins, Vol. 2

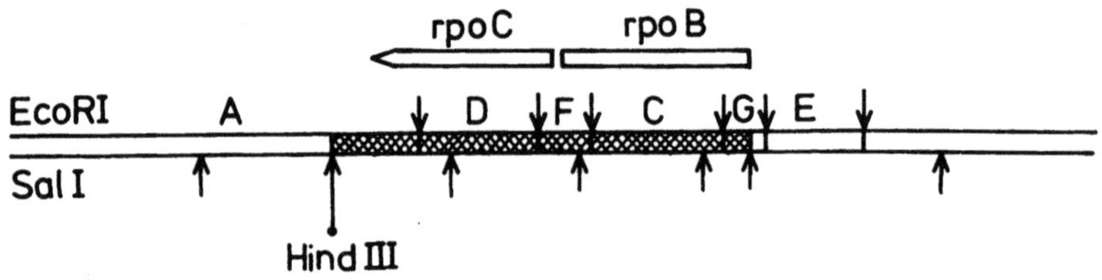

Fig.1. EcoRI and SalI restriction endonuclease cleavage map of the E.coli DNA region containing the structural genes (rpoB and rpoC) of the β and β'-subunits of RNA polymerase. Hind III cleavage region is also denoted in fragment EcoRI-A. The fragments, for which the primary structure is established, are hatched.

permitting investigation of its function.

Results and discussion

The rpoBC operon coding for β- and β'-subunits of E.coli RNA polymerase is located near 89 min on the genetic map of the bacterium. The positions of the splitting sites of several restriction endonucleases on the DNA in this region were established and given in Fig. 1. We determined primary structure of the EcoRI-G,EcoRI-C and EcoRI-F fragments containing the complete rpoB (β-subunit) gene and a proximal part of the rpoC (β'-subunit) gene /2/. For sequencing the rest part of the operon the fragments EcoRI-A - Sal I and EcoRI-D were cloned in pBR--322 plasmid. The recombinant plasmid, containing the first of them, was splitted with EcoRI and HindIII to obtain EcoRI-A - HindIII fragment. After isolation of this fragment and EcoRI-D fragment from the second recombinant plasmid each of them was splitted with several restriction endonucleases listed in Fig.2. The resulting mixtures of comparatively small subfragments were terminally labelled with ^{32}P by using T4 polynucleotide kinase and γ-^{32}P -ATP. Isolated individual subfragments were subjected to separation and then single strands were sequenced by the modified /5/ Maxam-Gilbert technique /6/. Practically the whole structure was determined from both complementary strands. The

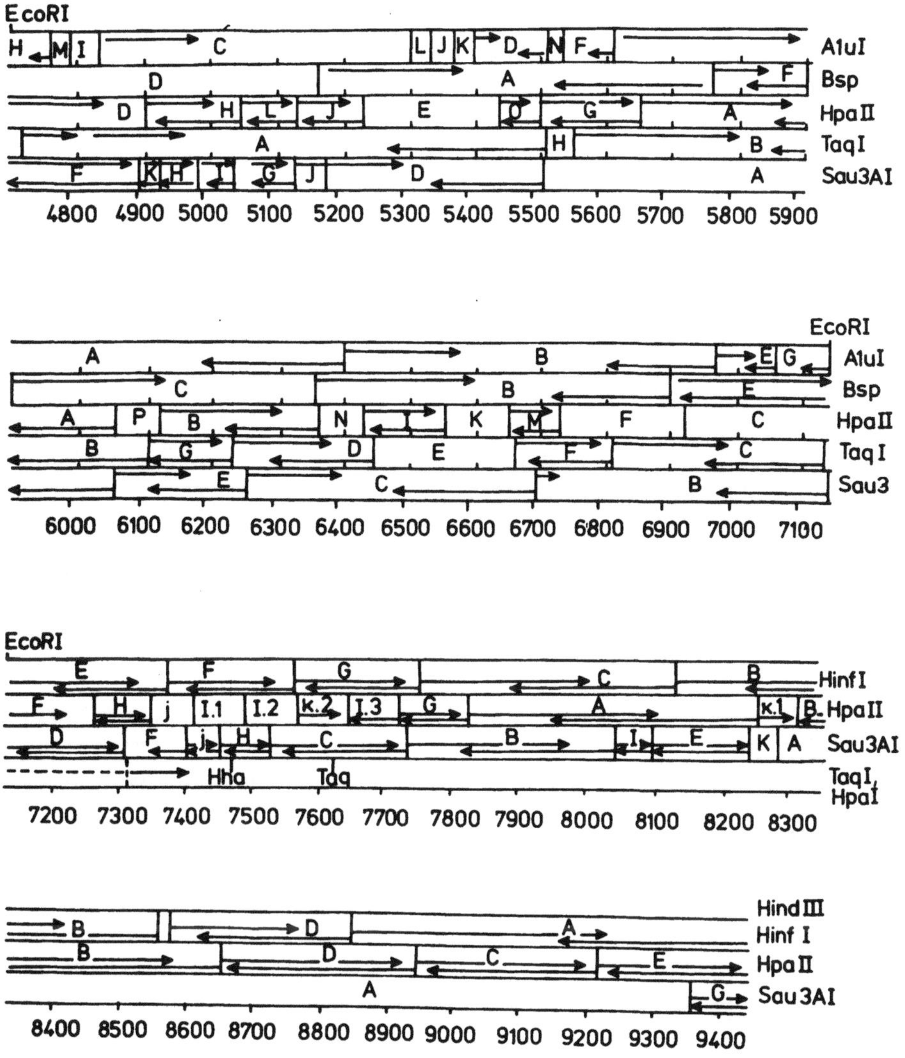

Fig. 2. The scheme for determining the sequences of fragments EcoRI-D and EcoRI-A - HindIII. The restriction subfragments are represented by rectangles. The arrows designate the lenghts of the determined subfragments of complementary chains.

complete structures of the fragments were deduced on the basis of overlaping of the subfragment sequences.

It should be noted that in the case of restriction endonuclease Taq I not all of the potential sites of splitting are

4130-4210	CTG CTG TCG GGT TAA AAC CCG GCA GCG GAT TGT GCT AAC TCC GAC GGG AGC AAA TCC GTG AAA GAT TTA TTA AAG TTT CTG
1-8	Met-Lys-Asp-Leu-Leu-Lys-Phe-Leu-
4211-4291	AAA GCG CAG ACT AAA ACC GAA GAG TTT GAT GCG ATC AAA ATT GCT CTG GCT TCG CCA GAC ATG ATC CGT TCA TGG TCT TTC
9-35	Lys-Ala-Gln-Thr-Lys-Thr-Glu-Glu-Phe-Asp-Ala-Ile-Lys-Ile-Ala-Leu-Ala-Ser-Pro-Asp-Met-Ile-Arg-Ser-Trp-Ser-Phe-
4292-4372	GGT GAA GTT AAA AAG CCG GAA ACC ATC AAC TAC CGT ACG TTC AAA CCA GAA CGT GAC GGC CTT TTC TGC GCC CGT ATC TTT
36-62	Gly-Glu-Val-Lys-Lys-Pro-Glu-Thr-Ile-Asn-Tyr-Arg-Thr-Phe-Lys-Pro-Glu-Arg-Asp-Gly-Leu-Phe-Cys-Ala-Arg-Ile-Phe-
4373-4453	GGG CCG GTA AAA GAT TAC GAG TGC CTG TGC GGT AAG TAC AAG CGC CTG AAA CAC CGT GGC GTC ATC TGT GAG AAG TGC GGC
63-89	Gly-Pro-Val-Lys-Asp-Tyr-Glu-Cys-Leu-Cys-Gly-Lys-Tyr-Lys-Arg-Leu-Lys-His-Arg-Gly-Val-Ile-Cys-Glu-Lys-Cys-Gly-
4454-4534	GTT GAA GTG ACC CAG ACT AAA GTA CGC CGT GAG CGT ATG GGC CAC ATC GAA CTG GCT TCC CCG ACT GCG CAC ATC TGG TTC
90-116	Val-Glu-Val-Thr-Gln-Thr-Lys-Val-Arg-Arg-Glu-Arg-Met-Gly-His-Ile-Glu-Leu-Ala-Ser-Pro-Thr-Ala-His-Ile-Trp-Phe-
4535-4615	CTG AAA TCG CTG CCG TCC CGT ATC GGT CTG CTG CTC GAT ATG CCG CTG CGC GAT ATC GAA CGC GTA CTG TAC TTT GAA TCC
117-143	Leu-Lis-Ser-Leu-Pro-Ser-Arg-Ile-Gly-Leu-Leu-Leu-Asp-Met-Pro-Leu-Arg-Asp-Ile-Glu-Arg-Val-Leu-Tyr-Phe-Glu-Ser-
4616-4696	TAT GTG GTT ATC GAA GGC GGT ATG ACC AAC *CTG GAA CGT CAG CAG ATC CTG ACT GAA GAG CAG TAT CTG GAC GCG CTG GAA
144-170	Tyr-Val-Val-Ile-Glu-Gly-Gly-Met-Thr-Asn-Leu-Glu-Arg-Gln-Gln-Ile-Leu-Thr-Glu-Glu-Gln-Tyr-Leu-Asp-Ala-Leu-Glu-
4697-4777	GAG TTC GGT GAC GAA TTC GAC GCG AAG ATG GGG GCG GAA GCA ATC ☆CAG GCT CTG CTG AAG AGC ATG GAT CTG GAG CAA GAG
171-197	Glu-Phe-Gly-Asp-Glu-Phe-Asp-Ala-Lys-Met-Gly-Ala-Glu-Ala-Ile-Gln-Ala-Leu-Leu-Lys-Ser-Met-Asp-Leu-Glu-Gln-Glu-
4778-4858	TGC GAA CAG CTG CGT GAA GAG CTG AAC GAA ACC AAC TCC GAA ACC AAG CGT AAA AAG CTG ACC AAG CGT ATC AAA CTG CTG
198-224	Cys-Glu-Gln-Leu-Arg-Glu-Glu-Leu-Asn-Glu-Thr-Asn-Ser-Glu-Thr-Lys-Arg-Lys-Lys-Leu-Thr-Lys-Arg-Ile-Lys-Leu-Leu-
4859-4939	GAA GCG TTC GTT CAG TCT GGT AAC AAA CCA GAG TGG ATG ATC CTG ACC GTT CTG CCG GTA CTG CCG CCA GAT CTG CGT CCG
225-251	Glu-Ala-Phe-Val-Gln-Ser-Gly-Asn-Lys-Pro-Glu-Trp-Met-Ile-Leu-Thr-Val-Leu-Pro-Val-Leu-Pro-Pro-Asp-Leu-Arg-Pro-
4940-5020	CTG GTT CCG CTG GAT GGT GGT CGT TTC GCG ACT TCT GAC CTG AAC GAT CTG TAT CGT CGC GTC ATT AAC CGT AAC AAC CGT
252-278	Leu-Val-Pro-Leu-Asp-Gly-Gly-Arg-Phe-Ala-Thr-Ser-Asp-Leu-Asn-Asp-Leu-Tyr-Arg-Arg-Val-Ile-Asn-Arg-Asn-Asn-Arg-
5021-5101	CTG AAA CGT CTG CTG GAT CTG GCT GCG CCG GAC ATC ATC GTA CGT AAC GAA AAA CGT ATG CTG CAG GAA GCG GTA GAC GCC
279-305	Leu-Lys-Arg-Leu-Leu-Asp-Leu-Ala-Ala-Pro-Asp-Ile-Ile-Val-Arg-Asn-Gly-Lys-Arg-Met-Leu-Gln-Glu-Ala-Val-Asp-Ala-
5102-5182	CTG CTG GAT AAC GGT CGT CGC GGT CGT GCG ATC ACC GGT TCT AAC AAG CGT CCT CTG AAA TCT TTG GCC GAC ATG ATC AAA
306-332	Leu-Leu-Asp-Asn-Gly-Arg-Arg-Gly-Arg-Ala-Ile-Thr-Gly-Ser-Asn-Lys-Arg-Pro-Leu-Lys-Ser-Leu-Ala-Asp-Met-Ile-Lys-
5183-5263	GGT AAA CAG GGT CGT TTC CGT CAG AAC CTG CTC GGT AAG CGT GTT GAC TAC TCC GGT CGT TCT GTA ATC ACC GTA GGT CCA
333-359	Gly-Lys-Gln-Gly-Arg-Phe-Arg-Gln-Asn-Leu-Leu-Gly-Lys-Arg-Val-Asp-Tyr-Ser-Gly-Arg-Ser-Val-Ile-Thr-Val-Gly-Pro-
5264-5344	TAC CTG CGT CTG CAT CAG TGC GGT CTG CCG AAG AAA ATG GCA CTG GAG CTG TTC AAA CCG TTC ATC TAC GGC AAG CTG GAA
360-386	Tyr-Leu-Arg-Leu-His-Gln-Cys-Gly-Leu-Pro-Lys-Lys-Met-Ala-Leu-Glu-Leu-Phe-Lys-Pro-Phe-Ile-Tyr-Gly-Lys-Leu-Glu-
5345-5425	CTG CGT GGT CTT GCT ACC ACC ATT AAA GCT GCG AAG AAA ATG GTT GAG CGC GAA GAA GCT GTC GTT TGG GAT ATC ☆CTG GAC
387-413	Leu-Arg-Gly-Leu-Ala-Thr-Thr-Ile-Lys-Ala-Ala-Lys-Lys-Met-Val-Glu-Arg-Glu-Glu-Ala-Val-Val-Trp-Asp-Ile-Leu-Asp-
5426-5506	GAA GTT ATC CGC GAA CAC CCG GTA CTG CTG AAC CGT GCA CCG ACT CTG CAC CGT CTG GGT ATC ☆CAG GCA TTT GAA CCG GTA
414-440	Glu-Val-Ile-Arg-Glu-His-Pro-Val-Leu-Leu-Asn-Arg-Ala-Pro-Thr-Leu-His-Arg-Leu-Gly-Ile-Gln-Ala-Phe-Glu-Pro-Val-
5507-5587	CTG ATC GAA GGT AAA GCT ATC CAG CTG CAC CCG CTG GTT TGT GCG GCA TAT AAC GCC GAC TTC GAT GGT GAC CAG ATG GCT
441-467	Leu-Ile-Glu-Gly-Lys-Ala-Ile-Gln-Leu-His-Pro-Leu-Val-Cys-Ala-Ala-Tyr-Asn-Ala-Asp-Phe-Asp-Gly-Asp-Gln-Met-Ala-
5588-5668	GTT CAC GTA CCG CTG ACG CTG GAA GCC CAG CTG GAA GCG CGT GCG CTG ATG ATG TCT ACC AAC AAC ATC CTG TCC CCG GCG
468-494	Val-His-Val-Pro-Leu-Thr-Leu-Glu-Ala-Gln-Leu-Glu-Ala-Arg-Ala-Leu-Met-Met-Ser-Thr-Asn-Asn-Ile-Leu-Ser-Pro-Ala-
5669-5749	AAC GGC GAA CCA ATC ATC GTT CCG TCT CAG GAC GTT GTA CTG GGT CTG TAC TAC ATG ACC CGT GAC TGT GTT AAC GCC AAA
495-521	Asn-Gly-Glu-Pro-Ile-Ile-Val-Pro-Ser-Gln-Asp-Val-Val-Leu-Gly-Leu-Tyr-Tyr-Met-Thr-Arg-Asp-Cys-Val-Asn-Ala-Lys-
5750-5830	GGC GAA GGC ATG GTG CTG ACT GGC CCG AAA GAA GCA GAA CGT CTG TAT CGC TCT GGT CTG GCT TCT CTG CAT GCG CGC GTT
522-548	Gly-Glu-Gly-Met-Val-Leu-Thr-Gly-Pro-Lys-Glu-Ala-Glu-Arg-Leu-Tyr-Arg-Ser-Gly-Leu-Ala-Ser-Leu-His-Ala-Arg-Val-
5831-5911	AAA GTG CGT ATC ACC GAG TAT GAA AAA GAT GCT AAC GGT GAA TTA GTA GCG AAA ACC AGC CTG AAA GAC ACG ACT GTT GGC
549-575	Lys-Val-Arg-Ile-Thr-Glu-Tyr-Glu-Lys-Asp-Ala-Asn-Gly-Glu-Leu-Val-Ala-Lys-Thr-Ser-Leu-Lys-Asp-Thr-Thr-Val-Gly-
5912-5992	CGT GCC ATT CTG TGG ATG ATT GTA CCG AAA GGT CTG CCT TAC TCC ATC GTC AAC ☆CAG GCG CTG GGT AAA AAA GCA ATC TCC
576-602	Arg-Ala-Ile-Leu-Trp-Met-Ile-Val-Pro-Lys-Gly-Leu-Pro-Tyr-Ser-Ile-Val-Asn-Gln-Ala-Leu-Gly-Lys-Lys-Ala-Ile-Ser-
5993-6073	AAA ATG CTG AAC ACC TGC TAC CGC ATT CTC GGT CTG AAA CCG ACC GTT ATT TTT GCG GAC CAG ATC ATG TAC ACC GGC TTC
603-629	Lys-Met-Leu-Asn-Thr-Cys-Tyr-Arg-Ile-Leu-Gly-Leu-Lys-Pro-Thr-Val-Ile-Phe-Ala-Asp-Gln-Ile-Met-Tyr-Thr-Gly-Phe-
6074-6154	GCC TAT GCA GCG CGT TCT GGT GCA TCT GTT GGT ATC GAT GAC ATG GTC ATC CCG GAG AAG AAA CAC GAA ATC ATC TCC GAG
630-656	Ala-Tyr-Ala-Ala-Arg-Ser-Gly-Ala-Ser-Val-Gly-Ile-Asp-Asp-Met-Val-Ile-Pro-Glu-Lys-Lys-His-Glu-Ile-Ile-Ser-Glu-
6155-6235	GCA GAA GCA GAA GTT GCT GAA ATT CAG GAG CAG TTC CAG TCT GGT CTG GTA ACT GCG GGC GAA CGC TAC AAC AAA GTT ATC
657-683	Ala-Glu-Ala-Glu-Val-Ala-Glu-Ile-Gln-Glu-Gln-Phe-Gln-Ser-Gly-Leu-Val-Thr-Ala-Gly-Glu-Arg-Tyr-Asn-Lys-Val-Ile-
6236-6316	GAT ATC TGG GCT GCG GCG AAC GAT CGT GTA TCC AAA GCG ATG ATG GAT AAC CTG CAA ACT GAA ACC GTG ATT AAC CGT GAC
684-710	Asp-Ile-Trp-Ala-Ala-Ala-Asn-Asp-Arg-Val-Ser-Lys-Ala-Met-Met-Asp-Asn-Leu-Gln-Thr-Glu-Thr-Val-Ile-Asn-Arg-Asp-
6317-6397	GGT CAG GAA GAG AAG CAG GTT TCC TTC AAC AGC ATC TAC ATG ATG GCC GAC TCC GGT GCG CGT GGT TCT GCG GCA CAG ATT
711-737	Gly-Gln-Glu-Glu-Lys-Gln-Val-Ser-Phe-Asn-Ser-Ile-Tyr-Met-Met-Ala-Asp-Ser-Gly-Ala-Arg-Gly-Ser-Ala-Ala-Gln-Ile-
6398-6478	CGT CAG CTT GCT GGT ATG CGT GGT CTG ATG GCG AAG CCG GAT GGC TCC ATC ATC GAA ACG CCA ATC ACC GCG AAC TTC CGT
738-764	Arg-Gln-Leu-Ala-Gly-Met-Arg-Gly-Leu-Met-Ala-Lys-Pro-Asp-Gly-Ser-Ile-Ile-Glu-Thr-Pro-Ile-Thr-Ala-Asn-Phe-Arg-
6479-6559	GAA GGT CTG AAC GTA CTC CAG TAC TTC ATC TCC ACC CAC GGT GCT CGT AAA GGT CTG GCG GAT ACC GCA CTG AAA ACT GCG
765-791	Glu-Gly-Leu-Asn-Val-Leu-Gln-Tyr-Phe-Ile-Ser-Thr-His-Gly-Ala-Arg-Lys-Gly-Leu-Ala-Asp-Thr-Ala-Leu-Lys-Thr-Ala-
6560-6640	AAC TCC GGT TAC CTG ACT CGT CGT CTG GTT GAC GTG GCG CAG GAC ☆CTG GTG GTT ACC GAA GAC GAT TGT GGT ACC CAT GAA
792-818	Asn-Ser-Gly-Tyr-Leu-Thr-Arg-Arg-Leu-Val-Asp-Val-Ala-Gln-Asp-Leu-Val-Val-Thr-Glu-Asp-Asp-Cys-Gly-Thr-His-Glu-
6641-6721	GGT ATC ATG ATG ACT CCG GTT ATC GAG GGT GGT GAC GTT AAA GAG CCG CTG CGC GAT CGC GTA CTG GGT CGT GTA ACT GCT
819-845	Gly-Ile-Met-Met-Thr-Pro-Val-Ile-Glu-Gly-Gly-Asp-Val-Lys-Glu-Pro-Leu-Arg-Asp-Arg-Val-Leu-Gly-Arg-Val-Thr-Ala-
6722-6802	GAA GAC GTT CTG AAG CCG GGT ACT GCT GAT ATC CTC GTT CCG CGC AAC ACG CTG CTG CAC GAA CAG TGG TGT GAC CTG CTG
846-872	Glu-Asp-Val-Leu-Lys-Pro-Gly-Tre-Ala-Asp-Ile-Leu-Val-Pro-Arg-Asn-Tre-Leu-Leu-His-Glu-Gln-Trp-Cys-Asp-Leu-Leu-
6803-6883	GAA GAG AAC TCT GTC GAC GCG GTT AAA GTA CGT TCT GTT GTA TCT TGT GAC ACC GAC TTT GGT GTA TGT GCG CAC TGC TAC
873-899	Glu-Glu-Asn-Ser-Val-Asp-Ala-Val-Lys-Val-Arg-Ser-Val-Val-Ser-Cys-Asp-Thr-Asp-Phe-Gly-Val-Cys-Ala-His-Cys-Tyr-
6884-6964	GGT CGT GAC *CTG GCG CGT GGC CAC ATC ATC AAC AAG GGT GAA GCA ATC GGT GTT ATC GCG GCA CAG TCC ATC GGT GAA CCG
900-926	Gly-Arg-Asp-Leu-Ala-Arg-Gly-His-Ile-Ile-Asn-Lys-Gly-Glu-Ala-Ile-Gly-Val-Ile-Ala-Ala-Gln-Ser-Ile-Gly-Glu-Pro-
6965-7045	GGT ACA CAG CTG ACC ATG CGT ACG TTC CAC ATC GGT GGT GCG GCA TCT CGT GCG GCT GCT GAA TCC AGC ATC CAA GTG AAA
927-953	Gly-Thr-Gln-Leu-Thr-Met-Arg-Thr-Phe-His-Ile-Gly-Gly-Ala-Ala-Ser-Arg-Ala-Ala-Ala-Glu-Ser-Ser-Ile-Gln-Val-Lys-

7046-7126	AAC AAA GGT AGC ATC AAG CTC AGC AAC GTG AAG TCG GTT GTG AAC TCC AGC GGT AAA CTG GTT ATC ACT TCC CGT AAT ACT
954-980	Asn-Lys-Gly-Ser-Ile-Lys-Leu-Ser-Asn-Val-Lys-Ser-Val-Val-Asn-Ser-Ser-Gly-Lys-Leu-Val-Ile-Thr-Ser-Arg-Asn-Thr-
7127-7207	GAA CTG AAA CTG ATC GAC GAA TTC GGT CGT ACT AAA GAA AGC TAC AAA GTA CCT TAC GGT GCG GTA CTG GCG AAA GGC GAT
981-1007	Glu-Leu-Lys-Leu-Ile-Asp-Glu-Phe-Gly-Arg-Thr-Lys-Glu-Ser-Tyr-Lys-Val-Pro-Tyr-Gly-Ala-Val-Leu-Ala-Lys-Gly-Asp-
7208-7288	GGC GAA CAG GTT GCT GGC GGC GAA ACC GTT GCA AAC TGG GAC CCG CAC ACC ATG CCG GTT ATC ACC GAA GTA AGC GGT TTT
1008-1034	Gly-Glu-Gln-Val-Ala-Gly-Gly-Glu-Thr-Val-Ala-Asn-Trp-Asp-Pro-His-Thr-Met-Pro-Val-Ile-Thr-Glu-Val-Ser-Gly-Phe-
7289-7369	GTA CGC TTT ACT GAC ATG ATC GAC GGC CAG ACC ATT ACG CGT CAG ACC GAC GAA CTG ACC GGT CTG TCT TCG CTG GTG GTT
1035-1061	Val-Arg-Phe-Thr-Asp-Met-Ile-Asp-Gly-Gln-Thr-Ile-Thr-Arg-Gln-Thr-Asp-Glu-Leu-Thr-Gly-Leu-Ser-Ser-Leu-Val-Val-
7370-7450	CTG GAT TCC GCA GAA CGT ACC GCA GGT GGT AAA GAT CTG CGT CCG GCA CTG AAA ATC GTT GAT GCT CAG GGT AAC GAC GTT
1062-1088	Leu-Asp-Ser-Ala-Glu-Arg-Thr-Ala-Gly-Gly-Lys-Asp-Leu-Arg-Pro-Ala-Leu-Lys-Ile-Val-Asp-Ala-Gln-Gly-Asn-Asp-Val-
7451-7531	CTG ATC CC*A GGT ACC GAT ATG CCA GCG CAG TAC TTC CTG CCG GGT AAA GCG ATT GTT CAG CTG GAA GAT GGC GTA CAG ATC
1089-1115	Leu-Ile-Pro-Gly-Thr-Asp-Met-Pro-Ala-Gln-Tyr-Phe-Leu-Pro-Gly-Lys-Ala-Ile-Val-Gln-Leu-Glu-Asp-Gly-Val-Gln-Ile-
7532-7612	AGC TCT GGT GAC ACC C*TG GCG CGT ATT CCG CAG GAA TCC GGC GGT ACC AAG GAC ATC ACC GGT GGT CTG CCG CGC GTT GCG
1116-1142	Ser-Ser-Gly-Asp-Thr-Leu-Ala-Arg-Ile-Pro-Gln-Glu-Ser-Gly-Gly-Thr-Lys-Asp-Ile-Thr-Gly-Gly-Leu-Pro-Arg-Val-Ala-
7613-7693	GAC CTG TTC GAA GCA CGT CGT CCG AAA GAG CCG GCA ATC C*TG GCT GAA ATC AGC GGT ATC GTT TCC TTC GGT AAA GAA ACC
1143-1169	Asp-Leu-Phe-Glu-Ala-Arg-Arg-Pro-Lys-Glu-Pro-Ala-Ile-Leu-Ala-Glu-Ile-Ser-Gly-Ile-Val-Ser-Phe-Gly-Lys-Glu-Thr-
7694-7774	AAA GGT AAA CGT CGT CTG GTT ATC ACC CCG GTA GAC GGT AGC GAT CCG TAC GAA GAG ATG ATT CCG AAA TGG CGT CAG CTC
1170-1196	Lys-Gly-Lys-Arg-Arg-Leu-Val-Ile-Thr-Pro-Val-Asp-Gly-Ser-Asp-Pro-Tyr-Glu-Glu-Met-Ile-Pro-Lys-Trp-Arg-Gln-Leu-
7775-7855	AAC GTG TTC GAA GGT GAA CGT GTA GAA CGT GGT GAC GTA ATT TCC GAC GGT CCG GAA GCG CCG CAC GAC ATT CTG CGT CTG
1197-1223	Asn-Val-Phe-Glu-Gly-Glu-Arg-Val-Glu-Arg-Gly-Asp-Val-Ile-Ser-Asp-Gly-Pro-Glu-Ala-Pro-His-Asp-Ile-Leu-Arg-Leu-
7856-7936	CGT GGT GTT CAT GCT GTT ACT CGT TAC ATC GTT AAC GAA GTA CAG GAC GTA TAC CGT CTG CAG GGC GTT AAG ATT AAC GAT
1224-1250	Arg-Gly-Val-His-Ala-Val-Thr-Arg-Tyr-Ile-Val-Asn-Glu-Val-Gln-Asp-Val-Tyr-Arg-Leu-Gln-Gly-Val-Lys-Ile-Asn-Asp-
7937-8017	AAA CAC ATC GAA GTT ATC GTT CGT CAG ATG CTG CGT AAA GCT ACC ATC GTT AAC GCG GGT AGC TCC GAC TTC C*TG GAA GGC
1251-1277	Lys-His-Ile-Glu-Val-Ile-Val-Arg-Gln-Met-Leu-Arg-Lys-Ala-Thr-Ile-Val-Asn-Ala-Gly-Ser-Ser-Asp-Phe-Leu-Glu-Gly-
8018-8098	GAA CAG GTT GAA TAC TCT CGC GTC AAG ATC GCA AAC CGC GAA CTG GAA GCG AAC GGC AAA GTG GGT GCA ACT TAC TCC CGC
1278-1304	Glu-Gln-Val-Glu-Tyr-Ser-Arg-Val-Lys-Ile-Ala-Asn-Arg-Glu-Leu-Glu-Ala-Asn-Gly-Lys-Val-Gly-Ala-Thr-Tyr-Ser-Arg-
8099-8179	GAT CTG CTG GGT ATC ACC AAA GCG TCT CTG GCA ACC GAG TCC TTC ATC TCC GCG GCA TCG TTC C*AG GAG ACC ACT CGC GTG
1305-1331	Asp-Leu-Leu-Gly-Ile-Thr-Lys-Ala-Ser-Leu-Ala-Thr-Glu-Ser-Phe-Ile-Ser-Ala-Ala-Ser-Phe-Gln-Glu-Thr-Thr-Arg-Val-
8180-8260	CTG ACC GAA GCA GCC GTT GCG GGC AAA CGC GAC GAA CTG CGC GGC CTG AAA GAG AAC GTT ATC GTG GGT CGT CTG ATC CCG
1332-1358	Leu-Thr-Glu-Ala-Ala-Val-Ala-Gly-Lys-Arg-Asp-Glu-Leu-Arg-Gly-Leu-Lys-Glu-Asn-Val-Ile-Val-Gly-Arg-Leu-Ile-Pro-
8261-8341	GCA GGT ACC GGT TAC GCG TAC CAC C*AG GAT CGT ATG CGT CGC CGT GCT GCG GGT GAA GCT CCG GCT GCA CCG CAG GTG ACT
1359-1385	Ala-Gly-Thr-Gly-Tyr-Ala-Tyr-His-Gln-Asp-Arg-Met-Arg-Arg-Arg-Ala-Ala-Gly-Glu-Ala-Pro-Ala-Ala-Pro-Gln-Val-Thr-
8342-8426	GCA GAA GAC GCA TCT GCC AGC C*TG GCA GAA CTG CTG AAC GCA GGT CTG GGC GGT TCT GAT AAC GAG TAA TCGTTAATCCGCAAA
1386-1407	Ala-Glu-Asp-Ala-Ser-Ala-Ser-Leu-Ala-Glu-Leu-Leu-Asn-Ala-Gly-Leu-Gly-Gly-Ser-Asp-Asn-Glu-Ter
8427-8532	TAACGTAAAAACCCGCTTCGGCGGGTTTTTTTATGGGGGGAGTTTAGGGAAAGAGCATTTGTCAGAATATTTAAGGAATTTCTGAATACTCATAATCAATGTAGAGA
8533-8639	TGACTAATATCCTGAAACTGACTGAACTAATTGAGTCAAACTCGGCAAGGATTCGATACTATTCCTGTGTAACTTTCTTAAGGAACGAGAATGAAACAGGAAGTGGA
8640-8746	AAAGTGGCGACCTTTTGGACATCCGGATGGTGATATTCGTGATTTATCATTTCTTGATGCTCATCAGGCTGTCTACGTTCAGCATCATGAGGGCAAAGAGCCTTTAG
8747-8853	AGTATCGCTTTTGGGTTACCTACTCTCTTCACTGCTTCACAAAAGATTATGAACATCAGACGAACGAAGAAAAACAATCGTTAATGTACCACGCGCCTAAAGAATCT
8854-8960	CGTCCCTTCTGCCAGCACCGTTATAACTTAGCGCGCACACACTTAAAAAGAACTATTTTGGCGCTGCCAGAAAGCAACGTTATTCATGCCGGGTATGGTAGCTATGC
8961-9067	CGTGATTGAGGTGGACTTAGACGGAGGAGATAAGGCATTTTACTTTGTTGCGTTCAGGGCTTTCAGGGAAAAGAAAAAACTCCGTTTGCATGTAACTAGCGCTTATC
9068-9174	CCATTTCTGAAAAACAGAAAGGTAAATCAGTGAAATTTTTCACCATTGCCTACAACTTATTGAGAAATAAGCAGCTTCCTCAGCCCTCAAAATAACAAAACCCACCT
9175-9281	TAAGGTGGGTTTCGCCAGAGAATTATCTCTGGTATTCAGAACGCCATTACCGGACTTTGCCTTGACCTTGCGATAATCGCAGGTTGCGGGATGTCTGAATTTCTTCA
9282-9388	GTCTGCTGCATCC*TGGAAGATGAGAACATGTGTTCTTATTTTCGTCTCTATCATAGTTGAGTATTTACTCTCTTACAATCAGATCTCTTTCATTGCTCAACAGGCGA
9389-9450	TGGCTTCAGACTTTGCATTACGGAATTTTTAAGAAAGGCAGGGCGAAACGAGGAAGAAGCTT

Fig. 3. The nucleotide sequence of the rpoBC operon segment containing the total structural rpoC gene and the total amino acid sequence of the β'-subunit of E.coli RNA polymerase. The nucleotide sequence of the complementary DNA chain, adequate to the sequence of mRNA, is given. Numeration of nucleotides corresponds to the complete rpoBC operon. The restriction EcoRI cleavage sites dividing fragments EcoRI-F, EcoRI-D and EcoRI-A - HindIII are situated between nucleotides 4709-4710 and 7145-7146. The underlined amino acid sequences are those the structures of which have been determined from analysis of corresponding peptides. C* - 5-methylcytosine. Inverted sequence repetitions entering into the proposed transcription terminator, are framed.

effective. There is the sequence TCGATC (7308-7311 b.p.) in the fragment which could not be digested with enzyme. The similar examples were already described /2/. Methylation of A residues in this sequence was proposed to be responsible for the effect.

For the protein structure analysis the isolated β'-subunit was cleavage with cyanogen bromide /7/. It was also digested with trypsin after the citraconic anhydride modification of lysine residues /8/. The determined peptides sequence comprises as much as 40% of the β'-subunit whole structure.

The N-terminal amino acid sequence Met-Lys-Asp-Leu-Leu-Lys-Phe-Leu of the β'-subunit was determined by means of automatic sequencing technique /7/.

The determined primary structure of the nucleic acid and the protein is presented in Fig. 3. The β'-subunit consists of 1407 amino acid residues, that corresponds to 155162,5 in mw. Its translation is initiated by GUG (4187-4189) codon. The Shine-Dalgarno sequence, GGAG (4176-4179), which is complementary to the 3'-end of 16S ribosomal RNA, is located at the distance of 9 nucleotides from the codon. Termination of the translation is brought about by amber codon TAA.

It is noteworthy that in interval 8433-8458 close to the terminating codon there is a thymidine cluster, which is preceeded by the sequence of hyphenated dyad symmetry. The structure of the site is strikingly similar to the known structures of the ρ-independent transcription terminator /9/. It was established recently /10/ that termination in the case of the rpoBC operon transcription really occurs close to the end of the rpoC gene. Therefore this sequence may be proposed to be a transcription terminator of the rpoBC operon.

The amino acid composition of the β'-subunit is Asp 81, Asn 48, Thr 77, Ser 71, Glu 109, Gln 50, Pro 57, Gly 115, Ala 124, Cys 15, Val 108, Met 36, Ile 92, Leu 139, Tyr 34, Phe 35, His 21, Lys 87, Arg 99, Trp 9. Thus, the β'-subunit is a basic protein, containing 207 basic and only 190 acidic amino acids. In the polypeptide strands of the β'- as well as β-subunits one can observe the regions of strong clustering of basic amino acids.

In the β'-subunit these sequences are located in intervals: 74-81, 213-222, 1167-1174 (Fig.3).It is of interest that the α-subunit does not contain similar clusters. Probably at least a few sequences participate directly in the contacts between RNA polymerase and DNA, and/or RNA.

Determination of the nucleotide sequence of the rpoC gene and amino acid sequence of the E.coli RNA polymerase β'-subunit completes the analysis of the core-enzyme primary structure.

References

1. Ovchinnikov,Yu.A., Lipkin,V.M., Modyanov,N.N., Chertov, O.Yu., Smirnov,Yu.V.: FEBS Lett. 76, 108-111 (1977).
2. Ovchinnikov,Yu.A., Monastyrskaya,G.S., Gubanov,V.V., Guryev,S.O., Chertov,O.Yu., Modyanov,N.N., Grinkevich,V.A., Makarova,I.A., Marchenko,T.V., Polovnikova,I.N., Lipkin, V.M., Sverdlov,E.D.: Eur. J. Biochem. 116,621-629 (1980).
3. Monastyrskaya,G.S., Guryev,S.O., Kalinina,N.F., Sorokin, A.V., Salomatina,I.S., Shuvaeva,T.M., Lipkin,V.M., Sverdlov,E.D., Ovchinnikov,Yu.A.: Bioorgan. Khim. 8 130-134 (1982).
4. Burton,Z., Burgess,R.R., Lin,J., Moore,D., Holder,S., Gross,C.A.: Nucl. Acids Res. 9, 2889-2898 (1981).
5. Ovchinnikov,Yu.A., Guryev,S.O., Krayev,A.S., Monastyrskaya, G.S., Skryabin,K.G., Sverdlov,E.D., Zakharyev,V.M., Bayev, A.A.: Gene 6, 235-249 (1979).
6. Maxam,A., Gilbert,W.: Proc. Nat. Acad. Sci.,USA 74, 560-564 (1977).
7. Shuvaeva,T.M., Lipkin,V.M., Nazimov,I.V., Modyanov,N.N., Ovchinnikov,Yu.A.: Bioorgan. Khim. 7, 1765-1777 (1981).
8. Atassi,M.Z., Habub,R.F.S.A. : Methods in Enzymology, S.P., Golovick, N.O.,Kaplan, Eds. N.Y., London: Academic Press, 25B, 546-553,1972.
9. Rosenberg,M., Court,D.: Ann. Rev. Genet. 13,319-353 (1972).
10. An,G., Frisen,J.D.: J. Bacteriol. 144, 904-916 (1980).

DETERMINATION OF AMINO ACID SEQUENCES OF LARGE HYDROPHOBIC PEPTIDES OF RHODOPSIN

Maria B.Kostina, Alexandre S.Zolotarev
Shemyakin Institute of Bioorganic Chemistry, USSR Academy of Sciences, Moscow, USSR

Introduction

The rhodopsin molecule consists of a single glycosylatet polypeptide chain, having a molecular weight 38 000. Rhodopsin is the integral membrane protein with a high proportion of hydrophobic amino acids (to 60%). Determination of the primary structure of such proteins is still a complicated task. During the course of rhodopsin sequencing we used methodological approaches which were worked out under structure-function investigation of bacteriorhodopsin (1). The basic methods for fragmentation of the polypeptide chain were chemical cleavages at methionine and tryptophan residues, enzymatic digestion of bleached membranes by chymotrypsin and trypsin and limited proteolysis of native disk membranes.
This report deals with some methodological aspects of determination of the primary structure of integral membrane proteins on the example of large hydrophobic fragments from different hydrolysates of rhodopsin.

Results

The most valuable information was obtained during the study of products resulting from cleavage of rhodopsin at the methionine residues. The cleavage was carried out with cyano-

Chemistry of Peptides and Proteins, Vol. 2

gen bromide according to standard procedures (2). The peptide mixture was separated by gel filtration on Bio-Gels in formic acid and by reversed-phase high-pressure liquid chromatography (h.p.l.c.) using the gradient system consisting of acetonitrile in a 0.1% trifluoroacetic acid and 10 mM ammonium acetate. Cyanogen bromide cleavage resulted in 15 peptides, 9 of which containing from 20 to58 amino acid residues. It was namely just these 9 peptides required three or more different subdigestions for each fragment to determine the complete amino acid sequences. The resulting complex mixtures of peptides were separated by gel filtration on Bio-Gels and by h.p.l.c. under the above described techniques. To analyse N-terminal amino acid sequences the Edman procedure was used, amino acids being identified as 1-dimethylaminonaphthalene-5-sulfonyl derivatives (3), phenylthiohydantoins (4) and 4-N,N-dimethylaminoazobenzene-4-thiohydantoins (5). The C-terminal amino acid sequences were established by analysis of carboxypeptidase A and B digestions (6) and by hydrazinolysis (7). The N-terminal sequences of large peptides were determined by automatic Edman degradation.

To illustrate the course of sequencing large CNBr peptides, we have chosen the most interesting examples, namely fragments CB5+CB6 and CB10.

Fragment CB5+CB6 (69 amino acid residues) was the largest of the cyanogen bromide cleavage and contained a Met-Thr bond, poorly cleaved with CNBr. For this fragment we attempted to run digestion with pepsin under acidic conditions at an enzyme:substrate ratio of 1:10 (w/w). The resulting peptic peptides covered the entire CB5+CB6 and their sequence determinations served as a basis for reconstruction of the cyanogen bromide fragment CB5+CB6.

A great deal of information was obtained from chymotryptic digestion of bleached membranes, the information being valuable not only for determination of the complete amino acid sequence of CB5+CB6 but for a number of other large hydrophobic fragments.

To break the intact membrane, the native disks were repeatedly frozen and thawed out, then bleached in the presence of 2 M hydroxylamine followed by dialysis against water. The preparative digestion with chymotrypsin was performed for 4 h in 0.1 N ammonium bicarbonate, pH 7.8, at an enzyme:substrate ratio of 1:20 (w/w). Separation of the chymotryptic peptides was achieved by gel filtration on Bio-Gels in formic acid, by cation-exchange chromatography on AGx4 in concentration and pH gradient of pyridine-acetate buffers and by h.p.l.c. according to previous procedure. A large set of peptides containing about 300 amino acid residues (85% of total structure) was generated by chymotryptic digestion. It was possible to use this overlapping set to our advantage in sequencing large hydrophobic fragments.
To obtain the missing overlaps within CB5+CB6 we used the cleavage of rhodopsin immobilized on the activated thiol support (CPG/Thiol, Pierce). Immobilized protein was subjected ,both, to proteolysis and chemical cleavage to obtain the peptides in question.
An attempt to digest fragment CB10 (46 amino acid residues) by pepsin was unsuccessful. The sequence of this fragment was solved mainly by cleavage with N-bromosuccinimide and by automatic Edman degradation. Fragment CB10 is of interest as its portion of the polypeptide chain is located at the cytoplasmic surface of the membrane and is prone to the action of a number of proteolytic enzymes (Fig.1). Sequencing membane-bound fragment F_2 (87 amino acid residues) resulting from limited proteolysis of native disk membranes via thermolysin provided important information on the C-terminal part of the molecule. The first step of sequencing this peptide was cyanogen bromide cleavage, fragment F_2 overlapping 6 CNBr peptides. The N-terminal cyanogen bromide peptide of F_2 was found to be derived from the C-terminal region of CB10. This circumstance allowed us to complete elucidation

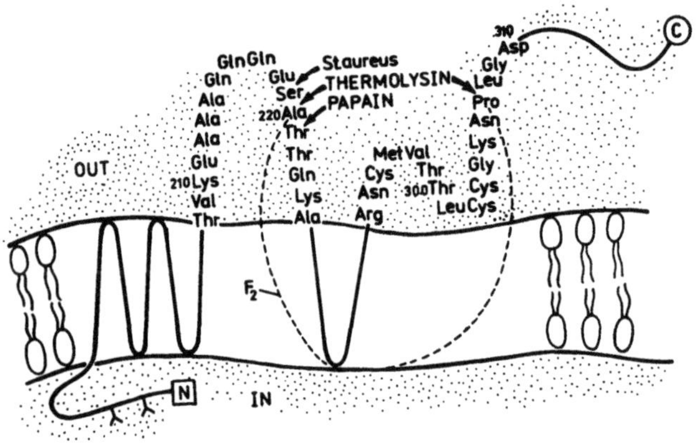

Fig.1. The action of proteolytic enzymes on the native disk membranes.

of amino acid sequence of cyanogen bromide fragment CB10. During the course of F_2 sequencing the elucidation of the structure of cyanogen bromide fragment CB12+CB13 proved to be the most difficult step. The fragment consisted of 51 amino acid residues and contained a Met-Thr bond, poorly cleaved with CNBr. The sequenator was widely used for structural study of CB12+CB13. Both, the intact fragment and the products resulting from its chemical cleavage were subjected

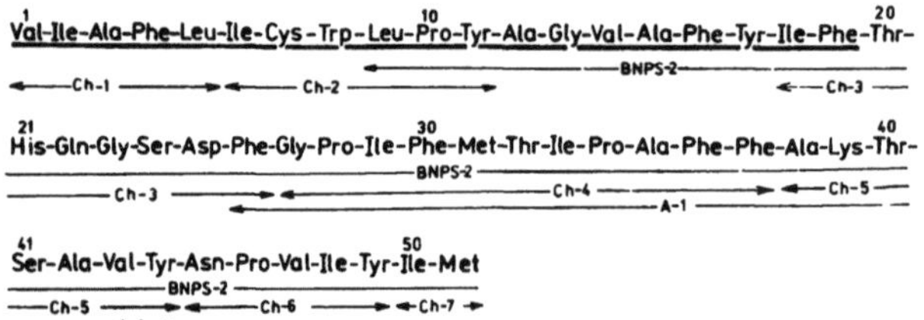

Fig.2. Amino acid sequence of the cyanogen bromide fragment CB12+CB13.

(——)-Amino acid sequence determined by automatic degradation; Ch-1 - Ch-7 -peptides resulting from digestion with chymotrypsin.

to automatic Edman degradation: peptides BNPS-2 and A-1 resulted from the cleavage of CB12+CB13 with BNPS-skatole and from partial acid hydrolysis, respectively. The reminder of the sequence was made available to analysis by supplementary cleavage of CB12+CB13 with chymotrypsin on suspension with stirring for 12 h at an enzyme:substrate ratio of 1:10(w/w). The data presented here has allowed us to determine the amino acid sequence of 166 amino acid residues (50% of total sequence).

References

1. Ovchinnikov, Yu.A., Abdulaev, N.G., Feigina, M.Yu., Kiselev, A.V., Lobanov, N.A.: FEBS Lett. 100, 219-224 (1979).
2. Steers, E., Graven, G.P., Anfinsen, C.B., Bethune, J.L.: J. Biol. Chem. 240, 2478-2484 (1965).
3. Grinkevich, V.A., Arzamazova, N.M., Potapenko, N.A., Grinkevich, Ch.A., Kravchenko, Z.B., Feigina, M.Yu., Aldanova, N.A.: Bioorg. Khim. 5, 1757-1774 (1979).
4. Chen, R.: Hoppe-Seyler's Z. Physiol. Chem. 357, 873-886 (1976).
5. Chang, J.Y., Brauer, D., Wittman-Liebold, B.: FEBS Lett. 93, 205-214 (1978).
6. Amber, R.P.: Methods in Enzymol. XXVB, 143-154, 262-272 Acad. Press, N.Y.-London 1972
7. Mesrob, B., Holeysovsky, V.: Czech. Chem. Commun. 32, 1976-1982 (1967).

THREE-DIMENSIONAL STRUCTURE OF ACTINOXANTHIN AT HIGH RESOLUTION

Alexander Kuzin, Sergey Trakhanov, Vladimir Pletnev

Shemyakin Institute of Bioorganic Chemistry, Academy of Sciences of the USSR, Moscow, USSR

Introduction

Due to the interest in biological properties of antitumor protein actinoxanthin (1) (isolated from Actinomyces globisporus, m.w. 10300) extensive structural and biochemical studies have been undertaken to establish the relationship between its conformation and action mechanism. The present paper describes the results of X-ray study of the three-dimensional structure of this protein in apoform at 2.0 Å resolution . Actinoxanthin crystallizes in space group $P2_12_12_1$ with cell dimensions: a=30.9 Å, b=48.8 Å, c=64.1 Å and Z=4. At the present stage the structure was refined by the restrained least-square procedure (2) to R-factor of 0.22.

Results

The actinoxanthin backbone folding drawn from C^{α} atom coordinates and the schematic diagram of the hydrogen bond system for the protein structure are shown in Fig.1 and 2, respectively. The actinoxanthin structure is characterized by the absence of α-helices and the presence of enhanced content of antiparallel β-structure. A distinctive feature of the molecular architecture is a cylindrical β-supersecondary structure formed by seven antiparallel β-strands (3). The topology of the arrangement of β-strands found in the protein structure is called "Greek key" (4). The protein structure shows strong

Chemistry of Peptides and Proteins, Vol. 2

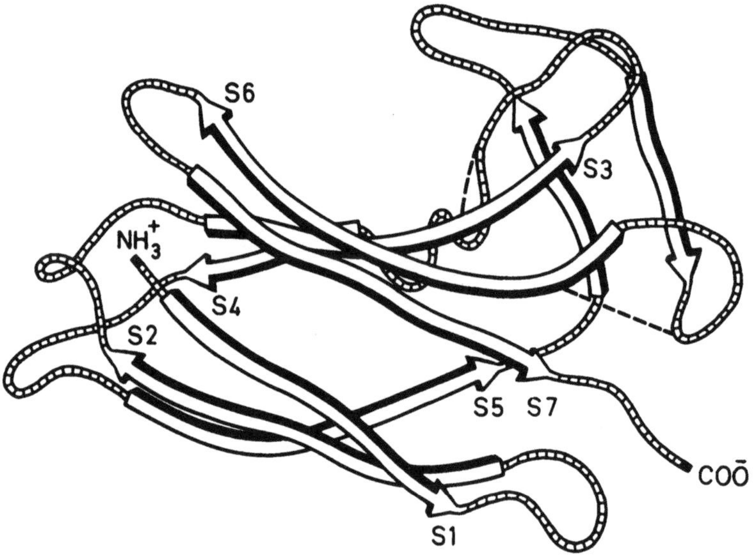

Fig.1. Topology of the actinoxanthin spatial folding. The arrows indicate β-strands, the strings indicate non-regular protein parts.

topological similarity with the supersecondary structure units in immunoglobulins, in the Bence-Jones protein Rhe, in superoxide dismutase and in Cu-proteins azurin and plastocyanin (3).

The inner part of the β-cylindrical barrel is tightly packed with hydrophobic side-chains oriented towards the axis: Ala (92), Val(6,18,20,64), Phe(4,51,61,63,106), Leu(90,99,104), Ile(31). Hydrophilic side-chains are arranged mainly on the external surface.

The schematic of the hydrogen bond system clearly indicates the extensions of interacting β-strands S_1-S_7 and β-segments of the hairpin shaped part 66-83 and also the location of

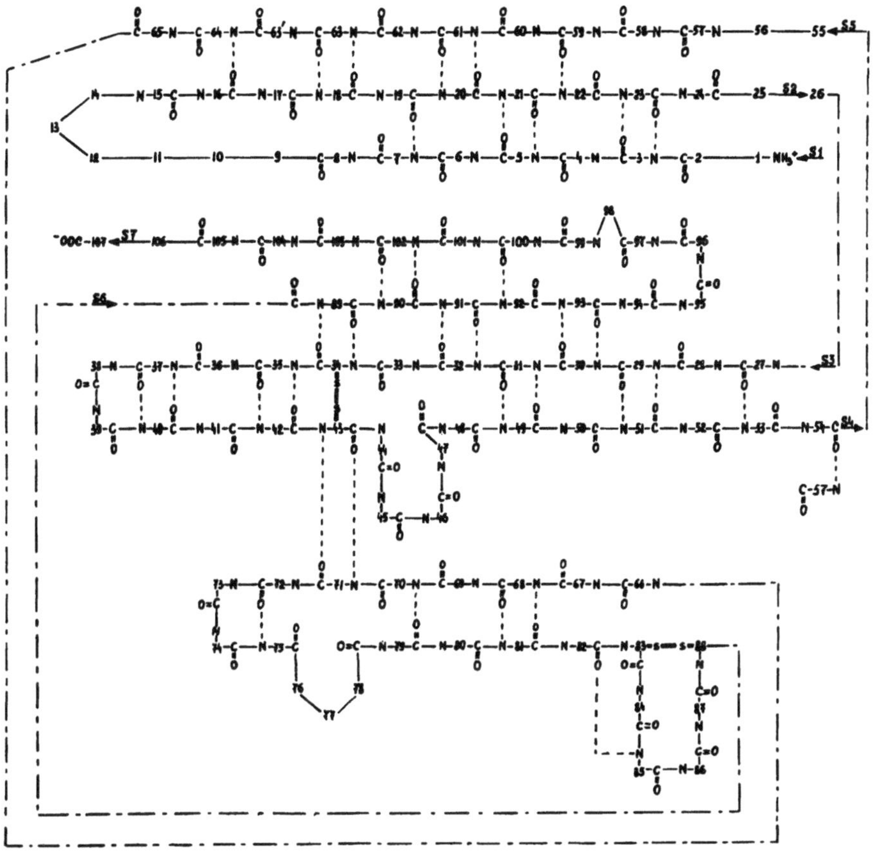

Fig.2. Schematic diagram of the hydrogen bond system in the actinoxanthin structure. The residue localized on the electron density map additionaly to amino acid sequence (5) is marked by number 63'.

β-turns. The distribution of the backbone angles φ and ψ on the Ala residue conformational map (which comprises the conformational states of all the other residues, except Gly) is shown in Fig.3; the high density of the φ,ψ points in the B-region provides additional evidence of the presence of extensive β-structure organization in the molecule.
The actinoxanthin structure found has a well-defined cavity, which is undoubtedly related to its biological activity. Pro-

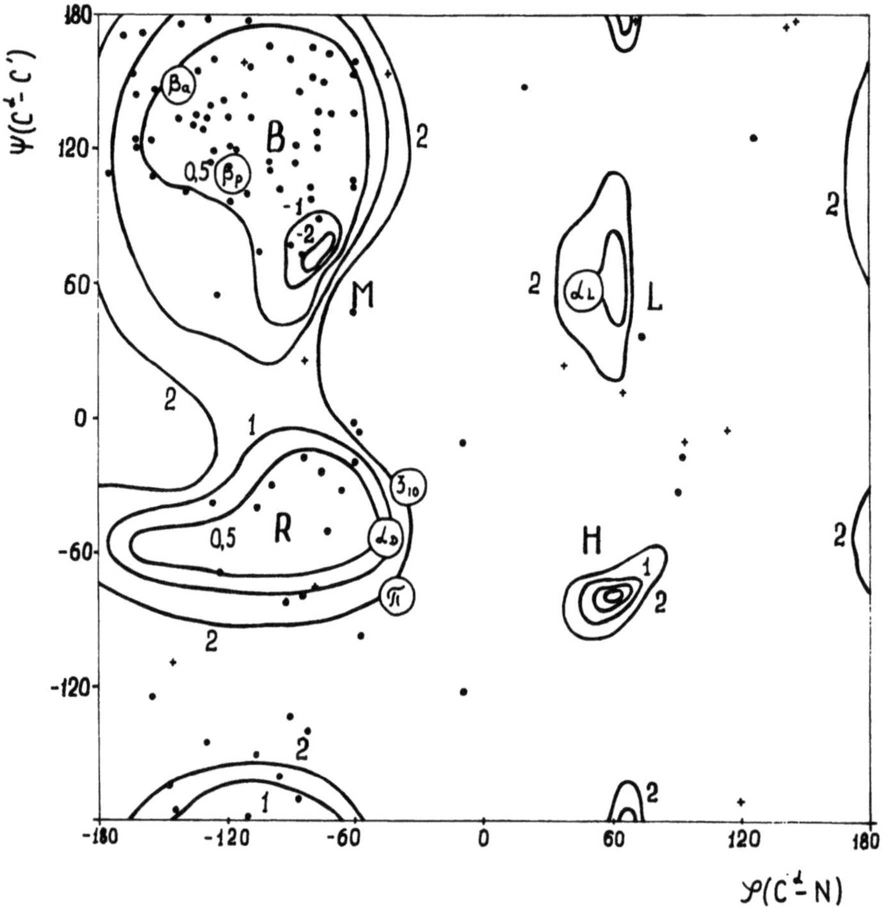

Fig.3. A plot of the backbone angles ϕ,ψ for actinoxanthin structure. (+) Gly residues; (.) other amino acid residues.

bably, it serves as a binding site for the low molecular weight component, the chromophore, in its active complex with the protein. The spectral studies of actinoxanthin in the active form have shown a marked resemblance of the non-protein chromophore group with that of the homologous antitumor antibiotic neocarzinostatin. There is a large number of noncharged residues as Gly,Ala and Pro and S-S bond 34-43 enclosing the internal and outlying parts of the molecular cavity with the main and side chains. Only three hydrophilic and one neutral side-chains of Asp(41), Asn(90,98) and His(101) respectively have a distinctive internal orientation. The are assumed to be the most probable candidates for binding the chromophore molecule. There is also an additional, very specific group of hydrophylic residues Gln(39), Thr(72,76), Ser(74) at the peripheral part of the cavity; this group is characterized by spatially adjacent tetrahedral arrangement of side-chains. The details of chromophore binding in the protein cavity are still unknown. This problem can in all probability be solved by X-ray analysis of actinoxanthin in an active holoform.

The authors express their acknowledgement to Dr. L.V.Malinina, Dr. A.G.Urzhumtsev and Dr. V.Yu.Lunin for their help in refining the actinoxanthin structure.

References

1. Khokhlov A.S., Cherches B.Z., Reshetov P.D., Smirnova G.M., Sorokina I.B., Koloditskaya T.A., Smirnov V.V., Navashin S.M., Fomina J.P.: J. Antibiot. 22, 541-544 (1969).
2. Agarval R.C.: Acta cryst. A34, 791-809 (1978).
3. Pletnev V.Z., Kuzin A.P., Trakhanov S.D., Kostetsky P.V. : Biopolymers 21, 287-300 (1982).
4. Richardson J.S.: Nature 268, 495-500 (1977).
5. Khoklov A.S., Reshetov P.D., Chupova L.A., Cherches B.Z., Zhigis L.S., Stoyachenko I.A.: J. Antibiot. 29, 1026-1034 (1976).

PRIMARY AND SECONDARY STRUCTURES OF PEPTIDES WHICH FORM VOLTAGE-DEPENDENT PORES IN LIPID-BILAYERS

Günther Jung, Günther Boheim, Roland Bosch, Hans Brückner, Peter Hartter, Ellen Katz, Wilfried A.König, Heribert Schmitt, Klaus-Peter Voges, Werner Winter

Institut für Organische Chemie der Universität Tübingen, D-74oo Tübingen, Germany

Introduction

The pore system of aggregated molecules of the polypeptide antibiotic alamethicin transports ions through lipid bilayer membranes (for reviews see ref. 1-5). These voltage-dependent pores can adopt different discrete conductance states depending on the pore size. The phenomena are best described by the " flip-flop gating model " (6,7), which is based on the dipole-dipole interactions of the α-helical alamethicin molecules under the influence of the applied membrane potential.

This model explains all electrochemical data of bilayer experiments,and it is consistent with the structural prerequisites found for pore formation by sequencing natural analogues, synthesis of membrane-active polypeptides and conformational studies. The following is a short summary,which includes recent results on sequence analysis, X-ray crystallo-

Chemistry of Peptides and Proteins, Vol. 2

```
ALAMETHICIN F 3o/5o
Ac-Aib-Pro-Aib-Ala-Aib-Ala-Gln-Aib-Val-Aib-Gly-Leu-Aib-Pro-Val-Aib-Aib-Glu-Gln-Pheol
              (Aib)(Ala)(Aib)    (Aib)                     (Val)(Val)(Gln)

TRICHOTOXIN A 4o/5o
    Ac-Aib-Gly-Aib-Leu-Aib-Gln-Aib-Aib-Aib-Ala-Aib-Aib-Pro-Leu-Aib-Iva-Glu-Valol
          (Ala)                    (Ala)                       (Aib)(Gln)

SUZUCACILLIN A/B (present state of sequence analysis)
Ac-Aib-Ala-Aib-Ala-Aib-Ala-Gln-Aib-Aib-Aib-Gly-Leu-Aib-Pro-Val-Aib-Iva-Glu-Gln-Pheol
  (Iva)               (Aib)      (Leu)                         (Aib)(Gln)
                                 (Iva)

SYNTHETIC NONADECAPEPTIDE
Boc-Aib-Ala-Aib-Ala-Aib-Ala-Aib-Ala-Aib-Ala-Gly-Ala-Aib-Pro-Ala-Aib-Aib-Glu-Gln-OMe
                                                                       (Bzl)

SYNTHETIC EICOSAPEPTIDE
Boc-(Ala-Aib-Ala-Aib-Ala)4-OMe

MELITTIN and MELITTIN 1-21                                                  (Ac)
For-Gly-Ile-Gly-Ala-Val-Leu-Lys-Val-Leu-Thr-Gly-Leu-Pro-Ala-Leu-Ile-Ser-Trp-Ile-Lys
 (H)                       (Ac)                         [Arg-Lys-Arg-Gln-Gln-NH2]
                                                             (Ac)
```

Figure 1. Sequences of natural and synthetic polypeptides which form potential dependent ion-conducting pores in lipid bilayer membranes.

graphy and conformational analysis by NMR and CD.

Sequences of membrane-modifying polypeptides

The unique voltage-dependent ion conducting phenomena were first detected on alamethicin, a polypeptide antibiotic produced by Trichoderma viride. Sequence analysis revealed an acetylated nonadecapeptide with a C-terminal aminoalcohol (for refs. see (2)). The isolated product was reported to be a mixture of analogues with amino acid exchanges at positions 6 (Ala/Aib) and 18 (Glu/Gln). Years later much more exchanges due to natural microheterogeneity were found using GLC-MS analysis of partial hydrolysates of alamethicin (8) (Fig.1). The production of these mixtures of analogues is common also to the trichotoxin-(2,9-12) and suzukacillin-(13) producing strains. These two antibiotics have no N-terminal proline. For sequence analysis and conformational studies they were selectively split by trifluoroacetic acid into helical N-termini and C-terminal N-prolyl-peptides.

Structure-activity studies on these natural analogues prompted us to design and synthesize a sequentially simplified nonadecapeptide maintaining still the conformation of alamethicin (14-16). Recently it was found that melittin and melittin 1-21, and their acetylated analogues exhibit also voltage-dependent single pore formation (7,17). Furthermore we could demonstrate that the C-terminal introduction of a dansyl group into trichotoxin A-4o increases the mean life time of pore states (17). And we recognized that the large dipole moment (18-2o) and lipophilicity of the α-helical parts constitute the main structural prerequisites for voltage-dependent pore formation. Therefore, we simplified further the sequence and build up α-helical deca-, pentadeca- and eicosapeptides Boc-(L-Ala-Aib-Ala-Aib-Ala)$_n$-OMe (n = 2-4). And, indeed, these helix dipoles exhibited alamethicin-like pore-formation (21). In this context it is noteworthy that HCO-(Ala)$_n$-OMe (n = 2o,24), HCO-(Ala-Ala-Gly)$_n$-OMe (n = 4,5) and Ac-(Leu-Ser(Bzl or free)-Leu-Gly)$_n$-OMe (n = 3,6) were found recently to produce discrete current fluctuations (22).

As a main result one can conclude that voltage-dependent membrane pore formation is entirely based on the dipole-dipole interactions within aggregates of α-helices. Therefore our findings are of general importance for the interactions between α-helical segments in membrane proteins.

X-ray studies on model peptides and conformation in solution

Alamethicin shows Cotton effects typical for α-helices in various solvents with highest ellipticities in alcoholic solvents and membrane-like environments (2,23). Similar results are found for suzukacillin (24) and trichotoxin (2) and N-terminal partial sequences with 12 respectively 13 residues (25). The α-helical conformation was also derived from the ^{13}C-NMR spectra of the natural peptides and of various helical model peptides (2,23-29).

X-ray analysis of some oligopeptides containing Aib residues revealed 3_{10}-helical structures, for example Tos-$(Aib)_5$-OMe (3o), Z-$(Aib)_5$-OBu^t (31) and Boc-Leu-Aib-Pro-Val-Aib-OMe (32). The relevance of these findings for the conformation of alamethicin was discussed in the recent literature.
In Tübingen the following X-ray analyses of Aib peptides were made so far: Boc-Aib-OH (33), Acetyl-D-isovaline (34), Boc-L-Ala-Aib-OH (35), the tripeptides Boc-Gly-L-Ala-Aib-OMe (36), Ac-L-Ala-Aib-Ala-OMe (37), the pentapeptide Boc-Aib-L-Ala-Aib-Ala-Aib-OMe (38) and the undecapeptide Boc-L-Ala-Aib-Ala-Aib-Ala-Glu(OBzl)-Ala-Aib-Ala-Aib-Ala-OMe (39,4o). It has been found, that the tripeptides adopt ß-turns type III, and that the pentapeptide shows a regular 3_{10}-helix. However, the undecapeptide model of the alamethicin helix adopts the α-helical structure with only small deviations from the ideal α-helix. This contradicts the speculations on the possible existence of a longer 3_{10}-helical segment in alamethicin. Only the C-terminal regions from the prolyl residues to the amino alcohols may adopt the 3_{10}-helical conformation. The C-termini of suzukacillin and trichotoxin A 4o were prepared in amounts sufficient for spectroscopic investigations and for comparison of the synthetic and the isolated material from hydrolysates of the natural peptides (24,41).
On comparison of the CD data of a larger number of α-helical

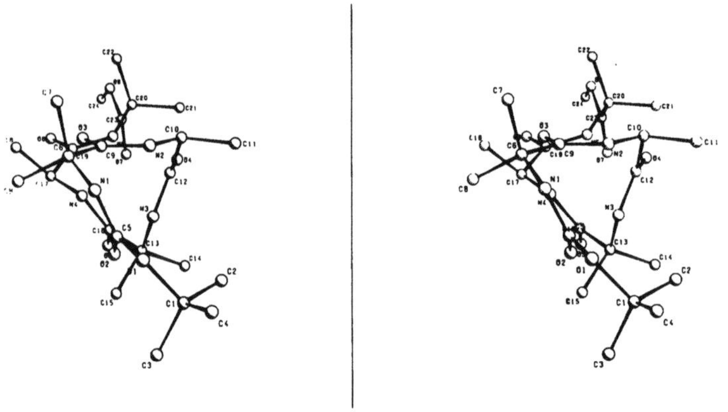

Fig.2. X-ray structure of the 3_{10}-helical pentapeptide Boc-Aib-L-Ala-Aib-Ala-Aib-OMe (stereoscopic view along the helix axis).

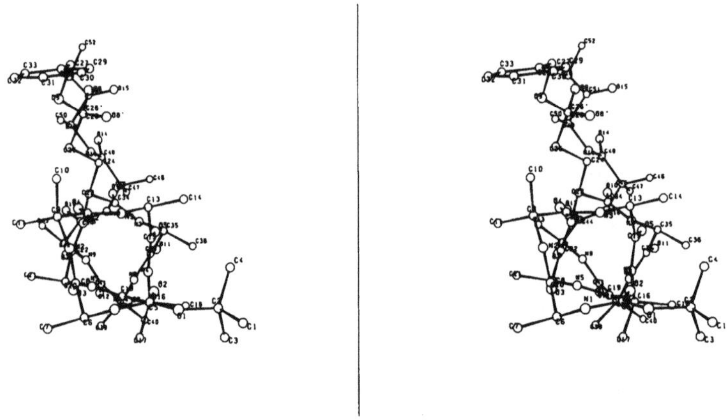

Fig.3. X-ray structure of the α-helical undecapeptide Boc-L-Ala-Aib-Ala-Aib-Ala-Glu(OBzl)-Ala-Aib-Ala-Aib-Ala-OMe (stereoscopic view along the helix axis).

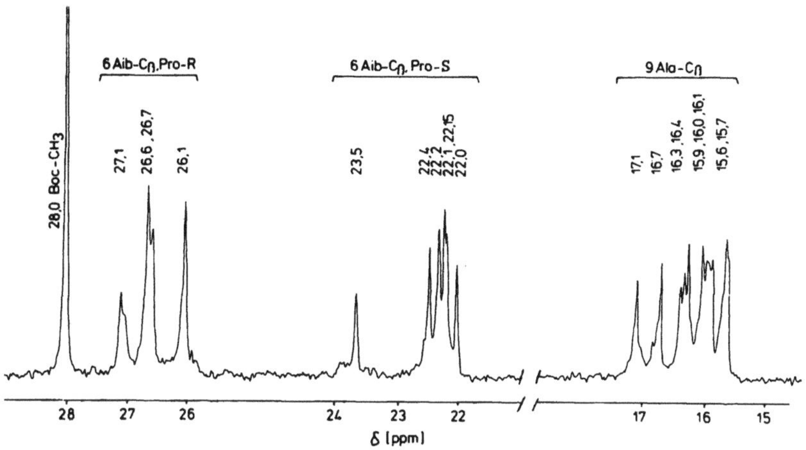

Fig.4. ^{13}C-NMR spectrum of the α-helical pentadecapeptide Boc-(L-Ala-Aib-Ala-Aib-Ala)$_3$-OMe (100.6 MHz ^{13}C; chloroform; C_β region).

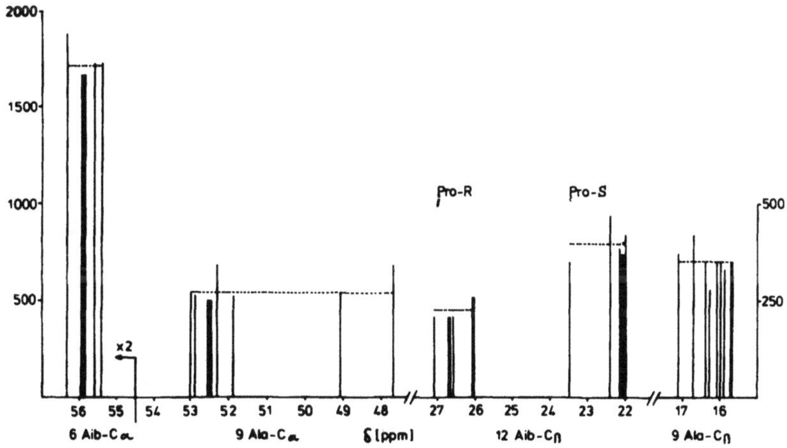

Fig.5. ^{13}C-NMR T_1-measurements (100.6 MHz) of Boc-(L-Ala-Aib-Ala-Aib-Ala)$_3$-OMe (solvent: chloroform; C_α and C_β carbon atoms).

polypeptides with different Aib contents, it was found, that the mean residue ellipticities increase with increasing Aib content. This is explained by the decrease in random-chain conformation respectively the increase in α-helix content. Thus, the particular role of the Aib residues is a stabilization of the α-helix dipole in lipophilic environment. The solubility of Aib containing polypeptides is usually excellent in most organic solvents (except ether or light petroleum). An almost complete assignment of the ^{1}H NMR spectrum of alamethicin was reported (42). Using ^{13}C NMR T_1 measurements (1oo.1 MHz for ^{13}C) on the pentadecapeptide Boc-(L-Ala-Aib-Ala-Aib-Ala)$_3$-OMe we could demonstrate recently that the pro-S methyl carbons of Aib residues are high-field shifted in α-helical environment and show almost the same T_1 values as the L-alanyl-C_β. On the other hand the pro-R Aib-C_β atoms absorb at lower field and relax faster due to stronger interaction with the peptide carbonyl groups. This confirms our earlier interpretations of the large magnetic non-equivalence in helical Aib model peptides and in the natural antibiotics (2,16,23,28,4o).

References

1. Boheim,G., Kolb,H.A.: J.Membrane Biol. **38**, 99-15o (1978).

2. Jung,G., Brückner,H., Schmitt,H.: in Structure and Activity of Natural Peptides, Voelter,W. and Weitzel,G.

(eds.), de Gruyter, Berlin, pp. 75-114 (1981).

3. Latorre,R., Alvarez,O.: Physiol.Reviews 61, 77-15o (1981).

4. Nagaraj,R., Balaram,P.: Acc.Chem.Res. 14, 356-362 (1981).

5. Balasubramanian,T.M., Kendrick,N.C.E., Taylor,M., Marshall,G.R., Hall,J.E., Vodyanoy,I., Reusser,F.: J.Am. Chem.Soc. 1o3,6127-6132 (1981).

6. Boheim,G., Hanke,W., Jung,G.: Biophys. Struct. Mech, in print.

7. Boheim,G., Hanke,W., Jung,G.: see these Proceedings (1982).

8. König,W.A., Aydin,M.: in Peptides 198o, Proc. 16th Eur. Pept.Symp. (Brunfeldt,K.,ed.), Scriptor, Kopenhagen, pp. 711-718 (198o).

9. Brückner,H., König,W.A., Greiner,M., Jung,G.: Angew. Chem. 91, 5o8-5o9 (1979); Angew.Chem.Int.Ed.Engl. 18, 476-477 (1979).

1o. König,W.A., Krohn,H., Greiner,M., Brückner,H., Jung,G.: Adv. in Mass Spectr. 8, 11o9-1115 (198o).

11. Aydin,M., Bloss,D.H., König,W.A., Brückner,H., Jung,G.: Biomed.Mass Spectr. 9, 39-42 (1982).

12. Brückner,H., Nicholson,G.J., Jung,G., Kruse,K., König, W.A.: Chromatographia 13, 2o9-214 (198o).

13. Katz,E., Jung,G., Aydin,M., König,W.A., manuscript submitted.

14. Oekonomopulos,R., Jung,G.: Liebigs Ann.Chem. 1979,1151-1172.

15. Oekonomopulos,R., Jung,G.: Biopolymers 19, 2o3-214(198o).

16. Oekonomopulos,R., Leibfritz,D., Jung,G.: Tetrahedron, in print.

17. Hanke,W.,Methfessel,C., Wilmsen,H.-U., Katz,E., Jung,G., Boheim,G.: Biochim.Biophys.A., in print.

18. Yantorno,R.E., Takashima,S., Mueller,P.: Biophys.J. 38, 1o5-11o (1982).

19. Schwarz,G., Savko,P.: Biophys.J. 39, 211-219 (1982).

2o. Schwarz,G., Savko,P., Jung,G.: Biochim.Biophys.A., in print.

21. Voges,K.-P., Boheim,G., Jung,G.: manuscript submitted.

22. Spach,G., Trudelle,Y., Heitz,F.: Abstr.Int.Symp. "Peptides, Polypeptides, and Proteins, Interactions and their Biological Implications", Galzignano, Italy (1982).

23. Jung,G., Dubischar,N., Leibfritz,D.: Eur.J.Biochem. 54, 395-4o9 (1975).

24. Katz,E., Jung,G.: to be published.

25. Mayr,W., Jung,G.: Liebigs Ann.Chem. 198o, 1489-15o6.

26. Jung,G., König,W.A., Leibfritz,D., Ooka,T., Janko,K., Boheim,G.: Biochim.Biophys.A. 433, 164-181 (1976).

27. Irmscher,G., Bovermann,G., Boheim,G., Jung,G.: Biochim. Biophys.A. 5o7, 47o-484 (1978).

28. Leibfritz,D., Mayr,W., Oekonomopulos,R., Jung,G.: Tetrahedron 34, 2o45-2o5o (1978).

29. Mayr,W., Oekonomopulos,R., Jung,G.: Biopolymers 18, 425-450 (1979).

30. Shamala,N., Nagaraj,R., Balaram,P.: J.Chem.Soc.Chem. Commun. 1978, 996-997.

31. Benedetti,E., Pedone,C., Toniolo,C.: in Peptides 1980, Proc.16th Eur.Pept.Symp., Brunfeldt,K., ed., Scriptor, Copenhagen, pp. 619-624 (1980).

32. Nagaraj,R., Balaram,P.: Acc.Chem.Res. 14, 356-362 (1981).

33. Mayr,W., Strähle,J., Jung,G.: Liebigs Ann.Chem. 1980, 715-724.

34. Bosch,R., Brückner,H., Jung,G., Winter,W.: Tetrahedron, in print.

35. Bosch,R., Voges,K.-P., Jung,G., Winter,W.: Acta Cryst., to be published.

36. Bosch,R., Jung,G., Winter,W.: Liebigs Ann.Chem. 1982, 1322-1329.

37. Schaal,H., Strähle,J., Winter,W., Brückner,H., Jung,G.: Liebigs Ann.Chem., to be published.

38. Bosch,R., Winter,W., Jung,G.: Acta Cryst., to be published.

39. Butters,T., Hütter,P., Jung,G., Pauls,P., Schmitt,H., Sheldrick,G.M., Winter,W.: Angew.Chem. 93, 904-905 (1981); Angew.Chem.Int.Ed.Engl. 20, 889-890 (1981).

40. Schmitt,H., Winter,W., Bosch,R., Jung,G.: Liebigs Ann. Chem. 1982, 1304-1321.

41. Brückner,H., Jung,G.: Liebigs Ann.Chem. 1982, in print.

42. Davis,D.G., Gisin,B.F. : FEBS Lett. 133, 247-251 (1981).

43. Jung,G., Bosch,R., Brückner,H., Katz,E., Schaal,H., Schmitt,H., Strähle,J., Voges,K.-P., Winter,W.: in Peptides 1982, Proc. 17th Eur.Pept.Symp. (Bláha,K., Malon,P., eds.), de Gruyter, Berlin, in print.

44. Jung,G., Bosch,R., Katz,E., Schmitt,H., Voges,K.-P., Winter,W.: Biopolymers, in print.

45. Jung,G., Bosch,R., Katz,E., Schmitt,H., Voges,K.-P., Winter,W.: in Physical Chemistry of Transmembrane Ion Motions, Elsevier Publ. Comp., in print.

PURIFICATION AND CHARACTERIZATION OF A SECOND RETINAL BINDING PROTEIN IN HALOBACTERIUM HALOBIUM

Peter Hegemann, Michael Steiner, Dieter Oesterhelt
Max-Planck-Institut für Biochemie
D-8033 Martinsried bei München, Germany

Introduction

Halobacterium halobium belongs to a class of bacteria which require a high concentration of NaCl for the maintenance of cell integrity and growth. The cells produce two retinal binding pigments: bacteriorhodopsin (BR) (1) which is present in large quantities and gives a purple colour to the cells, and halorhodopsin (HR) (2) which is present in much smaller quantities. Bacteriorhodopsin has been extensively studied but little is known about halorhodopsin. Characterization of this pigment has only been possible as a result of the isolation of mutants defective in the synthesis of bacteriorhodopsin (3). Upon excitation with light HR generates a membrane potential (negative inside). This potential is due either to the outward movement of sodium (4) or the inward movement of chloride (5). The chromophore has an excitation maximum of 588 nm and appears to have a photochemical cycle similar but not identical to bacteriorhodopsin (6,7). The large amounts of cytochrome present overlap the absorption spectrum of HR and have seriously hampered these studies. Thus, for more detailed analysis of the light induced changes in this chromophore, and for comparison with bacteriorhodopsin, purification of the pigment is necessary. Native HR is very unstable and may loose its chromophore if dissolved in detergent. Therefore, we decided to attempt to purify it in a denatured form and to renature it after purification in a way analogous to that which leads to the function-

Chemistry of Peptides and Proteins, Vol. 2

al reconstitution of bacteriorhodopsin in liposomes (8).

Results

Reconstitution and reduction of halorhodpsin. For purification of HR we used a strain L07 of H.halobium deficient in bacterio-opsin and in retinal synthesis but which produces halo-opsin. Halorhodopsin can be reconstituted in membrane vesicles prepared with high specific activity (6.4 Ci/nmole) by reduction of retinal with tritiated borohydride and reoxidation of the product with manganese dioxide in the same reaction vessel:

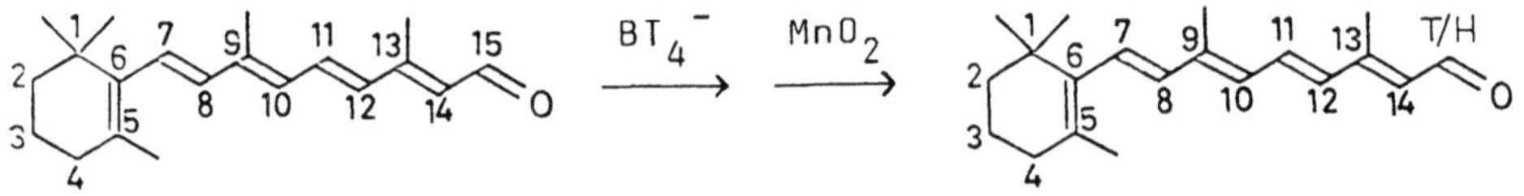

After filtration the solvent was evaporated and the residue was dissolved in methanol. This simple procedure yields pure all-trans C_{15}-^{3}H-retinal with only traces of the 13-cis isomer. Retinal-labelled HR was reconstituted by overnight incubation of membrane vesicles with substoichiometric amounts of this tritiated retinal at room temperature. The retinal protein linkage was stabilized by reduction with sodium cyanoborohydride at pH 5 for 4 hrs, in the dark, in the presence of 1% diethylether, a mild reversible membrane perturbant which facilitates the reduction. Gel electrophoresis in SDS-urea of lysates of these membranes showed only one protein component carrying the radioactive retinyl.

Purification procedure. As a first step in the isolation of halorhodopsin, membrane vesicles were lysed by dialysis against distilled water. The resulting membrane fraction was collected by centrifugation and washed with water. It was then dissolved in a buffer containing 10% SDS and 8M urea. The solubilized material was fractionated by gel filtration on AcA 44 Ultrogel (range: 10,000 to 130,000 M_V). Due to the very hydrophobic

nature of the proteins in this extract the monomeric proteins tend to form dimers and trimers as indicated by separation of analytical SDS gels. To avoid this we used 4% SDS without buffer as eluent for material on the column. A similar procedure had previously been reported for the separation of the subunits of cytochrome oxidase by Buse and Steffens (10).
Standard buffers such as Tris or phosphate have a strong influence on the micellar concentration of dodecyl sulfate and therefore on the loading of the protein with SDS. Under these conditions of low salt and high SDS retinyl protein elutes as a monomeric species before radioactive low molecular weight compounds, which arise from incomplete reconstitution.
The fractions from the column which contained the retinyl protein were purified further on a hydroxyl apatite column (Biogel HT equilibrated with 0.1% SDS at 22°C). The retinyl protein was applied directly to the column and eluted with a 500 ml gradient of sodium phosphate pH 6.4 (200 mM to 500 mM) and 0.1% SDS. This step does not give a very high factor of purification but it removes a number of proteins of a molecular weight very close to that of halorhodopsin, as well as lowering the SDS concentration from 4% to 0.1%. The radioactive fraction from this column was dialyzed against water to remove the phosphate and concentrated by ultrafiltration (Amicon YM 10). This concentrated sample was then applied to a preparative SDS-polyacrylamide gel electrophoresis. The gel contained 0.1% SDS and 3M urea and was otherwise prepared as described by Waehneldt (11). This technique allows the separation of protein mixtures of up to 10 mg per band.
This 3-step procedure yielded material which migrated as a single band of an apparent molecular weight of 25,000 Dalton on analytical SDS/urea polyacrylamide gels. The purification was about 400-fold with an overall yield of approximately 15%. From 15 l culture we obtained about 500 µg pure protein. Determination of the yield after each step is difficult, because

retinyl peptides are unstable and the specific radioactivity dropped during the purification procedure. Free radioactive retinal appears in all purification steps.

Characterization

The purified material was used in experiments to find homologies and differences between BR and the retinal binding component of HR; end group determination of this protein was not possible because it is blocked. We suppose that the end group is pyroglutamic acid as in BR. Treatment of purple membranes with pyroglutamase liberates the N-terminus, but it was impossible to carry out this reaction with either BR or HR in the presence of residual traces of SDS. The amino acid analyses of purified BR and HR are very similar.
Polyclonal antibodies against purified BR and HR were raised in rabbits. Each animal was injected initially with 500 µg of protein and then one month after the first injection four booster injections of 200 µg each at weekly intervals. Antigen-antibody reactions were carried out after gel electrophoresis of the proteins and transferred to a sheet of nitrocellulose (12). Although homologous antibodies gave strong reactions, cross reactions could not be detected indicating that different antigenic determinants must be present in both proteins. This should make it possible to use antibodies against HR on affinity columns for the isolation of the native pigment.
When both retinal proteins were partially digested with staphylococcus aureus-V8 protease the proteolytic pattern of the radioacitve retinyl peptides on SDS gels was similar but not identical. Digests of bacteriorhodopsin showed two extra peptides. Similar differences were also shown by digestion of the two proteins with subtilysin and fractionation of the digest by HPLC. To do this unlabelled carrier retinyl BR was mixed with ^{3}H-retinyl BR <u>or</u> ^{3}H-retinyl HR and digested for 4 hrs. The resulting fractions containing retinyl peptides were lyo-

philized, extracted with ethanol and fractionated on polar Si60 columns (13).

On the base of these observations we conclude that although these two retinal binding proteins have similar molecular weights and are similar in their primary sequences, they nevertheless must have differences in their primary structure. The strong similarities between BR and the retinal binding component of HR lead to the conclusion that these two proteins might work in the similar way and may both transport protons outward across the membrane. The sodium efflux and alkalisation of the medium could be produced of an additional protein tightly coupled to the retinal-binding component functioning as an H^+/ Na^+ antiporter to transport sodium out. The driving force for sodium efflux would be the chemiosmotic gradient created by the retinyl protein.

If HR functions as a chloride pump as suggested recently by Schobert and Lanyi (5) it could consist of the retinal-binding protein associated with an electrogenic Cl^-/OH^- antiporter. However, before these problems can be experimentally approached the primary sequence of the retinal protein from HR will have to be elucidated and reconstitution of the chromoprotein must be possible.

References

1. Oesterhelt, D. Stoeckenius, W.: Proc. Natl. Acad. Sci. USA 70, 2853-2857 (1973).
2. Mukohata, Y., Kaji, Y.: Arch. Biochem. 255, 243-250 (1981).
3. Wagner, G., Oesterhelt, D., Krippahl, G., Lanyi, J.K.: FEBS Lett. 131, 341-345 (1981).
4. Lindley, E.V., MacDonald, R.E.: Biochem. Biophys. Res. Comm. 88, 491-499 (1979).
5. Schobert, B., Lanyi, J.K.: J. Biol. Chem. (in press).
6. Lanyi, J.K., Weber, H.J.: J. Biol. Chem. 255, 243-250 (1981).

7. Weber, H.J., Bogomolni, R.A.: Photochem. Photobiol. 33, 601-608 (1981).

8. London, E., Khorana, H.G.: J. Biol. Chem. 257, 7003-7011 (1982).

9. Lanyi, J.K., Oesterhelt, D.: J. Biol. Chem. 257, 2674-2677 (1982).

10. Steffens, G., Buse, G.: Hoppe-Seyler's Z. Physiol. Chem. 357, 1125-1137 (1976).

11. Waehneldt, T.V.: Anal. Biochem. 43, 307-312 (1971).

12. Towbin, H., Staehlin, T., Gordon, J.: Proc. Natl. Acad. Sci. USA 76, 4350-4354 (1979).

13. Lemke, H.-D., Oesterhelt, D.: FEBS Lett. 128, 255-260 (1981).

HOMO- AND HETERONUCLEAR TWO-DIMENSIONAL NMR SPECTROSCOPY OF CYCLIC PEPTIDES[1)]

Horst Kessler, Regina Schuck
Institut für Organische Chemie der Universität, Niederurseler Hang, D-6000 Frankfurt, Germany

The potential of two-dimensional NMR methods (2D NMR) for conformational studies of peptides and small proteins is now well established (2). But all applications in this field are restricted to homonuclear techniques (Jeener, SECSY, NOESY). We will report here three examples where different heteronuclear and homonuclear techniques are used to analyze complicated peptide spectra.

The $CDCl_3$ solution of cyclo[Pro-NB.Gly_2] (NB.Gly = o-nitrobenzyl-glycine) exhibits a dominant crown (3) conformation (80 %) in slow equilibrium with 20 % boat (3) conformation (4).Eight AB-systems partially covered by proline α and δ signals yield a rather complicated spin pattern in the region between 3.5 and 5.5 ppm. Analysis by conventional, one-dimensional methods does not lead to an unambiguous assignment, whereas this problem is easily solved by two-dimensional methods.Geminal and long-range couplings have been analyzed by a 2D spin echo-correlated spectrum (SECSY). The exchange of each individual proton in one conformation into the specific position of the other conformation by ring inversion was followed by a 2D exchange spectrum at 331 K in $CDCl_3$ (Figure 1). The information from both 2D NMR spectra is summarized in Figure 2.

The ^1H NMR spectrum of the cyclic pentapeptide cyclo[Pro-Phe-Gly-Phe-Gly] and its Li complex was previously analyzed by usual temperature and solvent dependance studies of this molecule as well as of two specifically deuterated derivatives to obtain the sequential assignments. NOE difference spectroscopy and a SECSY spectrum yield these results

Chemistry of Peptides and Proteins, Vol. 2

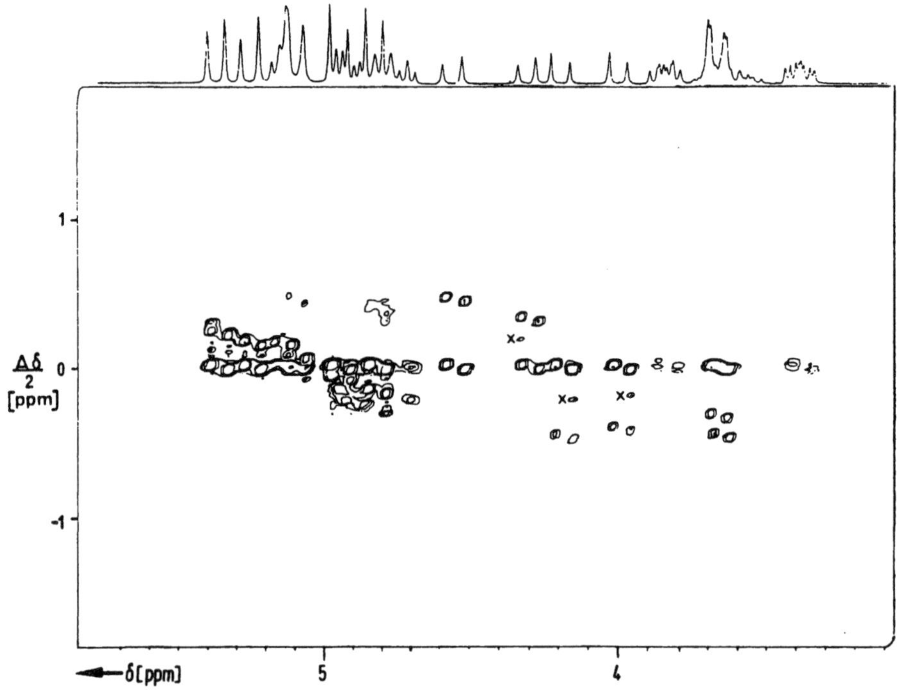

Figure 1: 2D Exchange spectrum of cyclo[Pro-NB.Gly_2] in $CDCl_3$ at 331 K.

spectroscopically in a much more convenient and less expensive way (5). The assignment of all C-13 signals of this peptide was possible by 2D-^1H-^{13}C shift correlation. The ^{13}C-chemical shift of the β- and γ-carbons of proline establish the trans conformation of the Gly^5-Pro^1-peptide bond (6). The observation of the nitrogen signals of the phenylalanines and the two glycines was performed via the INEPT technique (7). Again the assignment was done by 2D-^1H-^{15}N shift correlation (Figure 3). The interpretation of the NMR data results in the preferred conformation shown in Figure 3.

The third peptide is the naturally occuring cyclolinopeptide A (cyclo[Pro-Pro-Phe-Phe-Leu-Ile-Ile-Leu-Val]), where the synthesis of specifically deuterated derivatives was not yet possible. The assignment of amide protons as well as $C_\alpha\underline{H}$ and

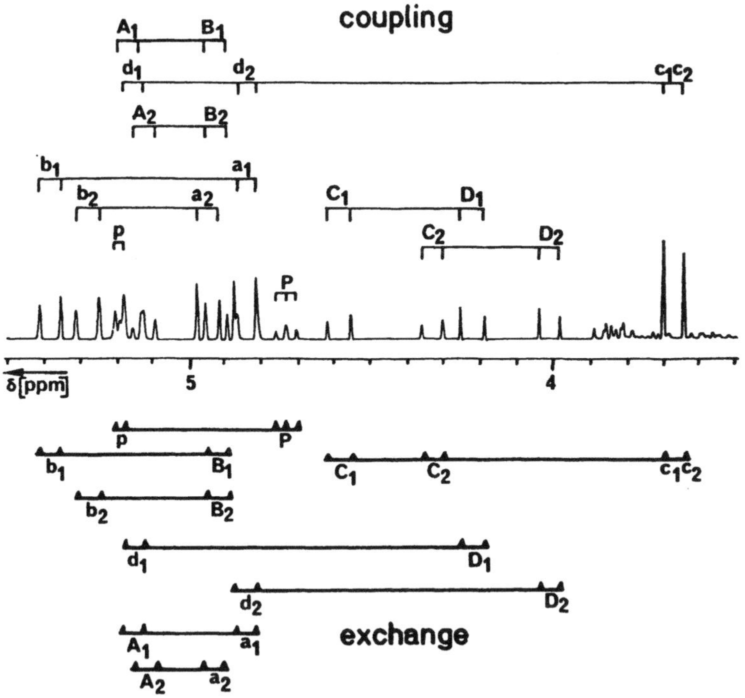

Figure 2. Coupling and exchange pattern of cyclo[Pro-NB.Gly_2] (A,B,a,b are benzylic, C,D,c,d Gly-C_α-protons).

and $C_\beta\underline{H}$ signals was performed by NOE difference measurements and a Jeener spectrum (8). The previous assignment (9) had to be revised.

The aliphatic part of the ^{13}C-NMR spectrum was assigned with the exception of four γ and all methyl carbons by means of a heteronuclear 2D-J-resolved NMR spectrum and a 2D-^{1}H-^{13}C shift correlation. The latter is shown in Figure 4 which also includes the assignment. Again the previous assigment (10) had to be revised.

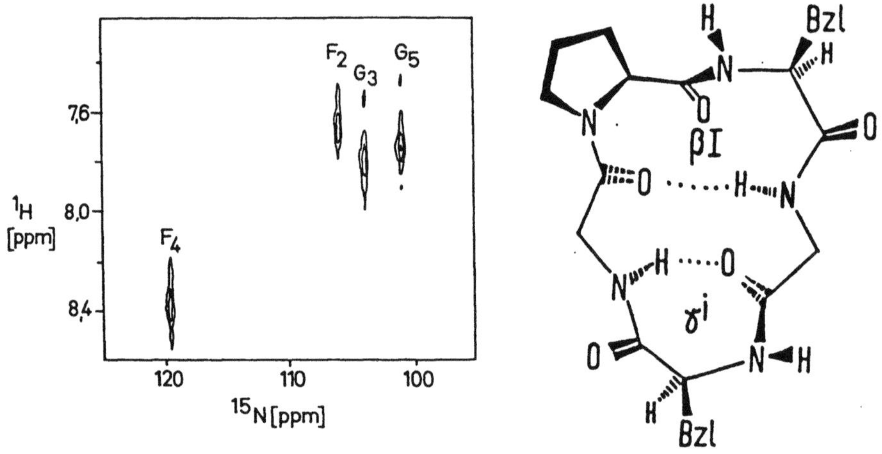

Figure 3. ^{1}H-^{15}N two-dimensional shift correlation of cyclo|Pro-Phe-Gly-Phe-Gly| and the conformation of the same molecule.

The interpretation of the spectral parameters thus obtained (NH-temperature gradients, solvent-induced shifts, coupling constants, NOE effects, ^{13}C chemical shifts) points to the fact that the molecule does not have a completely rigid backbone. However, in contrast to the literature (9,10,11) a cis Val^{9}-Pro^{1} peptide bond and an internal orientation of Leu-5 is certainly indicated.

In our opinion the combination of the different two-dimensional NMR techniques provides a powerful tool for spectral assignments. The recent developments in NMR technology allow the application of heteronuclear techniques even for peptides of medium size. Obviously these techniques only facilitate the extraction of the parameters and their assignments to molecular structure whereas the interpretation still requires care. However, even a careful interpretation of incorrectly assigned parameters leads to wrong results!

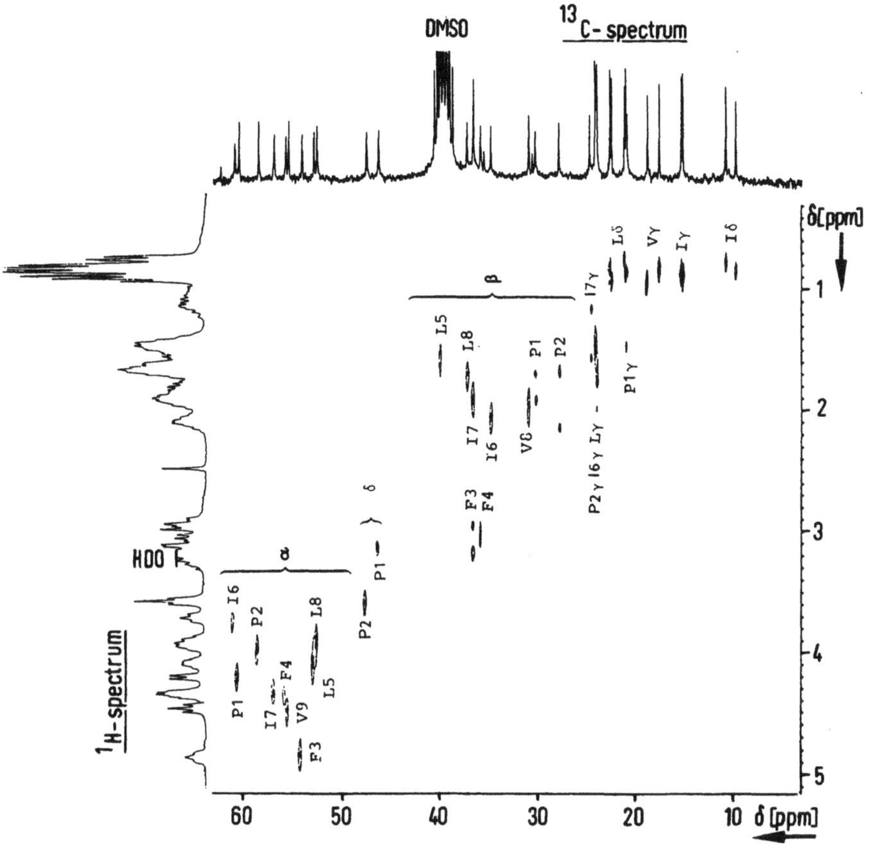

Figure 4. ^1H-^{13}C shift correlation of cyclolinopeptide A, cyclo|Pro-Pro-Phe-Phe-Leu-Ile-Ile-Leu-Val|, at 333 K in d_6-DMSO.

References

1. Peptide conformation, 21. For part 20 see:

   De Leeuw, F.A.A.M., Altona, C., Kessler, H., Bermel, W., Friedrich, A., Krack, G., Hull, W.E.: J.Amer.Chem.Soc, in press.

2. Nagayama, K., Wüthrich, K.: Eur.J.Biochem. 114, 365-374 (1981); Wagner, G., Kumar, A., Wüthrich, K.: Eur.J.Biochem. 114, 375-384 (1981).

3. Kessler, H.: in Stereodynamics of Molecular Systems, Ed. R.H.Sarma, Pergamon Press, New York, 1979, p. 187-196.

4. Kessler, H., Schuck, R., Siegmeier, R.: J.Amer.Chem.Soc.: in press.

5. Kessler, H., Hehlein, W., Schuck, R., J.Amer.Chem.Soc.: in press.

6. R.Deslauriers, I.C.P.Smith in: Topics in Carbon-13 NMR spectroscopy, Wiley, New York, 1976

7. Morris, G.A., Freeman, R.: J.Amer.Chem.Soc. 101, 760 (1979).

8. R.Schuck: Ph.D. thesis, Frankfurt, 1982.

9. Brewster, A.I., Bovey, F.A.: Proc.Nat.Acad.Sci. USA 68, 1199 (1971).

10. Siemion, I.Z., Klis, W.A., Sucharda-Sobczyk, A., Obermeier, R.: Ann.Soc.Chim.Polonorum 51, 1489 (1977).

11. Tonelli, A.E.: Proc.Nat.Acad.Sci. USA 68, 1203 (1971).

Miscellaneous and Biological Activity of Peptides and Proteins

BIOORGANIC CHEMISTRY OF RHODOPSINS

Yu.A.Ovchinnikov
Shemyakin Institute of Bioorganic Chemistry, USSR Academy of Sciences, 117988 Moscow, USSR

Bacteriorhodopsin and visual rhodopsin are integral membrane proteins containing retinal as a light-sensitive group /1,2/. Light absorption by these proteins results in generation of a considerable electric potential /1,3/. The energy of this potential is utilized for ATP synthesis and other vital functions in halophilic microorganisms, however, the ways of its utilization in the process of visual excitation are still obscure. Comparative study of structural organization of rhodopsins was a necessary stage for understanding their functioning.

The primary structure of bacteriorhodopsin /4,5/ has been earlier determined. The polypeptide chain of the protein is composed of seven α-helical rods /4, 6/ spanning the entire width of the purple membrane. Our experiments revealed that 27 amino acid residues of the bacteriorhodopsin polypeptide chain are exposed from the membrane and accessible to the action of proteolytic enzymes. Comparison of data on the primary structure and results of the action of proteolytic enzymes on the purple membranes enabled us to propose the model of bacteriorhodopsin arrangement in the membrane.

Bacteriorhodopsin investigation carried out in our laboratory stimulated the study of visual rhodopsin. Its structural analysis could clarify if this organization principle is general for all membrane proteins or only for bacteriorhodopsin because of its function.

To determine the primary structure of visual rhodopsin we used the approach that had been widely applied for the investigation of bacteriorhodopsin /4/.

Chemistry of Peptides and Proteins, Vol. 2

Our study of rhodopsin resulted in structural elucidation of its molecule, consisting of 347 amino acid residues. We used the method of limited proteolysis to find out in what fashion the rhodopsin polypeptide chain is arranged in the photoreceptor membrane. To localize the polypeptide chain regions exposed into cytoplasm, use was made of photoreceptor membranes of right-side orientation. Treatment of these membranes by papain, thermolysin and protease from *St.aureus* resulted in two membrane-bound fragments, being cleaved off the rhodopsin molecule. Their molecular weights, 25000 (F_I) and 12000 (F_2), were estimated by disc electrophoresis in polyacrylamide gel; the fragments (when membrane-bound) preserved spectra typical of the native molecule (λ_{max}=500nm) /7,8/. Fragment F_I is the N-terminal part of the protein molecule and contains polysaccharide chains; fragment F_2 belongs to the C-terminal region of the molecule and includes the lysine residue responsible for binding II-*cis* retinal via aldimine bond.Such noncovalent complex, containing fragments F_I and F_2, dissociated in a detergent upon illumination and that was used for purification and separation of the fragments by means of affinity chromatography on concanavalin-A-Sepharose. Analysis of the amino acid sequence of F_2 fragments obtained after proteolysis and elution from an affinity column by α -methylmannoside showed that all three enzymes attacked the rhodopsin molecule on the same site situated on the cytoplasmic membrane surface, that led to cleavage at bonds Ala-Thr (240-241) (papain), Ser-Ala (239-240) (thermolysin), Glu-Ser (238-239) (protease from *St.aureus)*. Besides, upon papain hydrolysis of the membranes the rhodopsin fragment Glu-Glu-Glu-Glu-Ser-Ala (235-240) was isolated from supernatant. Hydrolysis of the membrane-bound fragment F_I by carboxypeptidases A and B resulted in the amino acid sequence Lys-Glu-Ala-Ala-Ala-Gln (230-234) that also confirmed the side of papain cleavage. The C-terminal part of the rhodopsin molecule was also subjected to the action ot the proteases. According to data on limited proteolysis of the membrane by three enzymes, two regions exist in the rhodopsin polypeptide chain, the C-terminal

318-347 region and the 235-240 region, exposed into the cytoplasmic space. The 235-240 region forms the loop binding fragments F_I and F_2 and, apparently, it is functionally important for some photoreception processes.

At chymotrypsin action, the rhodopsin polypeptide chain in intact membranes is subjected to deeper hydrolysis comparing with the enzymes earlier used. Analysis of the obtained fragments showed that the membrane-bound part of rhodopsin preserving spectral characteristics of the native molecule, consists of four fragments F_I', F_2',F_2 and F_2'+F_2. The chymotryptic hydrolysate was separated by affinity chromatography on concanavalin-A-Sepharose and the fraction containing fragments F_2', F_2 and F_2'+F_2 was eluted after illumination. Fragment F_1 was eluted by α-methylmannoside solution. To characterize the latter fragment, cyanogen bromide hydrolysis was employed. The individual peptide of 23.000 molecular weight as well as the mixture of peptides of apparently the same molecular weights with N-terminal Arg and Lys, were isolated from the fraction obtained after illumination. Cyanogen bromide cleavage of the mixture of these peptides and separation of the obtained fractions allowed us to identify the regions of chymotryptic splitting of rhodopsin in the membrane using the following amino acid peptide sequences:

Arg-Phe-Gly-Gln-Asn-His-Ala-Ile-Met (147-155)

Lys-Ala-Gln-Lys-Gln-Val-Thr-Arg-Met (244-252)

One should note that upon chymotryptic hydrolysis the 317-335 and 336-347 fragments were cleaved off from the C-terminal part of the rhodopsin molecule.

The N-terminal and C-terminal amino acid sequences of the fragments, eluted from the column after illumination, were determined and we succeeded in locating two regions of the rhodopsin polypeptide chains situated on the cytoplasmic membrane surface and accessible to chymotrypsin action: Phe-Arg (146-147), Glu-Lys (243-244).

The data presented concern the membranes of native orientation. Intradiscal regions are inaccessible to the action of such nonpermeating agents as proteolytic enzymes. If rhodopsin is a transmembrane protein, its polypeptide chain must span the hydrophobic moiety of the membrane, i.e. some regions of the polypeptide chain are exposed into the intradiscal space. To study these regions, the native membranes were frozen and thawed for obtaining the vesicles of normal and reverse orientation. Chymotryptic hydrolysis of such membranes resulted in splitting off the peptides of sequences 1-10, 11-13, 14-24, 25-30 and 335-347. Determination of the amino acid sequence of the membrane-bound portion of the rhodopsin molecule gives the following sequence: Leu-Ala-Glu-Pro-Trp-Gln-Phe.

The analogues amino acid sequence was determined when analyzing the membrane-bound fragment after elution by α-methylmannoside after thermolysin hydrolysis. It testifies to the same site of the polypeptide chain of the rhodopsin molecule that was attacked by two enzymes. These findings show that 30 amino acid residues of the N-terminal part of rhodopsin are in intradiscal space.

In addition to the N-terminal part of the rhodopsin molecule, peptide bond 186-187 is also located on the intradiscal side of the membrane - as follows from the data on the limited papain proteolysis of unoriented membranes. From carboxymethylated hydrolysate the peptide of 4.000 molecular weight and amino acid sequence Cys-Gly-Ile-Asp-Tyr-Thr-Pro was isolated. Determination of the C-terminal amino acid sequence of this peptide, using carboxypeptidases A and B, gave the same result as in case of the F_1 from the papain hydrolysate of the normally oriented membranes. Analysis of the results on the limited enzymatic hydrolysis of rhodopsin in the membranes of right-side-out and inside-out orientation permits some conlusions concerining the arrangement of the polypeptide chain in the membrane. The N-terminal part of the molecule is exposed into the intradiscal space. In the vicinity of residue 31 the polypeptide chain enters the membrane. The C-terminal region (319-347) containing the phosphorylated serine and treonine residues is situated in cytoplasm. The polypeptide chain apparently spans the hydrophobic region of the lipid bilayer several times.

The important feature of the organization of the visual rhodopsin molecule in the photoreceptor membrane is the presence of the hydrophilic loop containing about 15 amino acids, on the cytoplasmic side between the fifth and the sixth α-helical rods. According to recent data this polypeptide chain is of the extreme importance as it provides binding of the first intermediate (GTP-binding protein) in a cascade of enzymatic reactions participating in signal amplification and light energy transformation into visual excitation.
Investigation of the protein complexes (transducin, phosphodiesterase, opsin kinase) started in our laboratory will probably shed light on the mechanism of function of the visual cell.
Location of the retinal residue in the protein globule is the next step in chemical studies of rhodopsins following elucidation of its structure and topography. Many investigators suppose that retinal of bacteriorhodopsin forms aldimine exclusively with the ε-amino group of Lys_{216} /9-11/, that was confirmed by experiments on bacteriorhodopsin reduction with complex hydrides in order to transform aldimine into a nonhydrolizable retinyl amine. If, however, the equilibrium occurs between two or more bacteriorhodopsin forms, differing in the aldimine position in the polypeptide chain, the reduction can lead to its shift to a more reactive form. In this case, retinyl groups can be distributed between several lysine residues, and as a limit they can be found on the lysine corresponding to the most easily reduced form. Is the existence of these forms possible in bacteriorhodopsin and under what conditions can it be realized?
If bacteriorhodopsin was preliminary subjected in purple membranes to chymotryptic cleavage into two fragments (the short one - residues 1-72 and the long one - residues 73-248) and then after the light or dark adaptation it was rapidly reduced in the dark in the presence of detergents (e.g. sodium dodecyl sulfate), then about one third of the retinyl residues appeared to be bound to short protein fragment. To establish the lysine residue in this fragment to which the retinal can be bound, bacteriorhodopsin with two completely dansylated lysine residues was used. In independent

experiments it was shown that both residues, Lys_{41} and Lys_{30}, were located in the short fragment. Upon reduction of dansylated and chymotrypsin-cleaved bacteriorhodopsin under the same conditions we observed practically analogous distribution of retinyl residues between short and long protein fragments. In this case Lys_{40} is the only carrier of the retinyl residue in the short fragment.

Does the amount of retinyl residues on Lys_{40} depend on the equilibrium shift at reduction or does it show a real ratio of various bacteriorhodopsin forms? To establish this it is necessary to determine if retinal is bound to the short or to the long protein fragment without any reduction. With this aim in view air-dried purple membranes preliminarily treated with chymotrypsin (or its glycerol suspension) were dissolved in neat of aqueous formic acid, and the fragments were separated under the conditions which rather effectively prevented aldimine hydrolysis (gel chromatography on Sephadex LH-60 in ethanol-formic acid mixture) /12/. It appeared that regardless of preliminary adaption to the light or to the dark about one third of retinal residues is bound to the short fragments. Accordance of the results obtained by both methods testified apparently to the presence of two bacteriorhodopsin forms differing in the location of the retinal residue in chymotrypsin-cleaved preparation, since the N-N migration of retinal seems unlikely at membrane solubilization in formic acid.

As shown earlier enzymatic cleavage of bacteriorhodopsin in purple membranes did not affect its spectral and functional properties /13/. The question arises if, indeed, such surrounding of the polyene chain and the mechanism of H^+-transport, which do not depend on whether retinal forms aldimine with Lys_{40} or Lys_{216}, are realized in bacteriorhodopsin.

Yet, sometimes the aldimine reduction apparently causes the shift of equilibrium between the forms differing in aldimine location. After hydride reduction of bacteriorhodopsin analogs containing residues of ω-arylpolyene aldehydes instead of retinal, the arylalkenyl residues were also found in short protein fragment, the shorter the polyene chain of these residues the larger their amount /14/. The reaction is rapid and cleavage of the peptide chain by hydrides did not occur /10/. Besides, upon separation of fragments of nonreduced proteins the corresponding aldehydic residues were observed exclusively in the long fragment of the protein where they bind to Lys_{216} /15/. This discrepancy can be explained by the fact that of two equilibrium aldimines formed by arylpolyenals with Lys_{40} and Lys_{216} the first one is reduced more rapidly.

Data in favour of retinal ability to form aldimine with the Lys_{40} residue in the native bacteriorhodopsin are not yet available. However, the obtained results undoubtedly point to the possibility of the retinal N-N migration and the question, concerning its possible role in functioning of bacteriorhodopsin and other retinal-containing proteins, arises once again.

References

1 Stoeckenius, W.: Acc.Chem.Res. 10, 337-344 (1980).

2. Uhl, R., Abrahamson, E.W.: Chem.Rev.81, 291-312 (1981).

3. Bolshakov, V.I., Kalamkarov, G.R., Ostrowsky, M.A.:Dokl. Acad. Nauk SSSR 248, 1485-1488 (1979).

4. Ovchinnikov, Yu.A., Abdulaev, N.G., Feigina, M.Yu., Kiselev, A.V., Lobanov, N.A.: FEBS Letters 100, 219-224 (1979).

5. Khorana, H.G., Gerber, G.E., Herlihy, W.C., Gray, C.P., Anderegg, R.J., Nihei, K., Biemann, K.: Proc.Natl.Acad. Sci. USA 76, 5046-5050 (1979).

6. Henderson, R., Unwin, P.N.T.: Nature 257, 28-32 (1975).

7. Saari, J.C: J.Cell Biol. 63, 480-491 (1974).

8. Pober, J.S., Stryer, L.: J.Mol.Biol. 95, 477-481 (1975).

9. Bayley, H., Huang, K.S., Radhakrishnan, R., Ross, A.H., Takagaki, Y., Khorana, H.G.: Proc.Natl.Acad.Sci. USA 78, 2225 (1981).

10. Lemke, H.D., Oesterhelt, D., FEBS Lett., 128, 255 (1981).

11. Muller E., Johnson, A.H., Akhtar, M.: FEBS Lett. 130, 187 (1981).

12. Gerber, G.E., Anderegg, R.J.,Herlihy, W.C., Gray, C.P., Biemann K., Khorana, H.G., Proc.Natl.Acad.Sci. USA 76, 227 (1979).

13. Abdulaev, N.G., Feigina, M.Yu., Kiselev, A.V., Ovchinnikov, Yu.A., Drachev, L.A., Kaulen, A.D., Khitrina L.V., Skulachev, V.P.: FEBS Lett., 90, 190 (1978).

14. Shkrob, A.M., Rodionov, A.V., Ovchinnikov, Yu.A.: Bioorgan. Khim. (USSR), 7, 1169 (1981).

15. Rodionov, A.V., Bairamashvili, D.I., Kudelin, A.B., Feigina, M.Yu., Shkrob, A.M., Ovchinnikov, Yu.A.: Bioorgan. Khim. (USSR), 7, 1328 (1981).

THE ROLE OF TYROSINE 26 AND OF TYROSINE 64 IN BACTERIORHODOPSIN FUNCTION

Horst-Dieter Lemke, Dieter Oesterhelt

Max-Planck-Institut für Biochemie
D-8033 Martinsried, Germany

Introduction

At present bacteriorhodopsin (BR) is one of the best studied integral membrane proteins. The amino acid sequence of BR is known (1,2) and a path of the polypeptide chain within the purple membrane (3) has been proposed. The retinal attachment site and the chromophore location within the protein have been determined (4-6). However, all this information together with electron diffraction data obtained at 7 Å resolution (7) do not allow an understanding of the molecular details of BR's action as a light driven proton pump.
Chemical modification is an appropriate approach for deducing the participation of single amino acid residues in the chromophore structure of the protein and in its catalytic mechanism. This report summarizes the results of experiments on tyrosine modification in BR.

Results

Treatment of purple membranes with tetranitromethane leads to the complete nitration of tyrosine 26 and 64, whereas tyrosines 131 and 133 only react in part (8). As a consequence of nitration, the chromophore of BR shifts in its absorption maximum from 570 to 532 nm and the extinction coefficient decreases from 63 000 $M^{-1}cm^{-1}$ to 45 000 $M^{-1}cm^{-1}$.

Chemistry of Peptides and Proteins, Vol. 2

After incorporation of the nitrated membrane into lipid vesicles the light-induced proton translocation activity is very much reduced compared to unmodified BR. A second technique for testing of BR functioning is to attach the membrane to a black lipid film. Function can be measured by recording the subsequent generation of a photocurrent on illumination (9,10). In this system only 10 % of the BR molecules are active after nitration (11). In addition, the kinetics for the formation of the photocurrent are significantly different from the kinetics measured with native purple membrane. This correlates qualitatively with an increased rate constant for formation and a decreased rate constant of decay of the $M_{412\ nm}$ intermediate in the photocycle upon nitration of BR (12,13).
The ionic reagent sodium dithionite was used to determine which of the four nitrated tyrosine residues account for the altered functional properties of the nitrated membrane. 3-Nitrotyrosines 64, 131 and 133 were reduced to their 3-aminotyrosine derivatives, without reactivation of the pump by this treatment. Therefore nitration of tyrosine 26 is assumed to be responsible for the changes in spectroscopic properties and activity of BR.
Nitration of a tyrosine lowers the pK of it's phenolic hydroxyl group considerably. Such a change must shift chromophore equilibria in BR if the tyrosine participates in the chromophoric structure (14,15). The blue shift of the chromophore upon nitration of tyrosine 26 allows one to assume that, in fact, this residue is close to the Schiff's base in the protein structure. Activity measurements revealed that a blue shifted chromophore does not mediate proton translocation (e.g. the 412 nm intermediate of the photocycle).
To elucidate the role of other tyrosine residues in BR, an ionic reagent, diazotized sulfanilic acid, was used for modification (16). After pre-treatment of purple membranes with methylacetimidate, which does not inactivate the proton pump but protects the lysine residues of the protein, it was found that

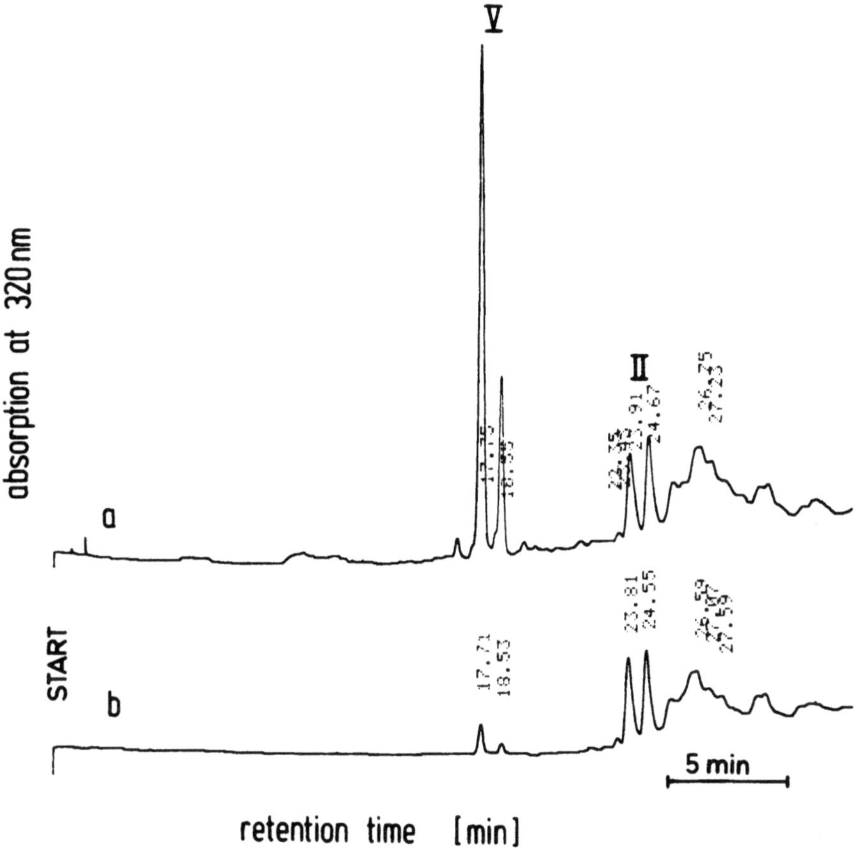

Figure 1. High performance liquid chromatography (HPLC) of CNBr- cleaved bacteriorhodopsin after azo-coupling (a) and subsequent reduction with sodium dithionite (b). Only fragments absorbing at 320 nm were detected.

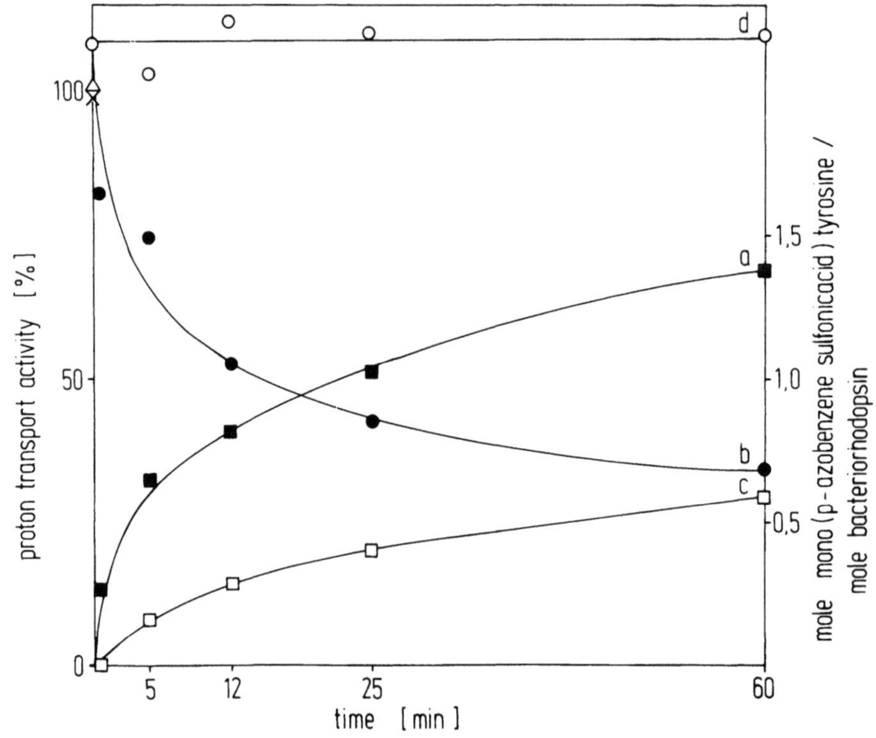

Figure 2. Reversible inhibition of bacteriorhodopsin upon diazo-labeling of tyrosine 64 and reduction with dithionite

a) time course of modification measured spectroscopically

b) proton pump activity of the modified samples

c) extent of modification after reduction

d) proton pump activity after reduction .

tyrosine 64 was predominantly labelled with diazotized sulfanilic acid. In contrast to nitration, diazo coupling of purple membrane neither changes the chromophore absorption nor the crystalline state of the membrane. Figure 1 shows the HPLC separation of modified and CNBr-cleaved BR. It demonstrates that diazo-coupled fragment V (trace a) carrying tyrosine 64 is predominantly labelled.

When diazo-coupled BR is treated with sodium dithionite only 3-azobenzene sulfonic acid tyrosine 64 is reduced to 3-amino-tyrosine 64 (trace b). The proton pump activity, which is inhibited upon diazo coupling (Fig.2, trace b) is restored upon addition of sodium dithionite (trace d). This correlates with the specific reduction of the diazo label on tyrosine 64 (Fig. 1, trace b, Fig.2, trace c). These results demonstrate that diazo-coupled tyrosine 64 blocks the proton translocation pathway.

To determine the relative location of tyrosines 26 and 64, purple membranes incorporated into lipid vesicles were reacted with dizotized sulfanilic acid. Tyrosine 26 is modified to the same extent in purple membrane sheets and in vesicles, whereas tyrosine 64 is modified in vesicles only to about 10 % of the value in purple membrane sheets. This indicates that the location of tyrosine 26 and of tyrosine 64 are on opposite sides of the membrane. The location of tyrosine 64 on the extracellular surface of the purple membrane can be demonstrated by diazo coupling and cleavage of the carboxyterminal part of BR by papain (16).

These results together clearly show that two tyrosine residues, tyrosine 26 close to the cytoplasmic surface and tyrosine 64 on the extracellular surface upon chemical modification prevent the proton translocating activity of BR.

References

1. Ovchinnikov, Yu.A., Abdulaev, N.G., Feigina, M.Yu., Kiselev, A.V., Lobanov, N.A.: FEBS Lett. 100, 219-224 (1979).

2. Gerber, G.E., Aderegg, R.J., Herlihy, W.C., Gray, C.P., Biemann, K., Khorana, H.G.: Proc. Natl. Acad. Sci. USA 76, 227-231 (1979).

3. Engelman, D.M., Henderson, R., McLachlan, A.D., Wallace, B.A.: Proc. Natl. Acad. Sci. USA 77, 2023-2027 (1980).

4. Bayley, H., Huang, K.S., Radhakrishnan, R., Ross, A.H., Takagaki, Y., Khorana, H.G.: Proc. Natl. Acad. Sci. USA 78, 2225-2229 (1981).

5. Lemke, H.-D., Oesterhelt, D.:FEBS Lett. 128, 255-260 (1981).

6. Mullen, E., Johnson, A.H., Akhtar, M.: FEBS Lett. 130, 187-193 (1981).

7. Unwin, P.N.T., Henderson, R.: J. Mol. Biol. 14, 425-440 (1975).

8. Lemke, H.-D., Oesterhelt, D.:Eur. J. Biochem. 115, 595-604 (1981).

9. Bamberg, E., Apell, H.J. Dencher, N.A., Sperling, W., Stieve, H., Läuger, H.: Biophys. Struct. Mechanism 5, 277-292 (1979).

10. Fahr, A., Läuger, P. Bamberg, E.: J. Memb. Biol. 60, 51-62 (1981).

11. Lemke, H.-D., Fahr, A., Oesterhelt, D., Bamberg, E. (in preparation).

12. Scherrer, P., Packer, L., Seltzer, S.: Arch. Biochem. Biophys. 212, 589-601 (1981).

13. Rosenbach, V., Goldberg, R., Gilon, C., Ottolenghi, M.: Photochem. Photobiol. 36, 197-201 (1982).

14. Fischer, U., Oesterhelt, D.: Biophys. J. 28, 211-230 (1979).

15. Fischer, U., Oesterhelt, D.: Biophys. J. 31, 139-146 (1980).

16. Lemke, H.-D., Bergmeyer, J., Straub, J., Oesterhelt, D.: J. Biol. Chem.,in press.

STRUCTURE-FUNCTION STUDIES ON GASTRIN RELATED PEPTIDES

Luis Moroder, Gianfranco Borin[+], Alessandro Lobbia, Jean-Pierre Bali[*] and Erich Wünsch

Max-Planck-Institut für Biochemie, Abteilung Peptidchemie, 8033 Martinsried bei München, BRD

[*]ER CNRS 228, Ecole Nationale Superieure de Chimie, Montpellier, France

Introduction

Extensive structure-function studies have provided convincing evidence that the main structural features responsible for the hormonal activity of the gastrins are located in a surprisingly limited sequence region, i.e. in the C-terminal Trp-Met-Asp-Phe-NH_2 (1,2). The related synthetic tetrapeptide amide had all the physiological properties of the natural hormone though in remarkable lower potency. The N-terminal extension of the gastrin molecule, particularly the characteristic penta-glutamic acid sequence could therefore possibly code for an additional functional information of this hormone.
To examine this question we have planned the synthesis of a series of gastrin analogues of increasing chain-length in the penta-glutamic acid sequence, whereby the methionine residue of the active center of the gastrins was substituted by norleucine in agreement with our recent findings concerning the usefulness of such replacement for biological and immunological studies (3).

[+]present address: Centro di Studi sui Biopolimeri, Via Marzolo 1, Universitá di Padova, Padova, Italy.

Chemistry of Peptides and Proteins, Vol. 2

Results

The syntheses of the gastrin analogues of the general formula Pyr-(Glu)$_x$-Ala-Tyr-Gly-Trp-Nle-Asp-Phe-NH_2, x = 0,1,2,3, and 4, were performed by the fragment condensation procedure taking advantage of our previous experiences in the synthesis of human little- and mini-gastrin analogues (4,5). The fully protected peptide derivatives were exposed to trifluoroacetic acid in the presence of scavengers. A screening of different cation trapping reagents clearly indicated 1,2-ethanedithiol as the most efficient to suppress at least partially tert-butylation at the tryptophan residues (6). Preparative high performance liquid chromatography (for experimental conditions see Table 1) of the resulting crude products led in good yields to the desired gastrin analoques at a high degree of purity as judged by thin layer chromatography in different solvent systems, by analytical high performance liquid chromatography on μ-Bondapak C18 and Rad Pak C18 columns, by amino acid analyses of the acid (6 M HCl, 110°C, 24 h, 2.5% thioglycolic acid) and enzymatic hydrolysates (aminopeptidase M digest; the N-terminus is cleaved by endopeptidases present in the commercial AP-M as di- and/or tripeptides) and by gas-chromatographic analyses of the derivatized acid hydrolysates as racemization test according to Frank et al (7). In the ^{1}H-nmr spectra of the gastrin analoques the typical sharp singulet of tert-butylated tryptophan was absent; additionally, the IR spectra showed no indication for the presence of reversed phase material (the products were repeatedly millipore-filtered using 0.2 μ filters).

This series of gastrin related peptides was then assayed for the gastric acid secretion potency by the experimental model of the reperfused stomach of anaesthetized fasted male wistar rats according to the method of Gosh and Schild (8) modified by Pham-Than-Chi et al (9). The responses are reported in Table 2 as micromoles of H^+ secreted at equimolar doses of the

$Pyr\text{-}[Glu(OBu^t)]_x\text{-}Ala\text{-}Tyr(Bu^t)\text{-}Gly\text{-}Trp\text{-}Nle\text{-}Asp(OBu^t)\text{-}Phe\text{-}NH_2$

↓ CF_3COOH, anisole/ethanedithiol

$Pyr\text{-}(Glu)_x\text{-}Ala\text{-}Tyr\text{-}Gly\text{-}Trp\text{-}Nle\text{-}Asp\text{-}Phe\text{-}NH_2$

(crude product)

Purification on HPLC (Jobin Yvon Prep 10)

x	column	eluent	flow rate	yield
0	Lichroprep RP 18, 25-40µm 40x4 cm	34% EtOH+66% 0.1M NH_4OAc,pH6.8	16 ml/min	85 %
1	Lichroprep RP 18, 25-40µm 40x4 cm	26% CH_3CN+74% 0.1M NH_4OAc,pH6.8	19 ml/min	71 %
2	Lichroprep RP 18, 25-40µm 40x4 cm	27% EtOH+73% 0.1M NH_4OAc,pH6.8	17 ml/min	50 %
3	Lichroprep RP 18, 25-40µm 40x4 cm	22.5% EtOH+77.5% 0.1M NH_4OAc, pH 6.8	19.5 ml/min	53 %
4	Lichroprep RP 18, 25-40µm 35x4 cm	21% EtOH+79% 0.1M NH_4OAc, pH 6.8	17.5 ml/min	70 %

Table 1. Deprotection and purification scheme of the gastrin analogues.

analogues injected intravenously, and as precent activity relative to that of 15-norleucine-human-little-gastrin (Nle^{15}-HG-[1-17]).

As expected the biological activity decreases by shortening the chain-length of the gastrin molecule; however, plotting the potencies of analoques against the chain-length, not a linear regression was obtained, but a sigmoidal graph (Fig. 1) with an abrupt increase of the activity from the lower levels typical for synthetic C-terminal gastrin fragments to the normal values of the naturally occuring mini- and little-

Gastrin peptides	Amount injected (nmol)	H^+ response (µmol)	% of Nle^{15}- HG-17 activity *
Nle^{15}-HG-(1-17)	0.25	15	100
Nle^{15}-HG-(5-17)	0.25	13	87
Nle^{15}-HG-(6-17)	0.25	10.9	72
Nle^{15}-HG-(7-17)	0.25	10	66
Nle^{15}-HG-(8-17)	0.25	4.9	34
Nle^{15}-HG-(9-17)	0.25	3	20
Nle^{15}-HG-(10-17)	0.25	2.5	17
Nle^{15}-HG-(14-17)	0.25	1.5	10

Table 2. Acid secretion potency of the gastrin analogues.
+) 15-Norleucine-human-little gastrin used as reference, was shown to be as potent as the parent synthetic methionine-peptide and 10% more active than the MRC-gastrin standard.

gastrin, upon incorporation of the third glutamic acid residue. A decreasing resistance to degradation in function of the penta-glutamic acid sequence should lead to a linear regression. Thus the results strongly imply for an important functional information coded in this particular region of the gastrin molecule.

In this context it is noteworthy that Abita et al (10) observed remarkable inhibitory effects on gastrin juice secretion by the activation peptide of trypsinogen. This peptide is liberated in large amounts during the duodenal activation of pancreatic zymogens and its primary structure is characterized by a surprising homology to the penta-glutamic acid sequence of the gastrins e.g. bovine trypsinogen activation peptide: H-Val-

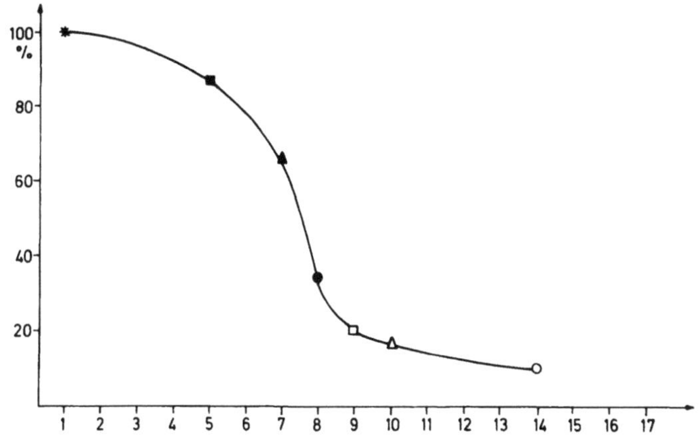

Fig. 1. Graphical correlation between biological activity and chain length of the analogues;
*) Nle^{15}-HG-(1-17), Nle^{15}-human-little gastrin;
■) Nle^{15}-HG-(5-17), Nle^{15}-human-minigastrin;
▲) Nle^{15}-HG-(7-17); ●) Nle^{15}-HG-(8-17); □) Nle^{15}-HG-(9-17); △) Nle^{15}-HG-(10-17); ○) Nle^{15}-HG-(14-17).

Asp-Asp-Asp-Asp-Lys-OH.
Furthermore as far as the affinity to specific antisera may mimic receptor recognition, anti-gastrin antisera exhibited an abrupt increase in binding of the gastrin analogues upon incorporation of a part of the penta-glutamic acid sequence as shown in Table 3.
To conclude, the acid sequence portion of the gastrins could carry the information for a specific transport mechanism of this hormone or constitute an additional recognition and thus, binding site for the receptor with consequent amplification of the hormonal activity.

Gastrin peptides	IC_{50} (femtomol)
Nle^{15}-HG-(5-17)	58.5
Nle^{15}-HG-(7-17)	68.5
Nle^{15}-HG-(8-17)	87.2
Nle^{15}-HG-(9-17)	216.7
Nle^{15}-HG-(10-17)	248.6
Nle^{15}-HG-(14-17)	>500

Table 3. Comparative cross-reactivities of the gastrin analogues versus ^{125}I-Nle^{11}-HG-[5-17] using antigastrin-antisera at a dilution of 1:50000; IC_{50}: femtomoles of gastrin analogue producing a 50% displacement of the tracer (12.5 pmole) from the antibody.

References

1. Tracy, H.J., Gregory, R.A.: Nature (London) 204, 935-938 (1964).
2. Gregory, R.A.: Bioorg. Chem. 8, 497-511 (1979).
3. Wünsch, E., Moroder, L., Gillessen, D., Soerensen, U.B., Bali, J.-P.: Hoppe-Seyler's Z.Physiol.Chem., in press.
4. Wünsch, E., Deimer, K.-H.:Hoppe-Seyler's Z.Physiol.Chem. 353, 1246-1254 (1972).
5. Moroder, L., Dress, F., Jaeger, E., Wünsch, E.: Hoppe-Seyler's Z.Physiol. Chem. 359, 147-154 (1978).
6. Wünsch, E., Jaeger, E., Kisfaludy, L., Löw, M.: Angew.Chem. 89, 330-331; Int. Ed. Engl. 16, 317-318 (1977).
7. Frank, H., Nicholson, G.J., Bayer, E.: J. Chromatog. Sci. 15, 174-176 (1977).
8. Ghosh, M.N., Schild, H.O.: J. Physiol. (London) 128, 35-36, (1955).
9. Pham-Tham-Chi, T.N., Bali, J.-P., Cayrol, B., Tubiana, M., Balmês, J.L., Boucard, M., Marignan, R.: Gastroenterol. Clin. Biol. (Paris) 2, 1031-1038 (1978).
10. Abita, J.P., Moulin, A., Lazdunski, M.: FEBS Lett. 34, 251-255 (1973).

ANALOGUES OF ANGIOTENSIN WITH VASODEPRESSOR ACTIVITY

Gunārs Chipens, Juris Ancāns, Daina Berga, Ilze Vosekalna, Natalya Mishlyakova, Andris Kriķis

Division of Peptide and Protein Bioregulators, Institute of Organic Synthesis, Latvian SSR Academy of Sciences, 226006 Rīga, USSR

Several groups of peptide bioregulators (hormones, kinins, toxins et al.), irrespective of their biological functions and origin, posses common structure fragments whose functions are to activate cell receptor systems and to determine specific quasi-cyclic shape of molecule [1,2]. In most cases, such quasi-cyclic structures are closed by ionic bonding involving a basic amino acid (arginine or lysine) in the common fragment and carboxyl group of the C-terminal amino acid (or amino dicarboxylic acid in the middle of peptide chain). Replacement of the ionic or salt bond for a covalent one or, in other terms, the transformation of quasi-cyclic structure into genuine cycle usually results in biologically active compounds with a selective and prolonged action [1,2]. As a typical example one can mention cycloanalogues of bradykinin, kallidin, corticotropin, tuftsin and others [3-5].

The method of total semi-empirical conformational calculation was also employed to determine the quasi-cyclic structures for angiotensin, however, the first attempts to obtain highly active cyclic analogues of this hormone were unsuccessful. The study of structure-function relationships in a number of bradykinin and kallidin cycloanalogues has demonstrated the essential role of the side chains of basic amino acids (arginine and lysine) for the manifestation of biological acitvity of cycloanalogues. These basic side chains are "lost" during cyclization. The ε-cycloanalogue of bradykinin lacks myotropic acti-

Chemistry of Peptides and Proteins, Vol. 2

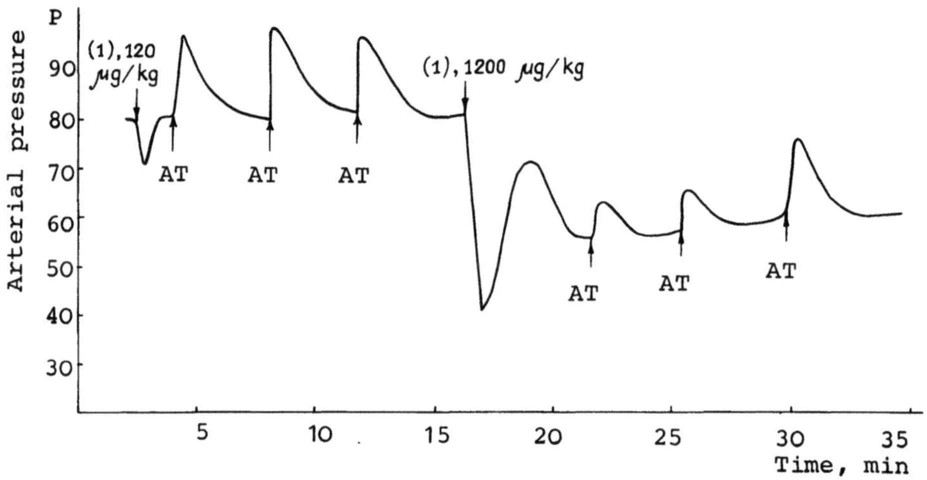

Fig.1. The effect of cyclo-[($N^{\varepsilon 1}$-Lys^1)-angiotensin] (1) on arterial blood pressure in anaesthetized rats and on the effect of angiotensin (AT), used in a 1 μg/kg dose.

vity (α=0), whereas ε-cyclokallidin shows the complete activity (α=1) of linear bradykinin [6].

With the above facts in mind, we have synthesized a novel cycloanalogue of angiotensin containing a basic amino acid (arginine) in the cyclic structure:

Lys-Arg-Val-Tyr-Ile-His-Pro-Phe (1).

The results of biological assays obtained with the cyclopeptide (1) were quite unexpected. The cycloanalogue (1) exerts a prolonged, dose-related depressor action on arterial blood pressure in anaesthetized rats (Fig.1). Intravenous administration of cyclo-[($N^{\varepsilon 1}$-Lys^1)-angiotensin] in 120,300,600 and 1200 μg/kg doses reduces arterial blood pressure in rats by 13,22,33 and 40 mm Hg, respectively, the duration of depressor effect increasing with increasing dose. Arterial pressure was not restored during 2 hours following cycloanalogue (1) injection (1200 μg/kg) and the pressor response to angiotensin was reduced in the presence of depressor effect during the first 10-15 min. Cyclopeptide (1) shows insignificant myotropic activity (α=0.39$\pm$0.03; pD_2=5.8$\pm$0.1) as judged by the in vitro experiments on the rat

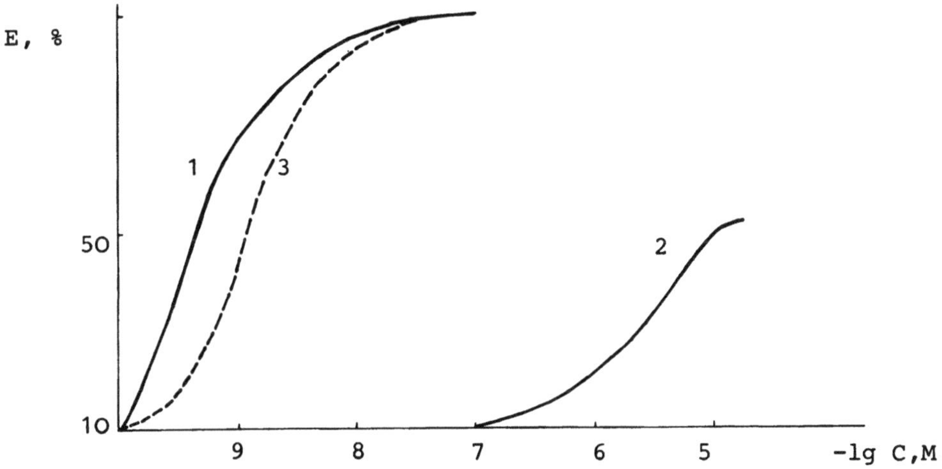

Fig.2. Cumulative dose-response curves of angiotensin (1), cyclo- [($N^{\varepsilon 1}$-Lys^1)-angiotensin] (2) and of angiotensin in the presence of cyclo-[($N^{\varepsilon 1}$-Lys^1)-angiotensin], (3, 10^{-8} M) applied to isolated rat ascending colon.

ascending colon, however the peptide inhibits the myotropic effect of angiotensin (Fig.2) when used in the 10^{-9}-10^{-7} M concentration range (pA_2=8.83$\pm$0.5). Thus, cyclo- [($N^{\varepsilon 1}$-Lys^1)-angiotensin] (1) shows a pronounced dualism of action - on the one hand, it exhibits the depressor effect characteristic of bradykinin, on the other hand, it inhibits the effects of angiotensin. The high pA_2 value indicates that cycloanalogue (1) practically does not differ from the native hormone in affinity to the rat ascending colon receptors. Therefore, it is possible that the biologically active structure of angiotensin, as in the case of other peptide bioregulators, is quasi-cyclic and the basic arginine residue of the common fragment makes an essential contribution to its biologic activity.

The prolonged depressor activity of cyclo- [($N^{\varepsilon 1}$-Lys^1)-angiotensin], characteristic of bradykinin cycloanalogues, suggests that during cyclization of angiotensin molecule some features of the biologically active space structure of bradykinin are generated which are sufficient for manifestation of activity. Some similar and presumably equifunctional elements in the primary structures of angiotensin and bradykinin

Asp-Arg-Val-Tyr-Ile-His-Pro-Phe

Arg-Pro-Pro-Gly-Phe-Ser-Pro-Phe-Arg

indicate that the specifity of their action is determined only by certain local structural differences. This prompted us to synthesize analogues of angiotensin and its C-terminal hexapeptide, containing an additional arginine residue in the C-end of the molecule (compounds III and IV in the Table).

PRIMARY STRUCTURES AND MYOTROPIC ACTIVITY OF FRAGMENTS AND ANALOGUES OF ANGIOTENSIN

Compounds		Rat ascending colon			Rat uterus		
		α	pD_2	pA_2*	α	pD_2	pA_2**
I	Lys-Arg-Val-Tyr-Ile-His-Pro-Phe	0.4	5.8	8.8			
II	Arg-Val-Tyr-Ile-His-Pro-Phe-Lys	1.0	5.5				
III	Asn-Arg-Val-Tyr-Ile-His-Pro-Phe-Arg	1.0	7.6	7.8	0.9	7.1	
IV	Val-Tyr-Ile-His-Pro-Phe-Arg	0.4	6.0	7.3	0.5	7.0	9.2
AT	Angiotensin (standard)	1.0	9.1				
BK	Bradykinin (standard)				1.0	9.3	

* Antagonistic effect to angiotensin

** Antagonistic effect to bradykinin, concentration in the case of compound (III) 10^{-9} M.

Nonapeptide (III), angiotensin analogue with an additional arginine residue in the carboxyl end, possessing 1% of the angiotensin pressor activity, has a myotropic effect on rat uterus as well as on rat ascending colon and also shows an appreciable antagonistic action to angiotensin when assayed on rat ascending colon. Nonapeptide (III) has no influence on the myotropic effect of bradykinin on rat uterus. Heptapeptide (IV), contrary to nonapeptide (III), has a depressor action in rats which is typical of bradykinin and amounts to 1% of the latter (Fig. 3). The myotropic activity of heptapeptide (IV) is essentially less pronounced as that of nonapeptide (III), the intrinsic activity on rat ascending colon and rat uterus being respectively only 0.4 and 0.5. Compound (IV) shows dualism of action: it is antagonistic to the myotropic effect of angiotensin at high

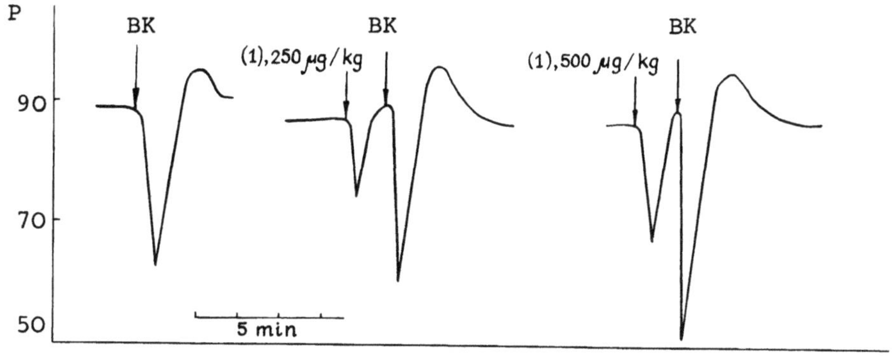

Fig.3. The effect of Val-Tyr-Ile-His-Pro-Phe-Arg on arterial blood pressure in anaesthetized rats and on depressor effect of bradykinin (BK) used in a 5 μg/kg dose.

concentrations (10^{-7}-10^{-6}M) on the rat ascending colon and in contrast to compound (III) tends to inhibit, at the same time, the myotropic effect of bradykinin, when tested at low concentrations on rat uterus.

The biological data demonstrate, that nonapeptide (III) exhibits neither bradykinin activity in vivo nor the effects of bradykinin in vitro, the latter, however, being observed in the case of heptapeptide (IV), despite the lack of N-terminal arginine residue, essential for the manifestation of bradykinin activity. The effects of compounds (III) and (IV) may be largely due to the products of their enzymatic splitting, especially in experiments in vivo. Kininase I which is mainly responsible for the rapid deactivation of bradykinin in blood plasma by removing the C-terminal residue of arginine, may also do so with compound (III) to give angiotensin (a potent pressor agent may be formed). In the case of compound (IV), the same enzyme may produce the C-terminal hexapeptide whose activity is only about 0.3% of that of angiotensin. So, the experimentally observed biological action of compounds (III) and (IV) may be the sum of several opposed effects. This is, in particular, indicated by the lack of pressor activity in the case of compound (I) which is resistant to kininase I.

In conclusion, it should be added that we do not claim the absolute validity of our interpretation. The final solution of problems concerned with the depressor activity discovered for the first time in the angiotensin series requires the synthesis of a number of linear and cyclic analogues.

References

1. Chipens, G.I., Krikis, A., Polevaya, L.K.: Biophysical and Biochemical Information Transfer in Recognition, Plenum Press, New York·London, 23-48 (1979).

2. Chipens, G.I., Polevaya, L.K., Veretennikova, N.I., Krikis, A.: Structure and Function of Low Molecular Peptides, Zinãtne, Riga, 11-124, 1980 (in Russian).

3. Chipens, G., Ancans, J., Afanasyeva, G., Balodis, J., Indulen, J., Kluša, V., Kudryashova, V., Liepins, E., Makarova, N., Mishlyakova, N.: Peptides 1976, Universite Bruxelles, Brussels, 353-360 (1976).

4. Chipens, G., Nikiforovich, G., Veretennikova, N., Vosekalna, I., Sosnov, A., Ancans, J., Mishlyakova, N., Liepins, E., Sekacis, I., Breslav, M.: Peptides. Structure and Biological Function, Pierce Chemical Company, Rockford, 567-570 (1979).

5. Romanovskis, P.J., Siskov, I.V., Liepkaula, I.K., Porunkevich, E.A., Ratkevich, M.P., Skujins, A.A., Chipens, G.I.: Peptides. Synthesis - Structure - Function, Pierce Chemical Company, Rockford, 229-332 (1981).

6. Chipens, G.I., Mutulis, F.K., Katayev, B.S., Kluša, V.E., Misins, I.P., Mishlyakova, N.V.: Int. J. Peptide Protein Res. 18, 302-311 (1981).

REGULATORY INFLUENCE OF PEPTIDE HORMONES ON NEURAL CELLS IN CULTURE

Bernd Hamprecht, Fridolin Löffler, Georg Reiser, Dietrich van Calker

Physiologisch-Chemisches Institut der Universität, Koellikerstr. 2, 8700 Würzburg, F.R.G.

Introduction

A still increasing number of peptide hormones is being discovered as common both to the brain and the alimentary system. In most cases very little or practically nothing is known on their function. Furthermore, the list of peptide hormones is growing that coexist in the same cell with classical neurotransmitters such as noradrenaline, dopamine and acetylcholine and with a functionally less well characterized biogenic amine such as serotonin (for review see ref. 1). Our ignorance of the function of such compounds is best illustrated by the range of terms used to classify them: Neurohormones, neurotransmitters, neuroeffectors, neuromodulators, neuroregulators, neurohumoral compounds, neuropeptides, neuroactive peptides. One way of approaching the problem of peptide hormone function in the nervous system is to ask the following sequence of questions: 1) Which cells carry receptors for such a peptide? 2) Which effector system(s) is (are) regulated by such a hormone-receptor complex? At the present time we do not know how many effector systems can be regulated by a given hormone-receptor complex, one or more. Examples for effector systems are: Adenylate and guanylate cyclases, ion channels, transport systems. 3) Which cellular processes are regulated as a consequence of the change in the effector system? Such processes may be of short-term (msec to min) or long-term (hours to days) duration.

Chemistry of Peptides and Proteins, Vol. 2

These questions could probably be answered best by using each of the cell types of the nervous system as a pure population. But what is a cell type? Neuron, glia, capillary endothelial cells? We know that there are neurons differing in location, size, morphology, content of neurohormones etc. But to ask how many different classes of neurons there are may be of little avail, since the answer may depend on the criteria applied and some differences may be quantitative rather than qualitative, or even transient. We also know there are different classes of glial cells: astrocytes, ependymal cells and oligodendrocytes. Again, it may well turn out to be necessary or useful to further subclassify these.

Evidence from the periphery indicates that different cells carry different sets of hormone receptors. If that were the case also with different types of brain cells, one might be able to separate cells by grabbing them at their specific receptors.

Therefore we set out to characterize the spectra of receptors present on the different cell types within the mixed population of cultured brain cells. For simplicity's sake we restricted our efforts primarily to receptors which interacted with adenylate cyclase as an effector system. In order to establish the methodology we first practiced with neuroblastoma x glioma hybrid cells (2), a permanent neuronal cell line, as a model system. Then we turned to primary cultures of brain cells from new-born mouse or rat. Such cultures mainly consist of astroblasts (immature astrocytes), oligodendroblasts, ependymal cells carrying beating ciliae, phagocytic cells (macrophages or microglia?), and probably capillary endothelial cells and fibroblasts (3). Finaly we carried out electrophysiological studies on the hormonal sensitivity of polyploid rat glioma cells (4) and of the hybrid cells.

Results

The level of cyclic AMP in neuroblastoma x glioma hybrid cells is regulated by two classes of hormones, those that raise it and those that inhibit this rise. Among the stimulating hormones are two peptides, secretin and glucagon, acting via separate receptors (5). The list of the inhibitory hormones also contains two types of peptides, opioids such as enkephalins and endorphins (6,7) and somatostatin (8). Each of the inhibitory hormones intervenes with the action of each of the stimulating ones.

Table 1. Hormones regulating the level of cyclic AMP in primary murine brain cell cultures (9).

Increase evoked by	Inhibition of increase evoked by
*noradrenaline (β-receptor)	*noradrenaline (α-receptor)
*secretin	*somatostatin
*vasoactive intestinal peptide	*adenosine (A_1-receptors)
*corticotropin, melanotropins	
*parathyrin	
*calcitonin	
(?)adenosine (A_2-receptors)	
(?)histamine (H_2-receptors)	
prostaglandins E_1,E_2	

*Hormones the receptors of which are most likely located on the same cell type. (?) Location of receptor in respect to those marked by an asterisk is unknown.

As in the hybrid cell model system the formation of cyclic AMP in primary cultures of newborn mouse or rat brain is either stimulated or inhibited, respectively, by two discrete sets of hormones including peptides. While secretin and vasoactive intestinal peptide (10), corticotropin and melanotropin (11), parathyrin and calcitonin (12) elevate the intracellular concentration of cyclic AMP, somatostatin (10) prevents this increase (Table 1).

The nonapeptide bradykinin does not influence the level of cyclic AMP in the hybrid cells and the polyploid rat glioma cells. However, it evokes a slow hyperpolarization of the plasma membranes of these cells, lasting for 10 - 30 s. Repetitive application of the peptide causes progressively lower responses and the onset of lag phases between peptide delivery and the beginning of the hyperpolarization response (13). Bradykinin also evokes, in the same cells, a transient increase in the intracellular level of cyclic GMP that follows a similar time course as the hyperpolarization response (14). Work is under way to investigate a possible causal relationship between changes in membrane potential, desensitization and lag period on the one side, and cyclic GMP-levels on the other side.

Discussion

The outcome of this work is severalfold:

The fact that one hormone inhibits the action of another may be best explained by the presence on the same cell of both types of receptors (15), and can be used to allocate a set of receptors to a single cell type within a mixture of cells. Such a set of receptors could subsequently be used as a fingerprint identity label for that cell type in the process of cell separation (15). Properly constructed probes fitting a given hormone receptor may be used in "fishing" a certain cell type out of a mixture of cells.

Glial cells previously not even thought of as possible targets of peptide hormones have now to be strongly considered as possible targets of neuropeptides.

Glial cells appear to carry receptors for hormones such as secretin and parathyrin which have not yet been clearly established to occur in brain. The present investigations might stimulate the search for the presence and production of these peptides in brain.

Besides the hormones listed in Table 1 no other hormone of a long list of peptides tested affected the level of cyclic AMP in the hybrid cells (5) and in primary brain cell culture (11). Therefore such cell systems could be used for discovering new hormonal agents, e.g., in extracts of nervous tissue (16).

These and other cell systems (17) could be employed in elucidating the question whether one receptor can regulate one or more effector systems. It is pertinent in this respect that the hybrid cells respond to opioids in several ways (18-21), although they appear to contain only one major class of opioid receptors (22,23).

References

1. Hökfelt, T., Lundberg, J.M., Schultzberg, M., Johansson, O., Ljungdahl, A., Rehfeld, J., in: Neural Peptides and Neuronal Communication (Costa, E., Trabucchi, M., eds.), pp. 1-23. Raven Press, New York 1980.
2. Hamprecht, B.: Int. Rev. Cytol. 49, 99-170 (1977).
3. Hallermayer, K., Harmening, C., Hamprecht, B.: J. Neurochem. 37, 43-52 (1982).
4. Heumann, R., Reiser, G., van Calker, D., Hamprecht, B.: Exp. Cell Res., in the press.
5. Propst, F., Moroder, L., Wünsch, E., Hamprecht, B.: J. Neurochem. 32, 1495-1500 (1979).
6. Brandt, M., Gullis, R.J., Fischer, K., Buchen, C., Hamprecht, B., Moroder, L., Wünsch, E.: Nature 262, 311-313 (1976).
7. Klee, W.A., Nirenberg, M.: Nature 263, 609-612.

8. Traber, J., Glaser, T., Brandt, M., Klebensberger, W., Hamprecht, B.: FEBS Lett. 81, 351-354 (1977).

9. van Calker, D., Hamprecht, B.: Adv. Cellul. Neurobiol. 1, 31-67 (1980).

10. van Calker, D., Müller, M., Hamprecht, B.: Proc. Natl. Acad. Sci. U.S.A. 77, 6907-6911 (1980).

11. van Calker, D., Löffler, F., Hamprecht, B.: J. Neurochem., submitted for publication.

12. Löffler, F., van Calker, D., Hamprecht, B.: EMBO J. 1, 297-302, 1982.

13. Reiser, G., Hamprecht, B.: Brain Res. 239, 191-199 (1982).

14. Reiser, G., Walter, U., Hamprecht, B.: Submitted for publication.

15. van Calker, D., Müller, M., Hamprecht, B.: J. Neurochem. 30, 713-718 (1978).

16. Glaser, T., Hübner, K., Hamprecht, B.: J. Neurochem. 39, 59-69 (1982).

17. Hamprecht, B., Amano, T., Kemper, W.: Brain Res. 101, 129-135 (1976).

18. Traber, J., Gullis, R., Hamprecht, B.: Life Sci. 16, 1863-1868 (1975).

19. Klee, W.A., Sharma, S.K., Nirenberg, M.: Life Sci. 16, 1869-1874 (1975).

20. Myers, P.R., Livengood, D.R., Shain, W.: Nature 257, 238-240 (1975).

21. Dawson, G.: J. biol. Chem. 254, 155-162 (1979).

22. Chang, K.J., Miller, R.J., Cuatrecasas, P.: Mol. Pharmacol. 14, 961-970 (1978).

23. Winter, A.: Diploma thesis, Univ. of Würzburg, 1982.

ENZYME IMMUNOASSAY OF HUMAN URINARY KALLIKREIN

Michael Franke, Susanne Rohrschneider and Reinhard Geiger
Abteilung für Klinische Chemie und Klinische Biochemie in der Chirurgischen Klinik der Universität München, Nussbaumstr. 20, D-8000 München 2, FRG
(Leitung: Prof. Dr. H. Fritz)

Introduction

Determination of tissue kallikrein (EC 3.4.21.35) in human urine and body fluids is of special interest because of its postulated physiological role in kidney function and blood pressure regulation, fertilization processes, glucose metabolism and intestinal absorption (1, 2).

Various methods for the determination of urinary kallikrein are available at present, but some assays such as the biological assays and the kinin-liberating assays (3) are cumbersome and not suitable for routine purpose.

Recently, a radioimmunoassay of HUK has been developed (4) and since that time this assay has become a common method for the determination of human tissue kallikrein. However, general problems of the radioimmunoassay such as long incubation periods, special and expensive equipment for radioactive measurements and the radioactive waste stimulated the development of alternatively but also highly sensitive immunoassays. This report describes an enzyme immunoassay for human tissue kallikrein allowing estimation of kallikrein levels as low as 0.5 µg/l in a reasonable short time (5 hours incubation time).

Chemistry of Peptides and Proteins, Vol. 2

Material and Methods

Human urinary kallikrein, used for the enzyme immunoassay as a standard, was purified as described (5). Horseradish peroxidase (grade 1) (EC 1.11.1.7) and bovine trypsin and chymotrypsin were products of Boehringer AG, Mannheim. Pig pancreatic kallikrein was a gift of Bayer AG, Wuppertal. Bovine serum albumin was purchased from Behringwerke AG, Marburg/Lahn. Tween 20 was a product of Serva AG. Microtiter plates M 129 B were purchased from Dynatech, Denkendorf, West-Germany.
Antibody production was performed as described by Mann et al. (4). IgG fractions were isolated according to Steinbuch and Audran (6).

Buffers

Buffer A	15 mmol/l Na_2CO_3, 350 mmol/l $NaHCO_3$, 0.2 g/l NaN_3, pH 9.6
Buffer B	1.5 mmol/l KH_2PO_4, 8 mmol/l Na_2HPO_4, 2.7 mmol/l KCl, 150 mmol/l NaCl, 0.05 % (w/v) Tween 20, pH 7.4
Buffer C	1.5 mmol/l KH_2PO_4, 8 mmol/l Na_2HPO_4, 2.7 mmol/l KCl, 150 mmol/l NACl, 0.05 % (w/v) Tween 20, 2 % (w/v) bovine serum albumin, pH 7.4
Buffer D	100 mmol/l citric acid, 100 mmol/l K_2HPO_4, pH 4.0

Coupling of peroxidase to IgG

Preparation of peroxidase IgG conjugate was performed as described by Nakane & Kawaoi (7). After gel filtration on Sephadex G-100 the anti-HUK-IgG peroxidase conjugate was diluted 1 : 50 with buffer C before applied to the enzyme immunoassay.

Peroxidase activity

Peroxidase activity was measured as described by Anaokar et al. (8).

Enzyme immunoassay conditions

Microtiter plates were coated with 0.2 ml per well of an IgG solution (10 µg IgG/ml buffer A) at 4^{0} C over night. The plates were then washed vigorously (3 times) in buffer B. Human urinary kallikrein standard samples and test samples were diluted with buffer C. Of these samples 0.2 ml were added to the wells and the plates were incubated at 37^{0} C for 3 h in order to bind the kallikrein to the solid-phase fixed antibodies. After incubation, the plates were washed (3 times) with buffer B and 0.2 ml of the anti-HUK-IgG peroxidase conjugate was added to each well and the plates incubated again at 37^{0} C for 2 h. After washing 0.2 ml substrate solution (0.5 ml ABTS (200 mg in 10 ml aqua dest.), 9.5 ml buffer D and 0.2 ml 0.06 % (v/v) H_2O_2) was added,and the plates were incubated this time for 30 min at room temperature. Thereafter, the reaction was stopped by addition of 0.05 ml sodium azid (10 g/l). The absorption was read photometrically at 405 nm with a Microelisa[R]-reader AM 115 from Dynatech, Denkendorf, West-Germany. Background generally ranged from 0.02 to 0.08 A_{405}. In each test a standard concentration curve with purified human urinary kallikrein was included. Standard curves were constructed by plotting the fraction of absorbance at 405 nm, $A - N/A_0 - N$ (A: absorbance, A_0: maximal absorbance obtained with excess HUK, N: absorbance of the blank) against the dose of human urinary kallikrein in a logit-log mode or in a linear-log mode. A blank contains buffer C instead of HUK.

Results

Concentration of the solid-phase IgG: The least amount of IgG required to coat the microtiter plates was determined from standard curves of data obtained with plates treated with concentrations of IgG ranging from 1 to 10 mg/l. The standard curves were identical for IgG concentrations between 10 and 5 mg/l, whereas standard curves obtained with IgG concentrations lower then 5 mg/l showed a decreased binding of HUK. Hence, we decided to use an IgG concentration of 10 mg/l for the assay.

Anti-HUK-IgG peroxidase conjugate dilution: The optimal dilution of the conjugate was determined from a standard curve obtained with microtiter plates coated with an IgG solution of 10 mg/l and conjugate diluted 100-, 50-, 20- and 10-fold with buffer C. As the 50-fold dilution was the highest dilution of the conjugate at which optimal absorbance at 405 nm occured, this conjugate dilution was chosen for the assay.

Time and temperature dependence of the antigen antibody reactions: To determine the optimal incubation conditions for the enzyme immunoassay the time and temperature dependence of the first incubation, the binding of human urinary kallikrein to the solid-phase fixed antibody, was studied at 22 and 37° C. The highest binding rate was observed at 37° C, where equilibrium was reached after 3 hours. The optimal time and temperature required for the second reaction, binding of anti-HUK-IgG peroxidase conjugate to solid-phase immunglobulin-fixed human urinary kallikrein, was similarily determined. This time equilibrium was reached after 2 hours at 37° C.

Standard curves and sensitivity: The standard curve was consistently linear for human urinary kallikrein concentrations between 0.25 and 60 µg/l under routine conditions (Figure). The sensitivity of an enzyme immunoassay can be defined as "the smallest amount of antigen giving a response which is distinguishable from the response in the absence of antigen" (9). The lowest concentration of human urinary kallikrein that produced a response greater than that caused in the absence of human urinary kallikrein was 0.25 µg/l, which corresponds to 50 pg per well.

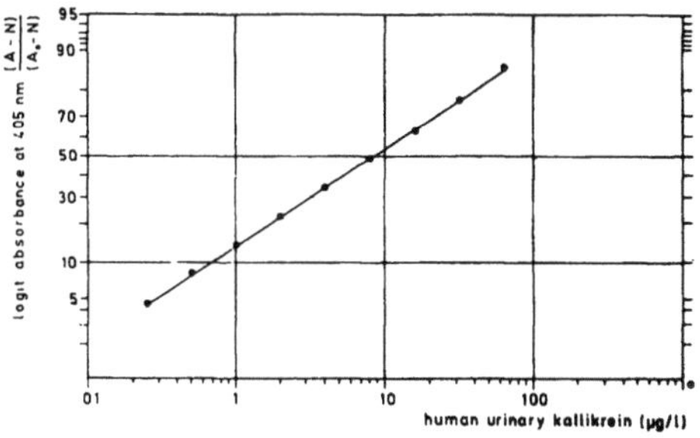

Figure. Dose response curve of purified human urinary kallikrein in the human urinary kallikrein enzyme immunoassay. Results are expressed in the logit-log transformation.

Precision: The intraassay coefficient of variation (N = 10) was 2 - 4 % for human urinary kallikrein (range 0.25 to 64 µg/l). The interassay coefficient of variation of identical samples (N = 10) containing 1,4,16,64 µg/l was 4 - 12 %.

Specificity: Cross-reaction of other serine proteinases with anti-HUK-IgG was tested under the given immunoassay conditions. A cross reaction was not observed with bovine trypsin and chymotrypsin, pig pancreatic kallikrein, boar acrosin and urine from rat, dog and rabbit. In human samples investigated immunologically active material was found in pancreatic juice, saliva, urine, serum and seminal plasma.

Recovery test: Recovery tests employing three different quantities of purified human urinary kallikrein to antigen-free rat urine showed a good precision of the assay. The data are listed in the Table.

Comparison of the enzyme immunoassay with other methods: In order to show the correlation of the enzyme immunoassay with other determination methods for human urinary kallikrein ,10 dialyzed urine test samples were measured by different kallikrein assays (10).

Table. Recovery test.

Human urinary kallikrein added (μg/l)	Human urinary kallikrein recovered (μg/l)	N	Recovery
20	19.0 ± 0.46	10	95 ± 2.3
10	10.5 ± 0.12	10	105 ± 1.2
5	5.1 ± 0.06	10	102 ± 1.2

Discussion

The sensitive and specific enzyme immunoassay developed for human urinary kallikrein should be especially suitable for further studies on the physiological and pathophysiological role of tissue kallikreins. As in humans the different tissue kallikreins are immunologically indistinguishable (11), this assay can be used to detect all of them.

The high specificy of the enzyme immunoassay could be demonstrated by employing various serine proteinases. Cross reactivity was not observed with bovine trypsin and chymotrypsin, boar acrosin and pig pancreatic kallikrein. Furtheron, in analogy to the results obtained by radioimmunoassay of human urinary kallikrein (4) we may be expect that human proteinases other than tissue kallikreins also do not interfere in the enzyme immunoassay.

A number of experiments were undertaken to compare the enzyme immunoassay with assays established already for the determination of human urinary kallikrein. For this purpose dialyzed urine samples from 10 different persons were employed.

The results obtained with the enzyme immunoassay were compared with those obtained by the dog blood pressure assay, a radioimmunoassay for human urinary kallikrein, a rat uterus assay after kallikrein-induced kinin liberation and assays with AcPheArgOEt and DValLeuArgNHNp as kallikrein substrates. The excellent correlation, however between the two immunoassays indicates the high validity of the enzyme immunoassay.

A considerable advantage of the enzyme immunoassay over the radioimmunoassay is among others the lack of radiation hazards, greater stability of the label, higher throughput and ease of automation. The assay procedure is simple, to perform in a relatively short time and requires only small quantities of kallikrein. The minimal detectable dose is 50 pg per well. Therefore, this assay offers a precise and feasible test for human urinary kallikrein and which can be used as an alternative to the radioimmunoassay.

Abbreviations

HUK: Human urinary kallikrein

IgG: immunoglobulin G

ABTS: 2,2'-azino-di(3-enthyl-benzthiazoline-6-sulfonate)

Acknowledgement

This work was supported by Deutsche Forschungsgemeinschaft, Sonderforschungsbereich 51, grant B-26, We wish to thank Prof. Dr. H. Fritz for many helpful discussions and comments during the work and for reading over the manuscript. We are indebted to Dr. Truscheit, Bayer AG, for the gift of porcine pancreatic kallikrein.

References

1. Schachter, M.:Pharmacol. Rev. 31, 1-17 (1980).

2. Geiger, R. In: Proteinases and their Inhibitors (Turk, V. and Vitale, Lj., eds.), Pergamon Press, Oxford, 353-376 (1981).

3. Geiger, R., Fritz, H.:Meth. Enzymol. 80, 466-492 (1982).

4. Mann, K., Göring, W., Lipp, W., Keipert, B., Karl, H. J., Geiger, R., Fink, E.: J. Clin. Chem. Clin. Biochem. 18, 395-401 (1980).

5. Geiger, R. Stuckstedte, U., Fritz, H.: Hoppe-Seyler's Z. Physiol. Chem. 361, 1003-1016 (1980).

6. Steinbuch, M., Audran, R.: Arch. Biochem. Biophys. 134, 279-284 (1969).

7. Nakane, P. K., Kawaoi, A.: J. Histochem. Cytochem. 22, 1084-1091 (1974).

8. Anaokar, S,, Garry, P. J., Standefer, J. C.: Clin. Chem. 25, 1426-1431 (1979).

9. Schurrs, A. W. W. M., Van Weemann, B. K.: Clin. Chim. Acta 81, 1-40 (1977).

10. Franke, M., Rohrschneider, S., Geiger, R.: J. Clin. Chem. Clin. Biochem, in press (1982).

11. Fritz, H., Fiedler, F., Warwas, M., Truscheit, E., Kolb, H. J., Mair, G., Tschesche, H.: In: Kininogenases - Kallikrein 4 (Haberland, G. L., Rohen, J. W., Suzuki, T. eds.) Schattauer Verlag, Stuttgart - New York, 15-28 (1977).

THE ALAMETHICIN PORE IS FORMED BY A VOLTAGE-GATED FLIP-FLOP OF α-HELIX DIPOLES

Günther Boheim[+], Wolfgang Hanke[+] and Günther Jung[++]
[+] Lehrbereich Zellphysiologie, Ruhr-Universität Bochum, Postfach 10 21 48, D-4630 Bochum, FRG
[++] Institut für Organische Chemie, Universität Tübingen, Auf der Morgenstelle 18, D-7400 Tübingen, FRG

Alamethicin, a nonadecapeptide antibiotic, forms voltage-dependent pores in planar lipid bilayers (1-3) and biological membranes (4) . Single alamethicin pores exhibit a peculiar sequence of conductance levels which are not integral multiples of a basic unit, but are consistent with oligomeric pores states of varying diameter ('barrel stave model') (3,5,6) . At progressively higher concentrations alamethicin produces a highly voltage-dependent conductance and finally lysis of the membrane. Next to alamethicin a number of similar peptide antibiotics with lytic activity have been described (7) . Suzukacillin A (8) , for example, also produces stable multilevel pores, while other analogues do not. Trichotoxin A 40, the primary structure of which is known (9) , induces a voltage-dependent conductance similar to that produced by alamethicin, but no stable pore states could be resolved (10) . Similar results were found by Vodyanoy et al. (11) for an unknown purified subfraction of natural alamethicin. A synthetic nonadecapeptide analogue also exhibits a voltage-dependent conductance without well defined conductance levels (10) . The chemical modification of trichotoxin A 40 at its C-terminal part by introduction of relatively small groups (-methylester, -hydrazide) did not change its pore forming behaviour. However, the introduction of the larger dansyl group led to a steeper current-voltage characteristic as compared to

Chemistry of Peptides and Proteins, Vol. 2

that of trichotoxin A 40 and to the observation of resolved multi-level pores (10).

Alamethicin and its analogues have pronounced amphiphilic character. The lipophilic N-terminal regions adopt an α-helical conformation in organic solvents and lipid vesicles (12). Peptide α-helices carry an intrinsic dipole moment, and values of 67 D (13) and 75 D (14) have been measured for that of the alamethicin molecule. In contrast the C-termini are more polar in character and do not adopt an α-helical conformation (12). The object of the present study is to demonstrate the significance of the structural features of this type of amphiphilic peptides for voltage-dependent pore formation. A model is proposed which describes the change in transmembrane orientation of antiparallel dipoles to a parallel array in response to an applied electric field. As a consequence of the electrostatic repulsion within an aggregate of parallel α-helix dipoles water-filled pores establish.

Melittin behaves similarly to alamethicin

The bee venom constituent melittin resembles alamethicin in amphiphatic nature, lytic activity and secondary structure, but is completely different in amino acid composition. Tosteson and Tosteson (15) reported stepwise current fluctuations due to the formation of melittin pores in planar bilayers. These pores showed low conductances (7-20 pS in 1 M NaCl) and lifetimes of several seconds. No multi-level behaviour was observed. Fig. 1 shows the current-voltage characteristics of planar bilayers modified by alamethicin (a) and melittin (b). Both exhibit a strong voltage-dependence of bilayer conductivity. Although this is more pronounced for alamethicin than for melittin, the similarity in behaviour is apparent.

Current fluctuation measurements at low levels of polypeptide induced membrane conductivity reveal the nature of pore formation. In Fig. 2 single pore current fluctuations are recorded

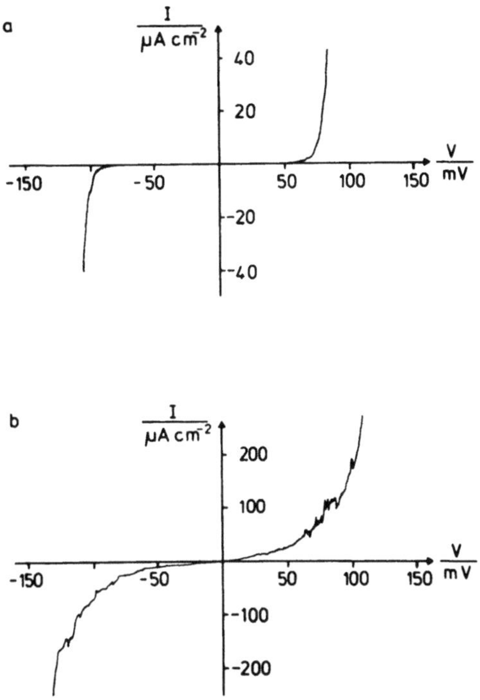

Fig. 1.

Current-voltage characteristics of alamethicin (a) and melittin (b), doped lipid bilayers showing voltage-dependence of peptide-induced membrane conductivity.

Bilayers were formed on the hole (area $\sim 2 \times 10^4$ μm^2) of a hexadecane pretreated sandwich teflon septum (18) from a solution of 2 mg lipid per ml solvent (hexane/ethanol 9:1). Lipid: 1-palmitoyl-2-oleoyl-glycero-3-phosphocholine (1,2-POPC; Avanti, Birmingham, USA); salt: 1 M KCl, pH$\sim$6; temperature: 20°C. Alamethicin: 0.2 μg cm^{-3} (a) and melittin: 2 μg cm^{-3} (b) were added to the cis-compartment only. Applied membrane potentials are referred to the trans-side of the bilayer as ground. A positive current thus flows from the cis-side to the trans-side.

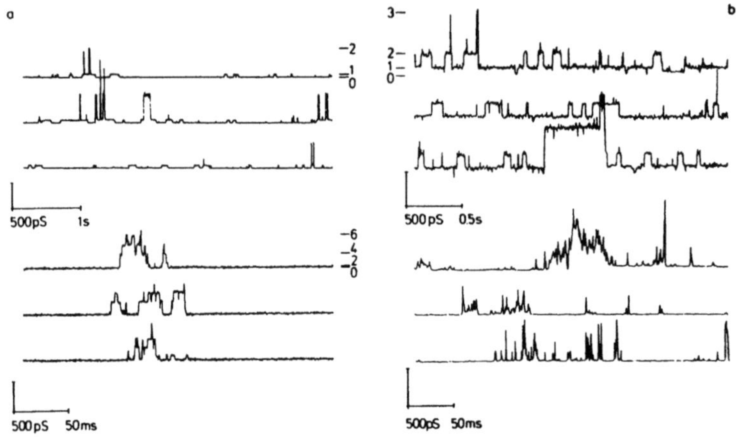

Fig. 2.

Single pore current fluctuations of alamethicin (a) and melittin (b) at two different ion concentrations.

The upper trace in (a) shows current fluctuations to occur mainly in pore states $\nu = 1$ and 2. At high ionic strength this is achieved by proper adjustment of the membrane voltage. The lower trace in (a) exhibits pore fluctuations in form of short-lived bursts. Under conditions of low ionic strength the pore mainly adopts the higher pore states $\nu = 4,5$ and 6. The corresponding current fluctuations induced by melittin are seen in (b). Whereas the upper trace shows long-lasting pore fluctuations between several non-integral conductance levels, the lower trace exhibits short-lived bursts. In the latter case pore level conductances could not be resolved.

Planar bilayers were formed using teflon sandwich septa as in Fig. 1 (a, upper trace) or across the tip of fire polished but otherwise untreated glass pipettes (19) (a, lower trace and b). Although the diameter of the pipette opening was 1-2 μm, the bilayer area estimated from capacitance measurements was ca. 100 μm^2, indicating that the bilayers were probably located some distance inside the pipette. In these experiments the trans-side corresponded to the inside of the pipette. Temperature: 20°C.

a. Upper trace - salt: 5 M NaCl, pH ~6; lipid: 50 % 1,2-POPC, 50 % bacterial phosphatidylethanolamine (bac PE, Avanti); alamethicin: 0.2 μg cm^{-3} (cis-side); voltage: 100 mV. Lower trace - salt: 50 mM NaCl, 2 mM Na, K-phosphate buffer, pH 7; lipid: 80 % 1,2-POPC, 20 % bac PE; alamethicin: 0.5 μg cm^{-3} (cis-side); voltage: 110 mV.
b. Upper trace - salt: 5 M NaCl, pH ~6; lipid: 1,2-POPC; melittin: 2 μg cm^{-3} (cis-side); voltage: 110 mV. Lower trace - salt: 50 mM NaCl, 2 mM Na,K-phosphate buffer, pH 7; lipid: 1,2-POPC; melittin: 2.5 μg cm^{-3} (cis-side); voltage: 150 mV.

---

at two different ion concentrations: 5 M NaCl (upper traces) and 50 mM NaCl (lower traces). Fig. 2a shows that the alamethicin pore adopts the lowest conductance states ν = 1 and ν = 2 over a long period of time at 5 M NaCl. At 50 mM NaCl short-lived burst-like events appear, and one finds the pore mainly in higher pore states (ν = 4,5 and 6). We conclude that electrostatic forces, whose effective radius depends on the ionic strength of the aqueous salt solution, determine pore formation and pore state transitions. Fig. 2 b shows the pore current fluctuations of melittin under identical ionic conditions. At 5 M NaCl stepwise current fluctuations are observed which exhibit multi-level behaviour of the melittin pore. At 50 mM NaCl short lived bursts appear but discrete conductance levels could not be resolved with melittin in contrast to alamethicin. Therefore we conclude that electrostatic forces influence alamethicin and melittin pore formation in the same way: pores are long-lived and stable at high ionic strength (small electrostatic forces), and they are short-lived at low ionic strength (large electrostatic forces). In view of a pore of variable diameter (3) one expects this behaviour to result

from repulsive forces.
The values of pore level conductances $\Lambda(\nu)$ from Fig. 2 are the following: alamethicin (5 M NaCl) — $\Lambda(1)$ = 53 pS, $\Lambda(2)$ = 550 pS, $\Lambda(3)$ = 2150 pS (not visible); melittin (5 M NaCl) — $\Lambda(1)$ = 83 pS, $\Lambda(2)$ = 290 pS, $\Lambda(3)$ = 1000 pS or $\Lambda(3)$ = 710 pS and $\Lambda(4)$ = 1210 pS; alamethicin (50 mM NaCl) — $\Lambda(2)$ = 15 pS, $\Lambda(3)$ = 80 pS, $\Lambda(4)$ = 230 pS, $\Lambda(5)$ = 390 pS and $\Lambda(6)$ = 560 pS. A comparison with the experimental data of Tosteson and Tosteson (15) shows that our lowest value of $\Lambda(1)$ = 83 pS (5 M NaCl, 1,2-POPC) is comparable to their value of Λ = 20 pS (1 M NaCl, asolectin bilayers). The conductance ratios of neighbouring conductance levels are different for alamethicin and melittin. Like the difference in pore selectivity (alamethicin: cation selective (16) , melittin: anion selective (15)) this seems to result from the unlike charges of the two peptides. The variation in melittin pore level conductances (e.g. $\Lambda(3)$) may be due to compositional heterogeneity of natural melittin.

The 'dipole flip-flop gating model'

The nature of pore formation by alamethicin-type antibiotics has long remained unclear. We consider the large intrinsic dipole moment of the α-helix to be the cause for voltage-dependent pore formation. The pore properties are completely explained by the repulsive and attractive electrostatic forces between parallel and antiparallel dipoles, respectively. Fig. 3 shows a schematic drawing of the pore formation process in terms of the dipole flip-flop gating model. Within the membrane, antiparallel dipoles form a stable preaggregate. Voltage application leads to reorientation of the dipoles parallel to the electrical field, and thus to formation of various pore states. The predictions of this model agree qualitatively and quantitatively with the experimental results of Boheim (3) and Boheim and Kolb (6) .

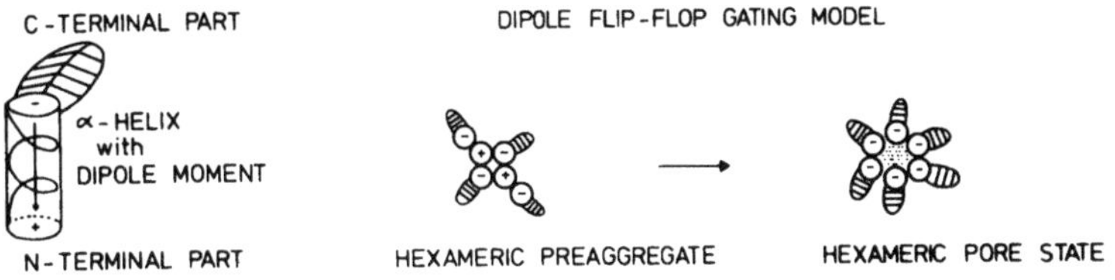

Fig. 3.

The dipole flip-flop gating model (schematic drawing). The elementary subunit of a pore is the single peptide molecule. Its structural features are (left side): a lipophilic α-helical part with intrinsic dipole moment and a hydrophilic part joined to the negatively charged terminus of the α-helix. We propose the following pore formation mechanism in agreement with the data of Boheim and Kolb (6) . A hexameric preaggregate is formed by the assembly of antiparallelly oriented α-helix dipoles. After appropriate voltage application two dipoles flip and orient themselves parallelly to the electrical field. A pore results from mutual repulsion of the parallel dipoles (hexameric pore state), and it is filled with water. Pore state transitions occur by the flip-flop of single molecules. This model encompasses the 'barrel stave model' (3,5,6) and in addition explains the voltage-dependence of pore formation.

Alamethicin and melittin, both, exhibit the same structural prerequisites: a lipophilic α-helix of adequate length and a hydrophilic C-terminal part. Both form multi-level pores. The peculiar amino acid α-aminoisobutyric acid in alamethicin is not essential for pore formation. Pore state distributions and pore state lifetimes of alamethicin and melittin are strongly dependent on ionic strength, which is consistent with the action of electrostatic forces. Dipole repulsion is reduced by the increase in dielectric constant due to water inflow into the pore interior (dielectric screening). It is then further reduced by the presence of a large number of ions (ionic screening).

Acknowledgement

A sample of the purified natural mixture of melittin was kindly provided by Dr. P. Hartter, Univ. Tübingen. We thank Dipl. Phys. C. Methfessel for helpful discussions. This work was supported by the Deutsche Forschungsgemeinschaft (SFB 114).

References

1. Mueller, P., Rudin, D.O.: Nature 217, 713-719 (1968).
2. Gordon, L.G.M., Haydon, D.A.: Biochim. Biophys. Acta 255, 1014-1018 (1972).
3. Boheim, G.: J. Membrane Biol. 19, 277-303 (1974).
4. Sakmann, B., Boheim, G.: Nature 282, 336-339 (1979).
5. Baumann, G., Mueller, P.: J. Supramolec. Struct. 2, 538-557 (1974).
6. Boheim, G., Kolb, H.A.: J. Membrane Biol. 38, 99-150 (1978).
7. Irmscher, G., Jung, G.: Eur. J. Biochem. 80, 165-174 (1977).
8. Boheim, G., Janko, K., Leibfritz, D., Ooka, T., König, W.A., Jung, G.: Biochim. Biophys. Acta 433, 182-199 (1976).
9. Brückner, H., König, W.A., Greiner, M., Jung, G.: Angew. Chem. Int. Ed. 18, 476-477 (1979).
10. Hanke, W., Methfessel, C., Wilmsen, H.-U., Katz, E., Jung, G., Boheim, G.: submitted for publication.
11. Vodyanoy, I., Hall, J.E., Balasubramanian, T.M., Marshall, G.R.: Biochim. Biophys. Acta 684, 53-58 (1982).

12. Jung, G., Brückner, H., Schmitt, H. in: Structure and Activity of Natural Peptides; W. Voelter and G. Weitzel, eds., de Gruyter, Berlin; pp. 75-114 (1981).

13. Yantorno, R.E., Takashima, S., Mueller, P.: Biophys. J. 38, 105-110 (1982).

14. Schwarz, G., Savko, P.: Biophys. J. (1982), in press.

15. Tosteson, M.T., Tosteson, D.C.: Biophys. J. 36, 109-116 (1981).

16. Hanke, W., Böheim, G.: Biochim. Biophys. Acta 596, 456-462 (1980).

17. Habermann, E.: Science 177, 314-322 (1972).

18. Schindler, H., Feher, G.: Biophys. J. 16, 1109-1113 (1976).

19. Neher, E., Sakmann, B., Steinbach, J.H.: Pflügers Arch. 375, 219-228 (1978).

CONFORMATIONS OF THE TRANSMEMBRANE CHANNEL FORMED BY GRAMICIDIN A

Sergei V.Sychev and Vadim T.Ivanov
Shemyakin Institute of Bioorganic Chemistry, USSR Academy of Sciences, 177988 Moscow, USSR

Gramicidin A (GA), the linear hydrophobic pentadecapeptide HCO-LVal-Gly-LAla-DLeu-LAla-DVal-LVal-LVal-DVal-LTrp-DLeu-LTrp-DLeu-L-Trp-DLeu-LTrp-DLeu-LTrp-$CONHCH_2CH_2OH$ /1/, forms cation channels in lipid bilayers and biological membranes /2-4/. According to Urry et al. /5,6/ the channel is formed by head-to-head association (in N-terminus to N-terminus fashion) of two π_{LD} helices. Veatch et al. /7,8/ suggested alternative dimer models of antiparallel ($\uparrow\downarrow\pi\pi_{LD}$) and parallel ($\uparrow\uparrow\pi\pi_{LD}$) double helices. The study of membrane conductivity allowed Bamberg et al. /9/ to exclude the $\uparrow\uparrow\pi\pi_{LD}$ helix from the possible structures of the active channel. Calculation of resonance splitting of the amide I band and comparison of the calculated and experimentally observed IR spectra showed /10/ that $\uparrow\downarrow\pi\pi_{LD}^{5.6}$ and $\downarrow\uparrow\pi\pi_{LD}^{7.2}$ helices are the dominant conformations of the antibiotic in solution and that the total fraction of $\vec{\pi}_{LD}^{4.4}$ $\overleftarrow{\pi}_{LD}^{4.4}$ and $\vec{\pi}_{LD}^{6.3}$ $\overleftarrow{\pi}_{LD}^{6.3}$ dimers is 25%.

In our resent work /11/ the conformational states of GA in dipalmitoylphosphatidylcholine (DPPC) and dimyristoylphosphatidylcholine (DMPC) liposomes at GA-lipid ratios from 1:10 to 1:350 were investigated by IR, CD (200-250 nm) and fluorescence spectroscopy. It was found that, depending on the GA-lipid ratio (Table 1) and the lipid type, GA can assume the conformation of a $\uparrow\downarrow\beta$-pleated sheet (state in which GA molecules forms a β-hairpin structure with one DL-turn), or a $\uparrow\downarrow\pi\pi_{LD}$ helix, or the mixture of both. Figure 1 presents the IR spectrum of GA incorporated in DPPC liposomes with an intensive phospholipid carbonyl band (1733cm^{-1}) and weak amide bands: amide I (1634 cm) and amide II (1550 cm^{-1}).

Chemistry of Peptides and Proteins, Vol. 2

Table 1.
Conformational States of GA in DPPC and DMPC Liposomes (20°C)[a].

Lipid	GA: lipid ratio (molar)	Predominant conformation
DPPC	I:I0-30	↑↓ππ$_{LD}$ helix + ↑↓β-sheet
	I:150	↑↓β-sheet
	I: >300	↑↓ππ$_{LD}$ helix
DMPC	I:35-350	↑↓β-sheet

a) Incorporation of GA into the lipid zone of liposomes was followed by fluorescence measurements (blue shift of the tryptophan emission).

Table 2.
Calculated Frequencies of the Amide I Band /10/.

Conformation	ν (cm^{-1})
↑↓ππ$_{LD}^{5.6}$	1634
↑↓ππ$_{LD}^{7.2}$	1636
$\vec{\pi}_{LD}^{4.4}$ $\overleftarrow{\pi}_{LD}^{4.4}$	1647
$\vec{\pi}_{LD}^{6.3}$ $\overleftarrow{\pi}_{LD}^{6.3}$	1656

Fig. 1b displays the amide I band obtained at a higher sensitivity and Fig. 1c - the same band in a molecular extinction scale after subtraction of the phospholipid absorption. Comparison of the observed frequency 1634 cm^{-1} with frequencies calculated for various structures of GA dimers /Table 2/ showed that at the GA-DPPC ratio 1:310 ↑↓ππ$_{LD}^{5.6}$ (or ↑↓ππ$_{LD}^{7.2}$) dimer is present in the bilayer. This result was in contrast with the data of Bamberg et.al. /12, 13/ on N-carboxyderivatives of GA with the -CO-$(CH_2)_n$-COOH group (n=2-6) instead of the formyl at the N-terminus in diphytanoylphosphatidylcholine membranes. In the presence of N-carboxygramicidins the membrane current decreased when pH changed from 4 to 6, i.e. within the titration range of carboxyl groups that strongly favours the $\vec{\pi}_{LD}$ $\overleftarrow{\pi}_{LD}$ dimers. To solve the mentioned contradiction we studied the N-carboxy-

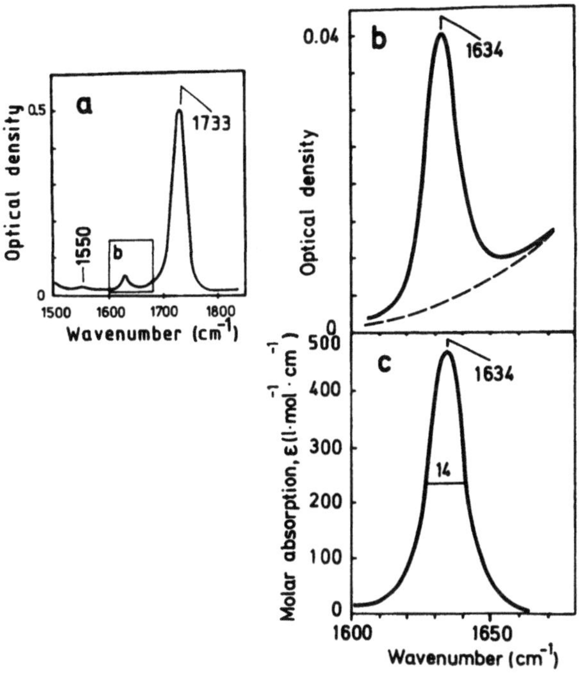

Fig. 1. IR spectrum of GA-DPPC liposomes (1:310 molar ratio). (a) Spectrum of a liposome suspension in D_2O. (b) Amide I band obtained with 20-fold higher sensitivity of the spectrophotometer (Perkin-Elmer- 180). Dashed line indicates the lipid absorption. (c) Amide I band after substraction of the lipid absorption.(a-c) Sonication and spectra measurements were made at 20°C.

derivatives of GA at pH 1.5 in DOPC and DPPC liposomes (Fig. 2, 3). Along with the $\uparrow\downarrow\pi\pi_{LD}^{5.6}$ helix (1634 cm^{-1}) the 1653 cm^{-1} component (Fig. 2b,c) was found in the spectra pointing to the presence of the $\overset{\rightarrow}{\pi}{}_{LD}^{6.3}\ \overset{\leftarrow}{\pi}{}_{LD}^{6.3}$ dimer (calculated frequency 1656 cm^{-1}). It is noteworthy that this component is also present in the spectrum of the antibiotic itself in DOPC but not in DPPC (cf. Fig 1c and 2). The relative concentration of the $\overset{\rightarrow}{\pi}{}_{LD}^{6.3}\ \overset{\leftarrow}{\pi}{}_{LD}^{6.3}$ dimer in case of GA is however quite low (∿5%) and for N-carboxygramici-

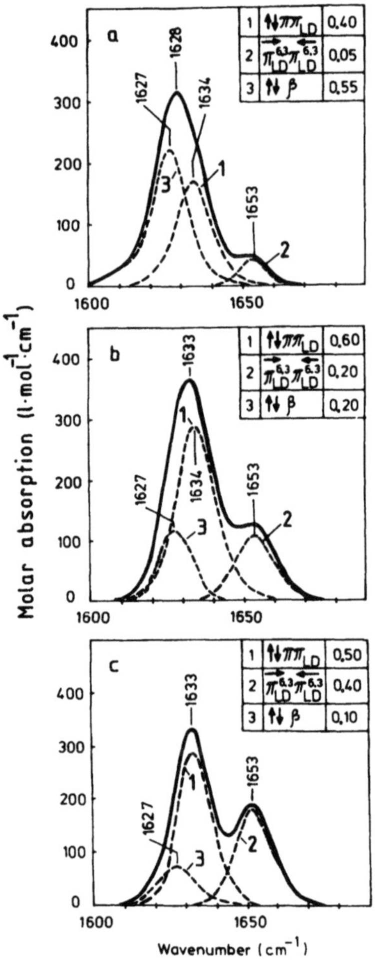

Fig. 2. IR amide I bands of GA and its N-carboxyderivatives in DOPC* liposomes (after substraction of the lipid absorption). Decomposition of the experimental profile into individual components is also shown. Suspension of liposomes (pD∿1.5) was prepared in D_2O+DCI. a) GA: lipid 1:100, T=20°C; (b) N-succinylgramicidin (n=2): lipid 1:50, T=20°; (c)N-suberylgramicidin(n=6): lipid 1:170, 1:340, T=45°C.

---

*dioleylphosphatidyl choline

dins it increases with the lengtheing of the $(CH_2)_n$ chain up to 40% at n=6 (Fig. 2).
The content of $\vec{\pi}_{LD}\ \overleftarrow{\pi}_{LD}$ dimers in DPPC is also the highest at n=6 (Fig.3). Thus one may conclude that (1) the $\uparrow\downarrow\pi\pi_{LD}$ helix is a dominating structure of GA in DOPC as well as in DPPC and (2) N-carboxyderivatives of GA show stronger tendency to form single stranded helical dimers than GA itself.
The data obtained allow a novel approach to the mechanisms of opening and closing of GA channels.
Bamberg et al. /12-14/ developed the dimerization hypothesis of channel functioning according to which the channel lifetime is determined by the lifetime of dimers. In this case the short lifetimes of N-carboxygramicidin channels (<0,5 sec /13/) must be due to rapid decay of $\vec{\pi}_{LD}\ \overleftarrow{\pi}_{LD}$ dimers. On the other hand the lifetime of the $\uparrow\downarrow$double helix of GA in nonpolar medium /8,10/ is $\sim 10^5$-times longer than the lifetimes of GA channels ($\sim$1 sec.). Therefore we believe that the process of GA channel closing is not due to decay of the double helix but is due to fluctuation of the membrane width and the ensuing conformational changes in the $\uparrow\downarrow\pi\pi_{LD}$ dimer: untwisting of the ends of the double helix or its partial "unwinding" (fig. 4) with formation of the terminal $\pi_{LD}^{4.4}$ helical turns (such helix was found in a monomeric state in a nonpolar medium /10/).
Using the relaxation method Bamberg et al. /13/ showed that kinetic properties of N-carboxyanalogs (pH4) also differ from those of G A. The experiments were interpreted according to the dimerization scheme, i.e. relaxation of the membrane current after the voltage jump was connected with a shift of equilibrium $A+A \underset{K_D}{\overset{K_R}{\rightleftharpoons}} A_2$ towards the dimer (K_R and K_D are the association and dissociation rate constants). Relaxation time τ is given by the equation /14/:

$$\frac{1}{\tau} = K_D + 4\sqrt{\frac{K_R\ K_D\ \lambda}{\Lambda L}}$$

where λ is the stationary conductance per unit area (after relaxation) which depends on GA concentration in the membrane, Λ

the single-channel conductance, and L Avogadro's number. The slope of the linear depence of $\frac{1}{\tau}$ on $\sqrt{\lambda}$ is determined by expression $b=\sqrt{\frac{K_R\ K_D}{\Lambda}}$.

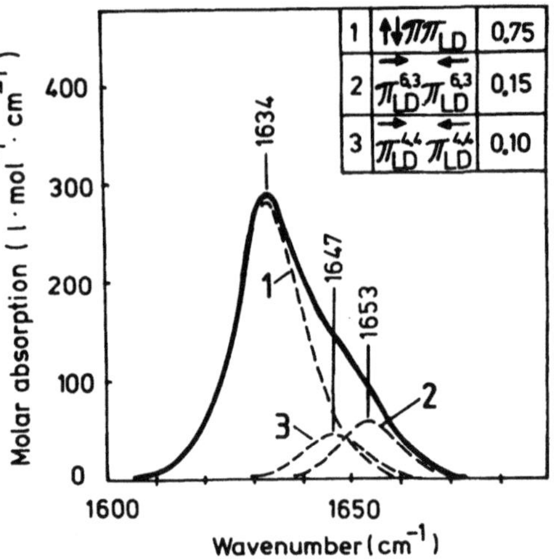

Fig. 3. IR amide I band of N-suberylgramicidin (n=6) in liposomes DPPC (after subtraction of the lipid absorption). Decomposition of the experimental profile into individual components is also shown. Suspension of liposomes (pD∿1.5) was prepared in D_2O+DCI, or (pD4) in 0.1 M D_2O+CD_3COOD/NaOD buffer. Peptide-lipid ratio is 1:180, T=20°C.

However, for the $\uparrow\downarrow\pi\pi_{LD}$ helix, opening and closure is a monomolecular reaction which therefore does not depend on GA concentration in the membrane, i.e. on λ. Thus the mentioned slope must depend on the ratio of conformers $\vec{\pi}_{LD}\overleftarrow{\pi}_{LD}$ and $\uparrow\downarrow\pi\pi_{LD}$ in the membrane: steeper slope refers to higher proportion of the $\vec{\pi}_{LD}\overleftarrow{\pi}_{LD}$ channels.

In the table 3 relaxation data on GA and its N-carboxyanalogs in the DOPC membrane are compared with our results on the conformational study of these compounds. The observed correlation between the steeper slope and the increased fraction of the $\vec{\pi}^{6.3}_{LD}\overleftarrow{\pi}^{6.3}_{LD}$ dimer in DOPC liposomes supports our conclusion that this channel type can dominate in N-carboxyderivatives of GA, whereas in the antibiotic itself the $\uparrow\downarrow\pi\pi_{LD}$ helical channel predominates. The low slope of $1/\tau$ *vs* $\sqrt{\lambda}$ dependence for GA in comparison with that

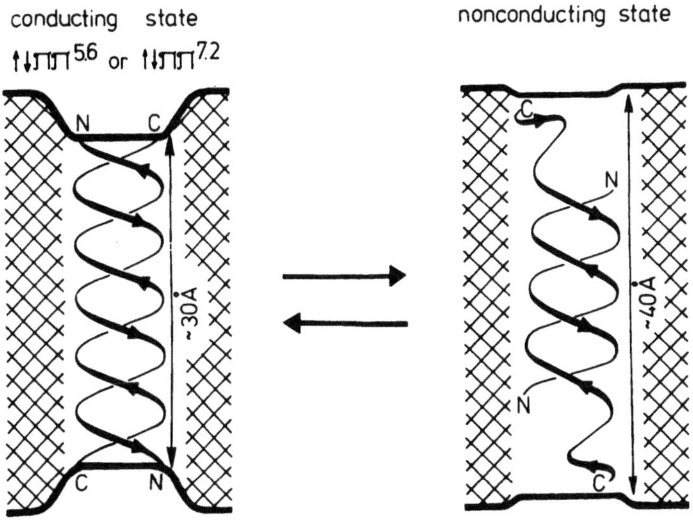

Fig.4. Switching on and off of the gramicidin A channel.

for N-carbohyderivatives is due to lower content of $\overrightarrow{\pi}^{6.3}_{LD}\overleftarrow{\pi}^{6.3}_{LD}$ conformation .According to the available data /14-16/ this slope (which was also obtained by autocorrelation analysis of membrane current noises /15/) in some membranes is even lower than in the DOPC membrane and, e.g., in the glycerylmonooleate membrane in n-decane it is close to zero /15/. The latter fact does not agree with the dimerization scheme. However, in accordance with our model it testifies to the complete absence of the Urry dimers in the membrane.

Table 3.

Slopes (b) of linear $1/\tau$ $vs\sqrt{\lambda}$ dependences for membranes modified with gramicidin and its N-carboxyl analogs /13/ a), and the $\vec{\pi}_{LD}\overleftarrow{\pi}_{LD}/\uparrow\downarrow\pi\pi_{LD}$ ratios for these compounds.

Compound[b)]	Lipid	$b=\sqrt{\frac{K_R\ K_D}{\Lambda}}$ $(M^{-1}\ s^{-1}\ S^{-6})$	b/b_{GA}	$\frac{[\vec{\pi}^{6.3}_{LD}\ \overleftarrow{\pi}^{6.3}_{LD}]}{[\uparrow\downarrow\pi\pi_{LD}]}$
gramicidin A	DPPC	6.96×10^{12}	1	0.12
N-succinylgramicidin (n=2)		1.39×10^{12}	20	0.33
N-glutarylgramicidin)n=3)		-	-	0.42
methyl ester of N-glutarylgramicidin		1.82×10^{14}	26	-
N-adipylgramicidin (n=4)		4.85×10^{11}	∿0,1	-
N-pimelylgramicidin (n=5)		2.24×10^{14}	32	-
N-suberylgramicidin (n=6)		1.44×10^{15}	210	0.80
gramicidin A	DOPC	-	-	0
N-suberylgramicidin[c)] (n=6)		-	-	0.33

a) 1M NaCL; V=150mV; pH 4,0; lipid in n-decane

b) Single-channel conductance of gramicidin from /15/

Slopes are calculated for the analogs with n=2,3,5 and are found graphically for analogs with n=4,6. For calculation the single-channel conductances in monoolein membrane /13/were used

c) the total contribution of $\vec{\pi}^{6.3}_{LD}\ \overleftarrow{\pi}^{6.3}_{LD}$ and $\vec{\pi}^{4.4}_{LD}\ \overleftarrow{\pi}^{4.4}_{LD}$ dimers was taken.

References

1. Sarges, R. and Witkop, B.: J.Am.Chem.Soc. 87, 2011-2027.
2. Pressman, B.C.::Proc.Natl.Acad.Sci.USA 53 1076-1080 (1965).
3. Muller, P. and Rudin D.O.: Biochim.Biophys.Res.Commun. 26, 398-404 (1967).
4. Hladky, S.B. and Haydon, D.A.: Nature 225, 451-451 (1970).

5. Urry, D.W.: Proc.Natl.Acad.Sci.USA 68, 672-676 (1971).

6. Urry, D.W., Goodall, M.C., Glickson, J.D., and Mayers, D.F.: Proc.Natl.Acad.Sci.USA 68, 1907-1911 (1971).

7. Veatch, W.R., Fossel, E.T., and Blout,E.R.: Biochemistry 13, 5249-5256 (1974).

8. Veatch, W.R. and Blout, E.R.: Biochemistry 13, 5257-5264 (1974).

9. Bamberg,E., Apell, H.J., and Alpes, H.: Proc.Natl.Acad.Sci. USA 74, 2402-2406 (1977).

10. Sychev, S.V., Nevskaya, N.A., Jordanov, St., Shepel, E.N., Miroshnikov, A.I., and Ivanov, V.T.: Bioorgan.Chem. 9, 121-151 (1980).

11. Sychev,S.V. and Ivanov, V.T.: Membranes and Transport Vol.2, A.N. Martonosi ed., pp.301-307, Plenum Corp., New York 1982

12. Bamberg, E., Alpes, H., Apell, H.-J., Bradley, R., Harter, B., Quelle, M.-J., and Urry, D.W.: J.Membrane Biol. 50, 257-270 (1979).

13. Apell, H.-J., Bamberg, E., and Alpes, H.: J.Membrane Biol. 50, 271-285 (1979).

14. Bamberg,E. and Lauger, P.: J.Membrane Biol. 11, 177-194 (1973).

15. Zingsheim, H.P. and Neher, E.: Biophys.Chem. 2, 197-207 (1974).

16. Kolb, H.-A. and Bamberg, E.: Biochim.Biophys.Acta 464,127-141 (1977).

STUDY OF VERATRIDINE RECEPTORS ASSOCIATED WITH SODIUM CHANNELS

Nickolai M. Soldatov, Tat'yana K. Prosolova, Nikita I. Kiyatkin and Eugenii V. Grishin

Shemyakin Institute of Bioorganic Chemistry, USSR Academy of Sciences, 117988 Moscow, USSR

Introduction

The sequential changes in membrane conductance to sodium and potassium ions lead to the generation of action potentials in electrically excitable membranes of nerve and muscle cells (1). The transport of these ions is very selective and is known to be carried out by different specified passways or "channels". Now the wide spectrum of biochemical approaches is used to study the fast *sodium channel*. This entity was proved to be a glycoprotein in nature (2) and possesses the affinity to a number of specific neurotoxins (3). Receptors for these toxins are located in different functional parts of the sodium channel and so far they can be subdivided into the following three entities (fig.1):

the *external mouth* of the channel with a carboxyl group capable to interact with sodium current inhibitor tetrodotoxin (4). Each sodium channel binds only one tetrodotoxin molecule (5), and for example, rat brain synaptosomes possess about 2 pmoles of receptor per mg of membrane protein (6).

The second part is a *gating structure* of the channel. It is composed of the potential sensitive inactivation gate freely moving in axoplasm (7) and the external membrane receptor for miniprotein toxins from *scorpion* (8) and *sea anemone* venoms (9).

The third functional part is a *selective filter* of the sodium channel. This fragment was proved to be located near the center of the membrane (10). It includes another carboxyl group (10)

Chemistry of Peptides and Proteins, Vol. 2

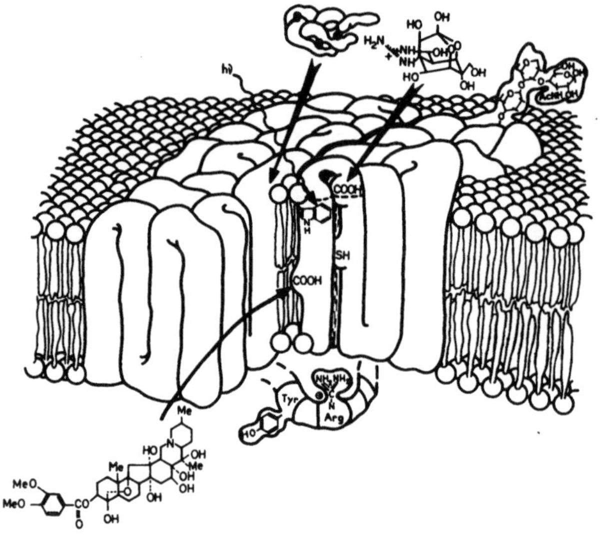

Fig.1. Schematic diagram of the sodium channel.

and is sensitive to neurotoxic alkaloids, *aconitine* (11) and *veratridine* (12).

All three receptors are independent but the interaction of binding sites with protein and alkaloid toxins was shown to exhibit positive heterotropic coorepativity (13).

The binding site of alkaloid toxins is apparently the most essential element of the channel because it regulates not only ionic selectivity but potential sensitivity as well (14). Up today this receptor is characterized very poorly. The one reason is relatively low affinity of the receptor to alkaloid neurotoxins which possess the apparent dissociation constants as high as about 10 µM.

On the other hand we failed to identify the synaptosomal receptor in direct radio-ligand assay with labelled analogs of,both, aconitine and veratridine which have been prepared by the electric discharge method in tritium gas. The result was attributed to dominance of nonspecific interaction of both toxins with membrane lipids (6,15).

Results

For the further study of the alkaloid neurotoxin receptor we used veratridine-Sepharose affinity chromatography. This matrix containing up to 5 μmoles of veratidine per ml was prepared by the reaction of hydroxyl groups of the toxin with epoxy-activated Sepharose 6B:

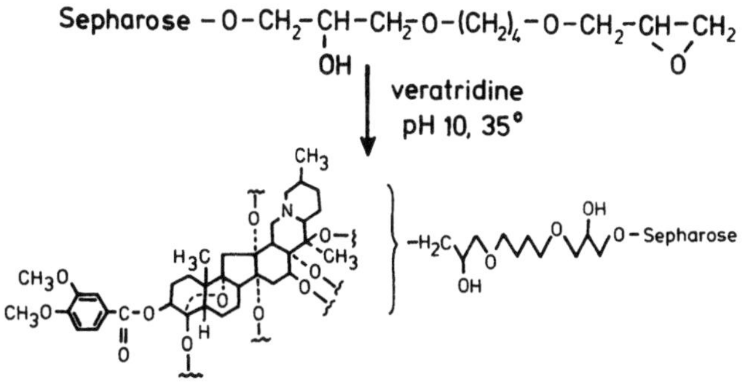

Chromatography of solubilized synaptosomal membranes was followed by the tetrodotoxin binding analysis. Thus it was shown that veratridine-Sepharose binds the receptor of tetrodotoxin when rat brain synaptosomal membranes solubilized in Lubrol PX are applied to the column. The matrix binding capacity is limited at the level about 20 pmoles of the receptor per ml that could be accounted for the low accessibility of the immobilized toxin molecules because the Sepharose 6B pore dimensions are large enough for toxin molecules being impermeable to solubilized receptor (Stokes radius ca. 80 Å (16)). On the other hand the esterification of some functionally important hydroxyl groups of veratridine could lead to the decrease of an essential quota of toxin molecules potentially reactive to the sodium channel binding site.

The specific elution of the tetrodotoxin receptor can be carried out by an excess of both aconitine and veratridine (fig. 2).

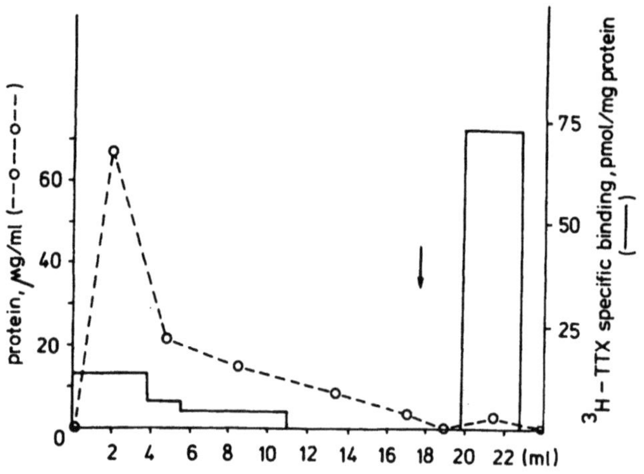

Fig. 2. Veratridine-Sepharose chromatography of rat brain synaptosomal membranes solubilized in Lubrol PX. Arrow shows a 2 ml impulse of 7 mM aconitine applied to the column (3 ml) at constant ionic strength and pH of elution buffer.

The single step of chromatography results in up to 50-fold enrichment of the receptor. This directly evidences in favour of the interaction of both alkaloids with the same binding site of the channel in spite of the difference in their chemical structures.

The spatial analysis of aconitine and veratridine structures (fig. 3) makes it possible to place a number of their oxygen functions in one-to-one correspondence with common environmental groups capable of hydrogen bonding. These countergroups may be assumed to represent oxygen atoms of the selective filter accrding to Hille model (17). Another intriguing particularity of the resulted arrangement is marked spatial proximity of aromatic rings which are also functionally important fragments of both neurotoxins. Probably these bulk groups are lo-

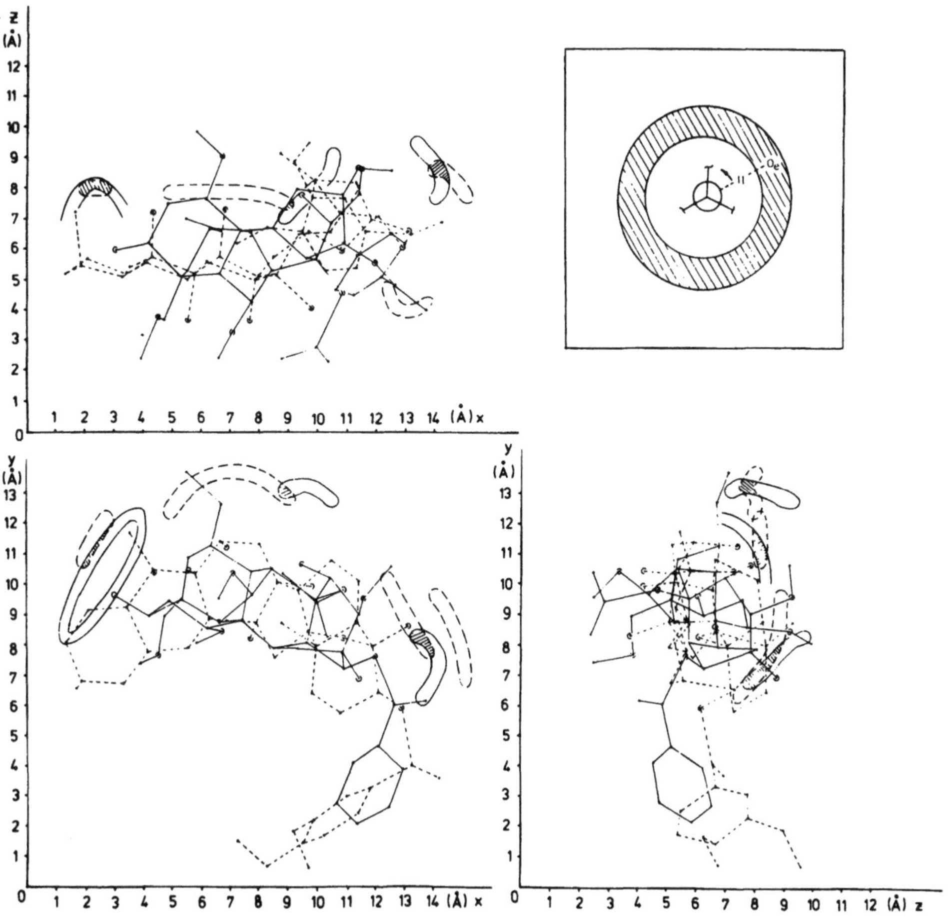

Fig. 3. The spatial structures of aconitine (—•—) and veratridine (--•--), oxygen atoms being signed by (——⊃). The dashed fields show the coordinates of receptor oxygen atoms (O_e) capable of hydrogen bonding. The insertion shows how coordinates for O_e were obtained. Here the dashed ring makes up a torque of the locus for O_e. It is situated at the distance of the hydrogen bond between O_e and a toxin molecule oxygen atom (mean value 2,5-3,0 Å (18)) and is its variation wide (ca. 0,5 Å).

cated in a hydrophobic loop of the channel protein or nearby in lipid phase. In the last case the protein chains of the ionic pore have to be situated on the periphery of the whole protein, being in contact with membrane lipids. That could explain why the selective filter so easily changes its dimensions upon interaction with alkaloid neurotoxins, and that may be a reason why the tetrodotoxin receptor in the external mouth of the channel is so sensitive to lipid environment (19). The adequacy of these possibilities is to be resolved in the forthcoming study of the sodium channel with the use of molecular probes as neurotoxins. The experimental data on veratridine-Sepharose affinity to the tetrodotoxin receptor presents a strong evidence that both alkaloid and tetrodotoxin binding sites belong to the same membrane macromolecule which does not dissociate in nonionic detergent. From this point of view it was intriguing to study the relation between the veratridine receptor and the binding site of scorpion neurotoxin. For this purpose we used a photosensitive and radioactive derivative of the toxin M_{10} from the Central Asian scorpion *Buthus eupeus* venom (20). This derivative interacts with two nerve membrane proteins of molecular weights 51 and 77 kD (21). Affinity chromatography of solubilized synaptosomal membranes labeled with this toxin derivative showed that after veratridine-Sepharose was thoroughly washed up, the excess of free veratridine induced the elution of only one of these membrane proteins of molecular weight about 73 kD. In the electropherogram of the respective fraction (fig.4) approximately 12 proteins were present of molecular weights less than 100 kD, and ca. 50-fold purification of the scorpion toxin binding protein was achieved.

So far the simultaneous use of the radio-ligand analysis, covalent labeling and affinity chromatography allowed us to find that in the presence of nonionic detergent all three groups of the receptors are capable to interact with immobilized veratridine, and the 73 kD protein is indeed a component of the sodium channel. It is very probable that all three receptors are elements of the same membrane protein because the veratridine binding site is intimately involved, both, in the functioning of the transmembrane pore of the channel and in the regulating mechanism of the gating system.

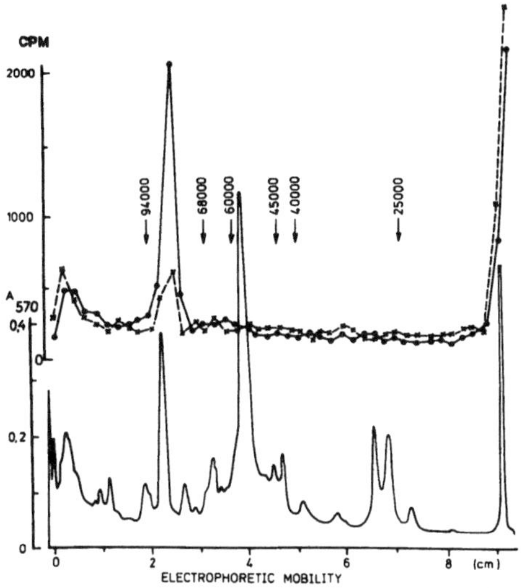

Fig. 4. SDS-electrophoretic analysis of the fraction that was eluted from veratridine-Sepharose by the impulse of 7 mM free veratridine upon chromatography of solubilized rat brain synaptosomes labeled by ^{125}I-DNPA-M_{10} in the abscence (—•—) or in the presence (—x—) of 50-fold molar excess of M_{10}. Arrows show the molecular weights of reference proteins. DNPA-dinitrophenylazide.

From the other side all three groups of the receptors are supposed to vary in sensitivity to encounted stages of separation. The further study of this aspect will disclose an outlook of the use of veratridine-Sepharose for the large-scale purification of the sodium channel. We believe that a combination of covalent labeling and biospecific methods of separation is indeed a prominent way for molecular characterization of the sodium channel.

Acknowledgments

We are very grateful to Prof. Yu.A.Ovchinnikov for support and advice during the course of these experiments, and Dr. P.Melnikov for help with the spatial analysis of toxins structures.

References

1. Hodgkin, A.L., Huxley, A.F.: J. Physiol. (London) 117, 500-544 (1952).
2. Cohen, S.A., Barchi, R.L.: Biochem. Biophys. Acta 645, 253-261 (1981).
3. Catterall, W.A.: Ann. Rev. Pharmacol.Toxicol. 20, 15-43 (1980).
4. Sigworth, F.J., Spalding, B.C.: Nature (London) 283, 293-295 (1980).
5. Cuervo, L.A., Adelman, W.J., jr.: J. Gen. Physiol. 55, 309-335 (1970).
6. Abita, J.-P., Chicheportiche, R., Schweitz, H., Lazdunski, M.: Biochemistry 16, 1838-1844 (1977).
7. Armstrong, C.M.: Quart. Rev. Biophys. 7,179-210 (1975).
8. Mozhayeva, G.N., Naumov, A.P., Nosyreva, E.D., Grishin, E.V.: Biochem. Biophys. Acta 597, 587-602 (1980).
9. Catterall, W.A., Beress, L.: J. Biol. Chem. 253,7393-7396 (1978).
10. Mozhayeva, G.N., Naumov, A.P., Negulyaev, Yu.A.: Biochem. Biophys. Acta 643, 251-255 (1981).
11. Mozhayeva, G.N., Naumov, A.P., Negulyaev, Yu.A., Nosyreva, E.D.: Biochem Biophys. Acta 466, 461-473 (1977).
12. Naumov, A.P., Negulyaev, Yu.A., Nosyreva, E.D.: Tsitologiya (USSR) 21, 692-696 (1979).
13. Catterall, W.A.: J. Biol. Chem. 252, 8669-8676 (1977).
14. Khodorov, B.I.: Neirofiziologiya (USSR) 12, 317-331 (1980).
15. Prosolova, T.K., Shamotienko, O.G., Grishin, E.V.: Abstracts of III USSR-Sweden Symposium on Physico-Chemical Biology, 239 (1981).
16. Hartshorne, R.P., Coopersmith, J., Catterall, W.A.: J. Biol. Chem. 225, 10572-10575 (1980).
17. Hille, B.: J. Gen. Physiol. 59,637-658 (1972).
18. The Hydrogen Bond. Recent developments in theory and experiments, vol. 2: Structure and Spectroscopy (eds P. Schuster, G. Zundel, and C.Sandorfy), Noth-Holland Publishing Co., Amsterdam·New York·Oxford, 1976
19. Agnew, W.S., Raftery, M.A.: Biochemistry 18, 1912-1919 (1979).
20. Grishin, E.V., Soldatov, N.M., Ovchinnikov, Yu.A., Mozhayeva, G.N., Naumov, A.P., Zubov, A.N., Nisman, B.Ch.: Bioorg.Khim. 6, 724-730 (1980).
21. Grishin, E.V., Soldatov, N.M., Ovchinnikov, Yu.A.: Bioorg. Khim. 6, 914-922 (1980).

THE ROLE OF NEUROTOXINS IN STUDYING SODIUM CHANNELS

Eugene V.Grishin
Shemyakin Institute of Bioorganic Chemistry, USSR Academy of Sciences, 117988 Moscow, USSR

Introduction

In recent years considerable progress was achieved towards our understanding of key moments of generation and transmission of nerve impulse. One should mention here the general theory of distribution of nerve excitation and the discovery of the determinig role of ion channels in this process (1). Electrophysiological experiments became the basis for the study of the molecular mechanism of signal transmission along the nerve fibers and first demonstrated application of neurotoxins as unique tools providing a deep insight into excitation phenomena (2). Neurotoxins interact quite specifically with definite structures of excitable membranes, disturbing the normal course of the nerve impulse transmission. Consequently the use of neurotoxin derivatives with various labels allows, in principle, to isolate and study in detail various functionally important components of a nerve fibre or a controllable cell.
The fast sodium channels are well known to play a decisive role during the generation of nerve impulse. The best way for the study of these membrane components is the utilization of neurotoxins that selectively affect the channel function. There are three groups of neurotoxins which in different way influence the sodium channel functioning (Fig.1). Tetrodotoxin and saxitoxin block the transmembrane current of sodium ions and apparently bind to the ion-conducting channel part. A number of alkaloid toxins: batrachotoxin, veratridine, aconitine and grayanotoxins activate sodium channels and change their ion selec-

Chemistry of Peptides and Proteins, Vol. 2

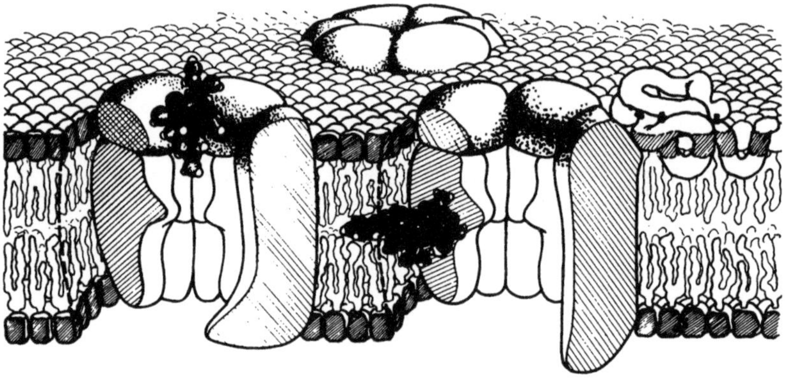

Fig. 1. Shematic representation of neurotoxin interactions with fast sodium channels.

tivity. It seems that these neurotoxins interact with the channel subunit carrying out the transport of sodium ions. Scorpion and sea anemone toxins affect the process of inactivation of fast sodium channels and it is quite likely that they bind to the membrane subunit, regulating the work of the ion-conducting channel part. Thus, the study of receptors for various neurotoxins which influence the functioning of sodium channels allows us to approach the understanding of their molecular nature. In the present paper aconitine, tetrodotoxin and scorpion neurotoxins were used to study the molecular organization of fast sodium channels in electrically excitable membranes.

Results

Investigation of the neurotoxin receptors is closely connected with preparation of toxin radioactive derivatives suitable for radio-ligand analysis. The radioactive analog of aconitine was prepared by the electric discharge method in tritium gaseous atmosphere. The obtained product was purified by thin-layer chromatography on Silica Gel. Specific radioactivity of biologically active derivative of aconitine is about 1,2 Ci/mmol and this compound could be used for the analysis of membrane receptors. Two plasma membrane fractions from crab axon and dog heart were prepared for the study of toxin binding. In binding experiments with $[^3H]$ aconitine (10^{-8} to 10^{-4}M) saturated regions were not found (Fig. 2).

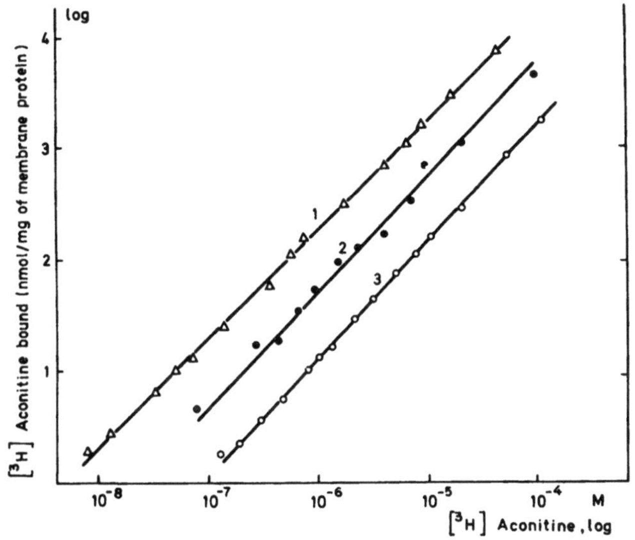

Fig. 2. Binding analysis of $[^3H]$ aconitine with dog heart plasma membrane (1,3) and solubilized crab nerve membrane (2).

It should be noted that the saturable binding of $[^3H]$ -aconitine was not also observed when axonal membranes were solubilized with Lubrol PX. Moreover, the interpretation of the binding process is quite difficult due to toxin aggregation in aqueous solutions and its association with membrane lipids. Aconitine seems not to be used for the radio-ligand analysis of sodium channel components in case of total membrane fractions.

To obtain the derivative of tetrodotoxin (TTX) with a high specific radioactivity use was made of the method of catalytic reduction of unsaturated substrates with gaseous tritium (3). At first the toxin was subjected to periodate oxidation (Fig.3).

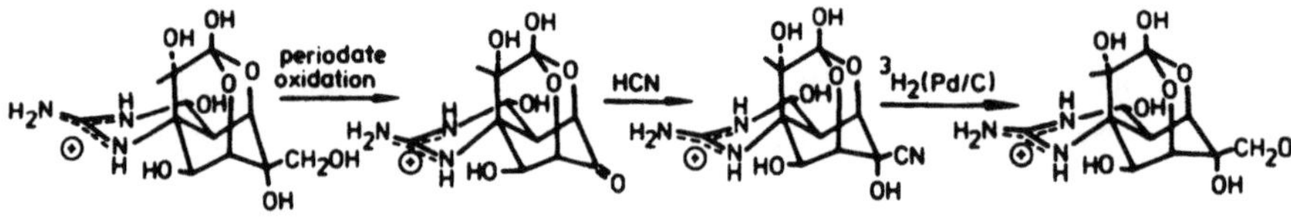

Fig. 3. Pathway of the synthesis of the TTX derivative with high specific radioactivity.

The formed ketoderivative was treated with HCN for the obtaining of the corresponding cyanohidrin. Then the catalytic reduction of the modified neurotoxin in the presence of 5% Pd/C with gaseous tritium was carried out. Here, the toxin analog containing an aldehyde group at C_{II} was, apparently, formed and upon the additional reduction it was transformed into the d,l form of tetrodotoxin. Purification of the radioactive toxin derivative was performed by chromatography on Biorex 70. According to the obtained data the specific radioactivity for various tetrodotoxin preparations was equal to 5-10 Ci /mmol and their yield was 4-6% relative to the original toxin. Binding studies of the radioactive tetrodotoxin analog with solubilized membranes from rat brain and crab axons testify to a high affinity of the modified toxin to the receptors (Fig.4).

The investigation of tetrodotoxin receptors from the membranes of crab axons and rat brain synaptosomes showed that they are quite stable in the membrane form but are rapidly invactivated upon solubilization with various detergents. Inactivation of the receptors can be partially prevented if tetrodotoxin and exogenic lipids are added to the solubilized membranes (4). At sedimentation in the sucrose density gradient it was established that TTX receptors from crab axonal membranes solubilized with Lubrol PX are present in two fractions with sedimentation coefficients 9,5 and 16S. This circumstance evidences in favour of the existence of two forms of tetrodotoxin receptors of molecular weights differing by about twice. TTX receptors from crab axonal membra-

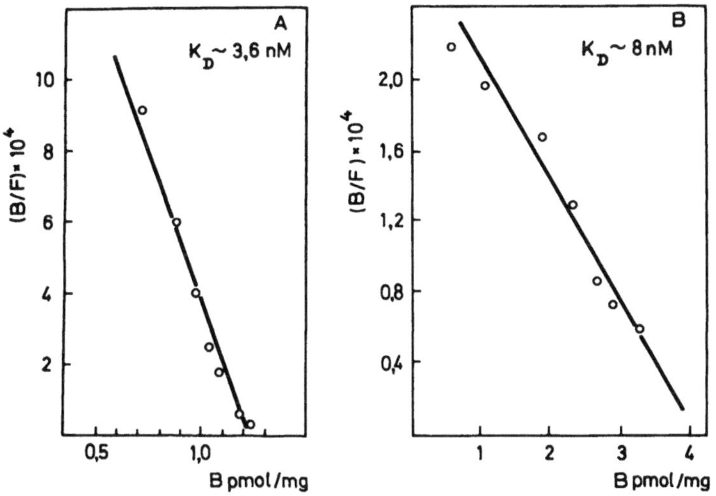

Fig.4 . Scatchard plots of [^{3}H]-TTX binding to solubilized rat brain membranes (A) and crab axonal membranes (B).

nes do not bind to concavaline and wheat germ lectin that, probably, points to the absence of hydrocarbon lectin-specific components in their molecules, but nevertheless does not exclude glycoprotein nature of the receptors.

Upon fractionation of axonal membranes solubilized with Lubrol PX by means of gel-chromatography, sedimentation in the sucrose gradient and ion exchange chromatography on DEAE-Sepharose the correlation between the tetrodoxin receptors and the proteins of 53000, 77000, 85000 and 230000 molecular weights was discovered (5). For unequivocal indentification of the TTX receptors their complete purification as well as their stabilization in the solubilized state are necessary.

Solubilization of synaptosomal membranes from rat brain with 0,5% Lubrol PX is accompanied by inactivation of tetrodotoxin receptors but to less degree than in case of crab axonal membranes. For isolation of the TTX receptors use was made of chromatography on DEAE-Sepharose, WGA-Sepharose and Sepharose CL-6B as well as of sedimentation in the sucrose density gradient. As a rule we succeeded in extraction of about 54% of the toxin

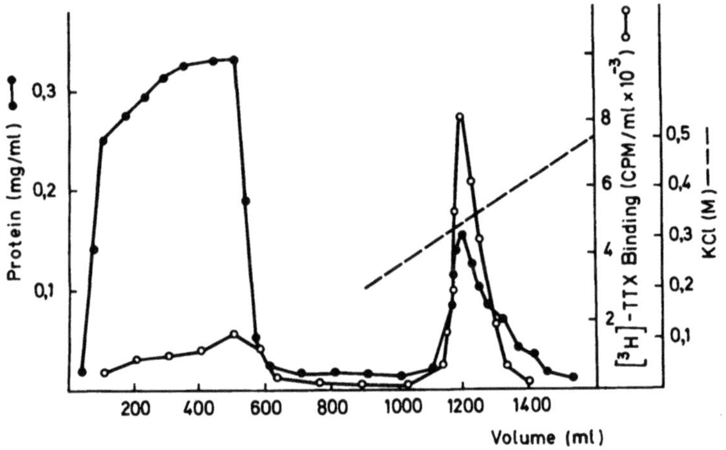

Fig. 5. Profile of elution of solubilized rat brain membranes from DEAE-Sepharose CL-6 B column.

receptor upon solubilization of rat brain membranes (620 mg of the protein, 930 pmol of the TTX receptor). With the aid of ion exchange chromatography on DEAE Sepharose CL-6B the 7-fold purification of the receptor was achieved (fig. 5). The same level of purification was observed when used WGA-Sepharose; in this case the binding activity of the receptor fraction was 225 pmol/mg of the protein (Fig.6). In the course of gel-chromatography on Sepharose CL-6B the receptor enrichment was even twice more (Fig.7). As a result the 340-fold purification of the TTX receptors was performed and the membrane fraction, binding more than 500 pmol of the toxin per mg of the protein, was obtained. It is noteworthy that calculation of the receptor activity was carried out without regard to the possible inactivation. The protein yield is about 0,05% relative to the original membrane preparation. The electrophoretic analysis of the purified receptor fraction showed that it is of relatively simple polypeptide composition. Mainly, the proteins of molecular weigts 250 000, 90000, 76000 and 53000 were found in the majority of experiments. Using the sedimentation analysis of the receptor

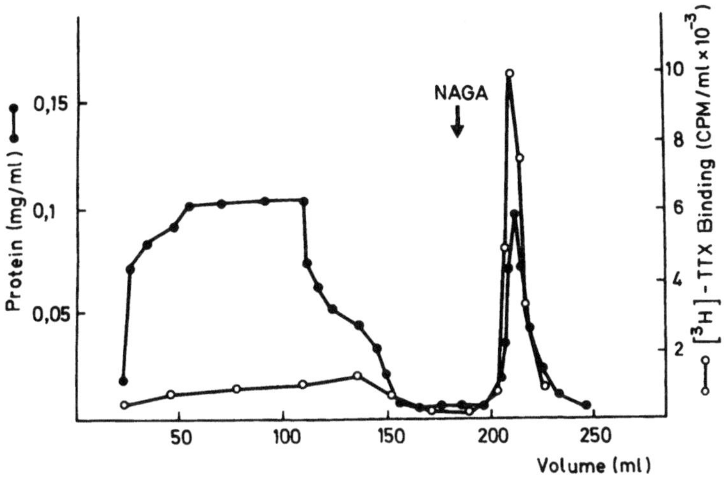

Fig. 6. Separation of the TTX receptor fraction from DEAE-Sepharose on WGA-Sepharose column. The arrow indicates the application of 20 mM N-acetylglucosamine.

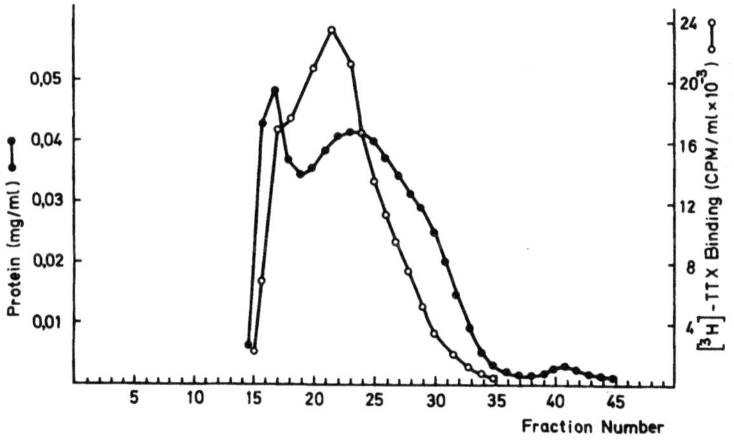

Fig. 7. Purification of the TTX receptor fraction from WGA-Sepharose on Sepharose CL-6B column.

fraction it was established that its sedimentation coefficient is about 11S.

Neurotoxins of the scorpion venom selectively act on the gating structure of sodium channels thus slowing down their inactivation rate (6). This fact allows to assume that the receptors of scorpion toxins are the functionally important components of the system providing transmembrane transfer of sodium ions across the electroexcitable membrane. And that is the reason why investigations of scorpion toxin receptors are of undoubt interest though connected with considerable difficulties due to dependence of neurotoxin binding on the membrane potential value as well as the low receptor content in the membrane (7).

In our Institute from the venom of scorpion *Buthus eupeus* 17 individual polypeptide neurotoxins were isolated and characterized, complete amino acid sequences being established for 4 of them (8). To elucidate the nature of receptors for scorpion neurotoxins use was made of a toxin photosensitive derivative preserving the biological activity. Upon photoactivation the modified neurotoxin analog was shown to be specifically bound by two protein components of electrically excitable membranes of 76000 and 51000 molecular weights (9).

Binding of scorpion neurotoxin by preparations of electroexcitable membranes is characterized by the high level of nonspecific sorption apparently due to interaction of the toxin molecule with the lipid phase of the membrane. Such assumption suggests the existence of the typical hydrophobic region in the toxin molecule. This region is localized in the 46-56 fragment of the neurotoxin M_{10}. It seems very important that this fragment contains almost half of the conserved amino acid residues. Analysis of the tertiary structure of the variant-3 toxin from scorpion *Centruroides sculpturatus* showed that practically all conserved residues are clustered together on the surface of the molecule, forming a large continuous patch, and the hydrophobic region is situated in the vicinity of this patch (10). In the neurotoxin M_{10} Lys-57 is situated near the similar patch of conserved and hydrophobic amino acid residues. Consequently, at the moment of interaction with sodium channels the photosensitive toxin analog labeled at Lys-57 must

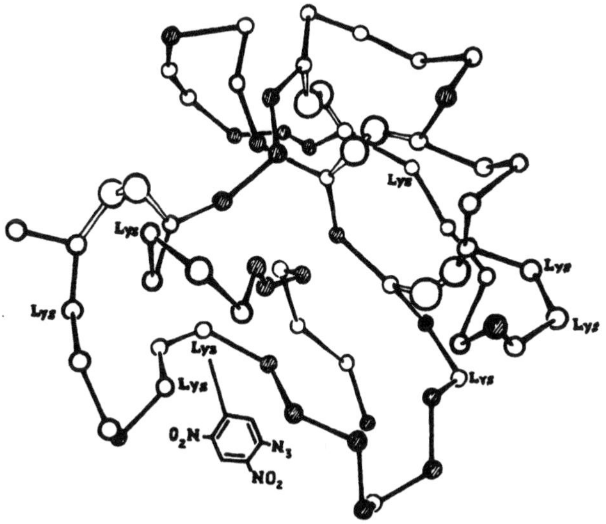

Fig. 8. Probable polypeptide chain arrangement of modified neurotoxin M_{10}. Hydrophobic amino acid residues are hatched.

modify predominantly the receptor components of electroexitable membranes. Fig. 8 shows a possible scheme of the arrangement of the polypeptide chain of the neurotoxin M_{10} designed on the basis of the data on X-ray analysis of the toxin-3 from the scorpion *Centruroides sculpturatus* venom.

Two proteins of 270000 and 38000 mol. weights were found to be modified by the photosensitive analog of the toxin from the scorpion *Leiurus quinquestriatus* upon its interaction with rat brain synaptosomes (11). The reason of these differences in characteristics of scorpion toxin receptors is still obscure. Probably it is mainly connected with variety of features of the used toxins, isolated from venoms of two different scorpion species. It is noteworthy that scorpion neurotoxins can possess the principally different modes of action (12,13), and for the toxin from *Leiurus quinquestriatus* possibility of interaction even with the choline receptor was shown (14). Further investigations of the receptors of various neurotoxins will shed light on the detail structure and mode of action of fast sodium channels.

References

1. Hodgkin, A.L., Huxley, A.F.: J. Physiol.London 117, 500-544 (1952).

2. Narahashi, T.: Physiol.Rev. 54, 813-889 (1974).

3. Kovalenko, V.A., Pashkov, V.N., Grishin, E.V., Ovchinnikov, Yu.A., Shevchenko, V.P., Myasoedov, N.F.: Bioorg.Khim. (USSR) 8, 719-721 (1982).

4. Agnew, W.S., Raftery, M.A.: Biochemistry 18, 1912-1819 (1979).

5. Kovalenko, V.A., Pashkov, V.N., Grishin, E.V.: Bioorg.Khim. (USSR) 7, 1828-1837 (1981).

6. Mozhayeva, G.N., Naumov, A.P., Soldatov, N.M., Grishin, E.V.: Biofizika 24, 235-241 (1979).

7. Mozhayeva, G.N., Naumov, A.P., Nosyreva, E.D., Grishin, E.V.: Biochim.Biophys.Acta 597, 587-602 (1980).

8. Grishin, E.V.: Int.J.Quantum.Chem. 19, 291-298 (1981).

9. Grishin, E.V., Soldatov, N.M., Ovchinnikov, Yu.A.: Bioorg. Khim. (USSR) 6, 914-922 (1980).

10. Fontecilla-Camps, J.C., Almassy, R.J., Suddath, F.L., Watt, D.D., Bugg, C.E.: Proc.Natl.Acad.Sci.USA 77,6496-6500 (1980).

11. Catterall, W.A., Hartshorne, R.P., Beneski, D.A.: Toxicon 20, 27-40 (1982).

12. Gillespie, J.I., Meves, H.: J.Physiol.London 308, 479-487 (1980).

13. Carbone, E., Wanke, E., Prestipino, G., Possani, L.D., Maelicke, A.: Nature 296, 90-91 (1982).

14. Culp, W.J., McKenzie, D.T.: Proc.Natl.Acad.Sci. USA 78, 7171-7175 (1981).

STRUCTURE AND MECHANISM OF ACTION OF NEUROTOXINS FROM THE VENOM OF LATRODECTUS SPIDERS

Shavkat Salikhov, Tatyana Slavnova, Mukhrodj Tashmukhamedov, Marat Adylbekov, Alexander Korneyev, Janna Abdurakhmanova, Abid Sadykov
Institute of Bioorganic Chemistry, Tashkent, USSR

Introduction

Presynaptically acting neurotoxins occur rare in nature and they are not enough studied, that is why the investigation of their structure and mechanism of action , their usage as "tools" will give more information concerning presynaptic membrane and contribute to better understanding of molecular mechanisms in nerve-muscular transmission.
According to the mechanism of action and the effect produced presynaptically acting neurotoxins may be divided into the following groups:

1) Neurotoxins isolated from the venom of Elapidae snakes. They are considered to be the proteins of basic nature possessing phospholipase activity.
2) Neurotoxins produced by microorganisms Clostridium botulinum and Clostridium tetani.
3) Neurotoxins isolated from the venom of Latrodectus spiders.

Results

Neurotoxic phospholipases and toxins produced by Clostridium microorganisms selectively act on cholinergic system by blocking acetylcholine release while the main effect of

Chemistry of Peptides and Proteins, Vol. 2

Latrodectus spider toxins is expressed in inducing massive release of mediators from various nerve terminals(1-2). Regretfully we don't have information enough for detailed discussion of relationship between the structure and function of toxins isolated from various species of Latrodectus genus. To study the structure and mechanism of action of the main neurotoxins of Latrodectus spiders we isolated and identified this toxin from the venom of various species of Latrodectus genus, namely, from the venom of Latrodectus mactans tredecimguttatus widely spread in Central Asia and from the rare species Latrodectus pallidus pavlovski Charit and from the venom of spiders Latrodectus Dahli Levi bred from the cocoons in an insectarium.

When studying the whole venoms by electrophoresis in polyacrylamide gels 15 dyed protein bands have been observed. The protein components with various m.w. have been seen(10,000--200000), some of them are common for all species of spiders. The fractionation technique was the same for all species. It included the following steps: the treatment by ammonium sulphate, gel chromatography, ion-exchange chromatography.The procedure resulted in 10 fractions from the venom of each species. The fraction possessing maximal activity eluted at 0,28 M NaCl from the column containing cellulose D-52.Electrophoresis in PAAG at pH=8,9 in the presence of SDS gave only one band in anode region. During N-terminal analysis the amino group of toxin N-terminal amino acid appeared to be blocked. Antiserum for homogeneous toxin of Latrodectus tredecimguttatus formed only one precipitation line with the toxins of various spider species.

The data obtained in the process of isolation and purification given earlier (3), identify the quarternary structure of the toxin molecule. At neutral and weak alkaline pH the structure is presented as biologically active dimer and tetramer being capable to cleave into non-active monomer at low pH values in the presence of detergents (fig.1).

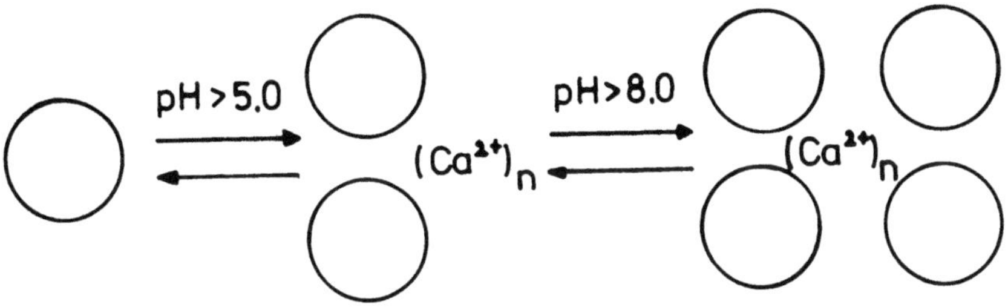

Figure 1. Various states of neurotoxin depending on the medium.

Amino acid composition shows marked similarity in protein molecules of various spider species(Table 1).

Table 1. Amino acid composition of neurotoxins from the venom of Latrodectus spiders.

Amino acid	Number of residues L.tredecimguttatus	Number of residues L.pallidus Pavlovsky	Number of residues L.Dahli Levi
Asp	80	74	53
Thr	29	25	38
Ser	33	28	36
Glu	67	74	79
Pro	34	29	31
Gly	38	30	42
Ala	40	30	34
Val	49	32	38
Met	12	13	16
Ile	38	29	34
Leu	59	51	47
Tyr	25	29	32
Phe	34	37	31
His	23	30	33
Lys	36	34	46
Arg	37	34	50
Trp	4	not det.	not det.
All in all	638	579	640

In spite of the considerable similarity in functional activity and in physico-chemical parameters the final conclusion may be drawn after finding the analogy in amino acid sequences. We are just working hard on this problem.

The mechanism of action of the neurotoxins of Latrodectus spiders has been studied electrophisiologically on frog neuro-muscular preparation. The toxin has been added to the bathing solution in concentrations $5 \times 10^{-7} - 1 \times 10^{-5}$ g/ml. In case the toxin has been added to Ringer solution containing Ca^{2+} and Mg^{2+} the increase in mepp frequency has been observed after some latent period, the duration of which depended on the toxin concentration. The frequency increased 100-200 times when the toxin showed its maximal effect. The removal of the toxin followed 10 minutes contact with the preparation did not prevent the development of its effect. Toxin addition to the medium with Ca^{2+} concentration maintained at 10^{-8}M did not induce the typical effect expressed in mepp frequency increase. However, the change of the medium for that one containing Ca^{2+}, Mg^{2+} or Sr^{2+} ions immediately resulted in the rise in mepp frequency. This toxin effect can be eliminated by changing the medium for the starting one.without bivalent cations(fig.2).

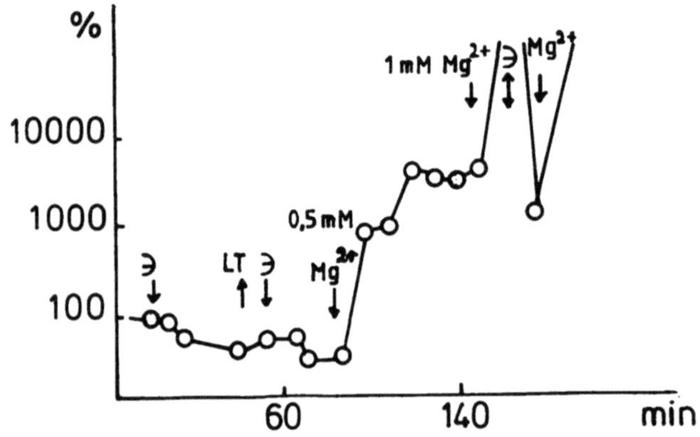

Figure 2. Activation of α-latrotoxin effect by Mg^{2+} and Sr^{2+} ions.

Thus, the effect of toxin can be controled by adding or removing of bivalent cations from the medium. The threshold concentrations of Ca^{2+} and Mg^{2+} are equal to 0,5 mM, but the maximal effect is observed only in the presence of 1 mM Mg^{2+} or Ca^{2+} or 4 mM Sr^{2+}.

There exist several points of view concerning the molecular nature of the activating effect of bivalent cations. The venom of Latrodectus spider (5) and its neurotoxin have been shown to increase the model phospholipid membrane permeability for K^+, Na^+ and Ca^{2+} ions by forming the conductivity channels. The same phenomenon has been observed in our experiments, too. If the toxin can form additional ion channels in presynaptic membrane (4,5), then the penetration of bivalent cations into axoplasma may result in the increased discharge of mediator quanta. However, the typical effect of the toxin may be also observed in the absence of bivalent cations, in case we increase the toxity of the bathing solution by addding 120 mM of saccharose or by raising the temperature of the medium (6,7). This finding may be considered to be the evidence of the necessity for the presynaptic membrane or the toxin molecule itself to be in a certain state to ensure the proper effect. To verify the last supposition the experiments have been carried out with the toxin treated by various concentrations of chelate-forming agents, EGTA and EDTA, before adding it to the bathing solution. The preincubation of the toxin for 2 hours with 1-10 mM EGTA or 10 mM EDTA appeared to prevent the development of the typical effect in spite of the fact that the toxin incubated first in Ca^{2+}-free medium was then added to the medium containing 1,8 mM Ca^{2+}.

If the amount of the native toxin is added to the medium with the toxin previously treated by EGTA the mepp frequency is immediately increased.

The loss of activity caused by the effect of the chelate-forming agents may be explained by the necessity to have bivalent cations for maintaining the active form of the toxin molecule, i.e. its dimer and tetramer structure. This fact is evidenced by the experiments to restore the toxin activity lost after the treatment by EGTA. This can be done by adding for 15 min. the toxin previously incubated with EGTA to the bath with the preparation and then washing it by normal Ringer, containing 1,8 mM Ca^{2+}. Similar to the previous experiments no increase in mepp frequency has been observed. But if Ca^{2+} concentration was raised up to 5 mM, mepp frequency jumped up to 490±100%/ /4 experiments (fig.3).

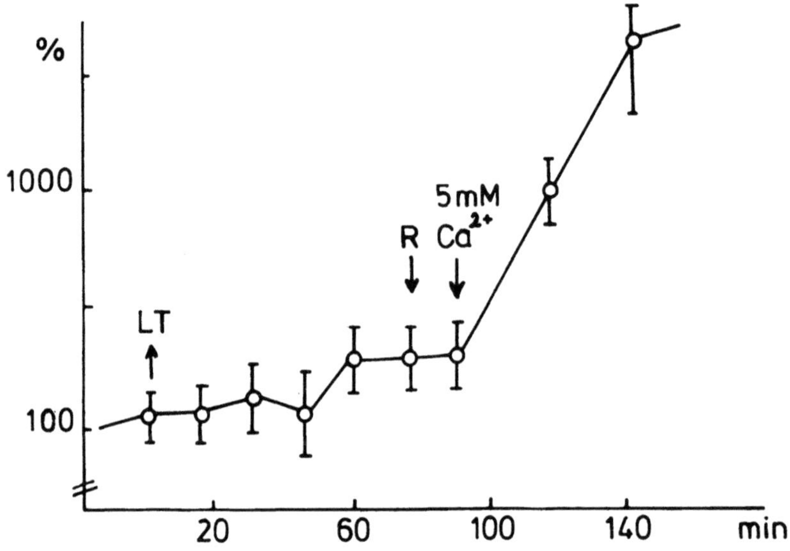

Figure 3. Reactivating influence of increased Ca^{2+} concentration on the effect of 1 x 10^{-5} g/ml of α-latrotoxin previously treated by 5mM EGTA.

It appears that the changes in the toxin molecule induced by incubation in the presence of EGTA and EDTA may be eliminated by the use of higher Ca^{2+} concentration even in case the non--active toxin had been already strongly absorbed on presynaptic membrane. As it has been already shown that α-latrotoxin takes the form of monomer (3,4) at pH lower than 5,0 we carried out the experiments at pH as low as 4,8 using citrate-phosphate buffer (3 mM). The addition of the toxin to the bath with normal Ringer (Ca^{2+}=1,8 mM, pH=7,4) resulted in mepp frequency rise within 5-10 minutes. The change of this solution for a similar one but at pH=4,8 rapidly decreased mepp frequency. This effect can be reversed by rising pH up to 7,4(fig.4).

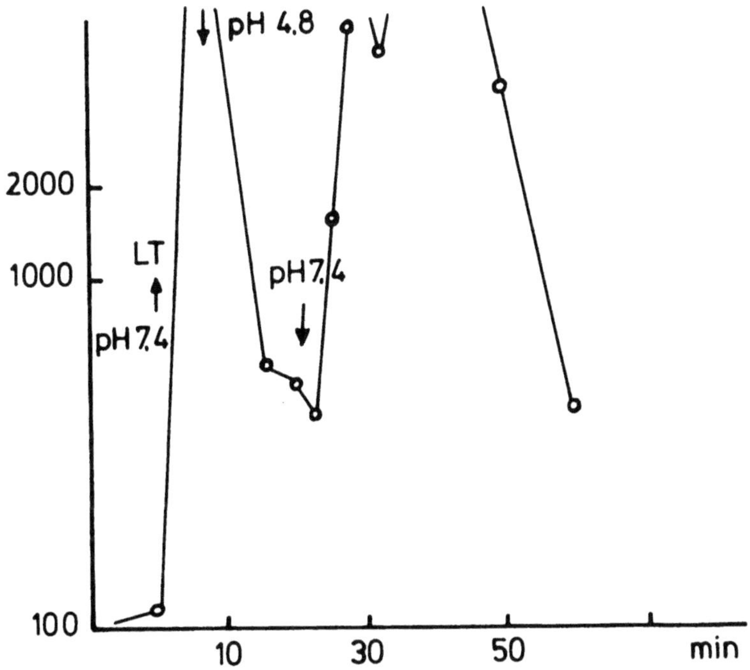

Figure 4. The influence of the medium acidification on α-latrotoxin effect.

The efficiency of α-latrotoxin action on bilayer phospholipid membranes depends also on pH of the medium, it is considerably increased in pH range 5,0-8,0.Thus, the effect of the toxin can be controled by changing pH as well as by changing the concentration of bivalent cations.

The findings given above show that the conditions leading to the transition of the toxin into its monomer form deprive this toxin of the ability to act specifically on presynaptic membrane, but they probably do not influence its irreversable binding.

References

1. Howard, B.D., Gundersen, C.B. Jr.: Ann.Rev.Pharmacol. Toxicol. 20, 307-336 (1980).

2. Tzeng, M.C., Siekevitz, P.: Advances in Cytopharmacology, Vol. 3, ed. B.Ceccarelli, F.Clementi, Raven Press, New York 1979

3. Salikhov, Sh., Tashmukhamedov, M., Adylbekov, M., Korneyev, A., Sadykov, A.: DAN SSSR 262, 2, 485-488 (1982).

4. Finkelstein, A., Rubin, L.L., Tzeng, M.C.: Science 193, 1009-1011 (1976).

5. Krasilnikov, O.V., Ternovsky, V.I., Tashmukhamedov, B.A.: Biofizika 27, 72-75 (1982).

6. Misler, S., Hurlbuf, W.P.: Proc. Nat. Acad. Sci. USA 76, 991-995 (1979).

7. Gario, A., Mauro, A.: J. Gen. Physiol. 73, 245-263 (1979).

TOPOGRAPHY OF THE ACETYLCHOLINE RECEPTOR - NEUROTOXIN BINDING SITES PROBED BY SPECTROSCOPY AND PHOTOLABELING

Victor I. Tsetlin, Kirill A. Pluzhnikov and Vadim T. Ivanov
Shemyakin Institute of Bioorganic Chemistry, USSR Academy of Sciences, 117988 Moscow, USSR

Introduction

In recent years much interest has been focused on cobra venom neurotoxins (see reviews (1,2). Firstly, these proteins of relatively low molecular weight (~7,000-8,000) possess a stable conformation and are convenient objects to develop new methods of protein chemistry. Secondly a numerous family of homologous neurotoxins provides wide opportunities for structure-functional studies. However, comparison of toxicity for various neurotoxins and their chemically modified analogs did not allow to delineate a limited number of functional groups which could be regarded as a neurotoxin "active center". On the contrary, it seems that highly specific neurotoxin action - binding to the acetylcholine receptor of postsynaptic membrane (AchR) and concomitant blocking the neurotransmission - results from involvement of many charged and hydrophobic groupings (1).

Our purpose was to identify those regions of the neurotoxin molecule which are directly implicated in AchR binding. To this end, we monitored by means of EPR and fluorescence spectroscopy the AchR binding of a series of neurotoxins with grafted spin or fluorescence labels (3-5). In addition, photoinduced cross-linking was studied between AchR and various neurotoxin II *Naja naja oxiana* (NT-II) photoactivable derivatives (6). The present report deals with the topography of the AchR and neurotoxin interacting sites described on the totality of EPR, fluorescence and photolabeling data.

Chemistry of Peptides and Proteins, Vol. 2

Results

Several photoactivable derivatives of postsynaptically acting neurotoxins have been described in literature (7-9). However, a single labeled product seems to be isolated only in (8). Our task was to prepare a group of NT-II biologically active derivatives, each comprising one photoactivable label in different regions of the neurotoxin surface. The convenient sites to incorporate such labels are the amino groups (Leu 1, Lys 15, Lys 25, Lys 26, Lys 44 and Lys 46) situated in the three out of for disulfide-confined loops of NT-II. We used the reaction with N-hydroxysuccinimidyl p-azidobenzoate, based on the earlier found conditions (4,10) for NT-II modification with spin-labeled N-hydroxysuccinimides.

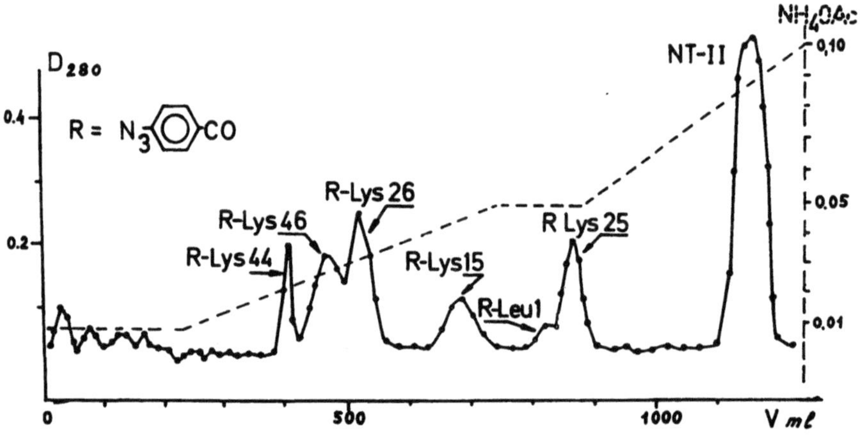

Fig.1 Separation of NT-II p-azidobenzoyl derivatives by Bio Rex 70 column chromatography (6).

Although many proteins with covalently bound photosensitive labels were described in literature, there are only a few exceptions when the label position was determined (and at that relying upon the indirect methods). In the modified conditions of

(11) we managed to reduce the azide group and S-S bridges avoiding intramolecular crosslinking. The reduced and carboxymethylated derivatives were subjected to trypsinolysis, the hydrolysates being separated on Sephadex G-25 sf column. The position of the modified residues (see Fig.1) was established judging from the presence of a label in the corresponding peptide and prevention of the trypsin cleavage at the modified lysine residues.

Binding studies were performed with the AchR isolated from the *Torpedo marmorata* electric organs using affinity chromatography on a neurotoxin-Sepharose in Triton X-100. (The AchR activity ranged from 7 to 10 nmole α-toxin binding sites per 1 mg receptor protein).

Either non-covalent or covalent binding of the NT-II photoactivable derivatives to AchR was assessed from their capacity to compete with ^{3}H -acetylated NT-II or toxin 3 *Naja naja siamensis*. The experiments demonstrated high efficacy of a non-covalent complex formation for the all 6 photoactivable derivatives. It means that the incorporation of p-azidobenzoyl groups at the specified positions does not preclude the NT-II binding to AchR (similarly to earlier described acetylation or spin labeling(3)).

Photolysis of the AchR complexes with the NT-II photoactivable derivatives was performed using a 250 W halogen lamp without cutting off the short-wave irradiation. The control experiments showed that, under the adopted procedure, there was virtually no change in the spectral characteristics of NT-II, AchR, or their complex, and no loss in the AchR specific activity. The extent of crosslinking was determined from the decline in the complex capacity to bind a radioactive neurotoxin after illumination. With one exception, all the tested compounds form extremely effectively (70-80%) the photoinduced crosslinks with AchR. The derivative with the label at Lys 44, which binds to AchR as tightly as the rest of analogs, but manifests no crosslinking on illumination, may be regarded as an "internal reference" for assessing the crosslinking specificity.

These results demonstrate quite clearly that the photoactivable labels placed at specified sites of the NT-II molecule are covalently bound to AchR as a result of illumination of the respective neurotoxin - AchR complexes. This allows to reliably identify the region of contacts between the neurotoxin and AchR. The question might arise : are the data on photolabeling in accord with the earlier EPR and fluorescence studies? Analysis of the changes in the spin label mobility and accessibility for external media which accompany the complexation with AchR, allowed the following conclusion to be made: the spin labels attached to Leu 1, Glu 2, Lys 15, Lys 26, His 31 and Lys 46 residues are bound to AchR with greater or lesser efficacy, while that at Lys 44 is exposed in solution in free neurotoxin as well as in its AchR complex (3-5).

Fluorescence data on the dansylated NT-II derivatives corroborated the above conclusion in relation to the Lys 25, Lys 26 and Lys 46 residues. However, they provided evidence against the contacts between the AchR surface and dansyl groups at the Leu 1 α-amino group or Lys 15 ε-amino group (5). In view of the photoaffinity modification, it is reasonable to believe that Leu 1 and Lys 15 amino groups still can interact with AchR. Apparently there are extended regions of the tight contacts between AchR and neurotoxin(for example in vicinity of bound Lys 26, His 31 and Lys 46 side chains (5), along with weaker contacts (possibly, of lesser functional importance). The latter seems to be true for the sites involving Leu 1 and Lys 15 residues: the incorporated label, depending on its size and character, may be either attached or detached from the AchR surface.

In general, photolabeling data are in accord with the results of spectroscopic approaches (see Fig.2) and substantiate the previously formulated ideas on the topography of neurotoxin and AchR interacting sites. This concerns, in particular, the evaluation of the binding site dimensions in AchR and the presence of the AchR functional groupings near the bound fragments of NT-II (see 5,10).

In view of a great similarity in the spatial structure and action of NT-II and other related short-chain neurotoxins, it

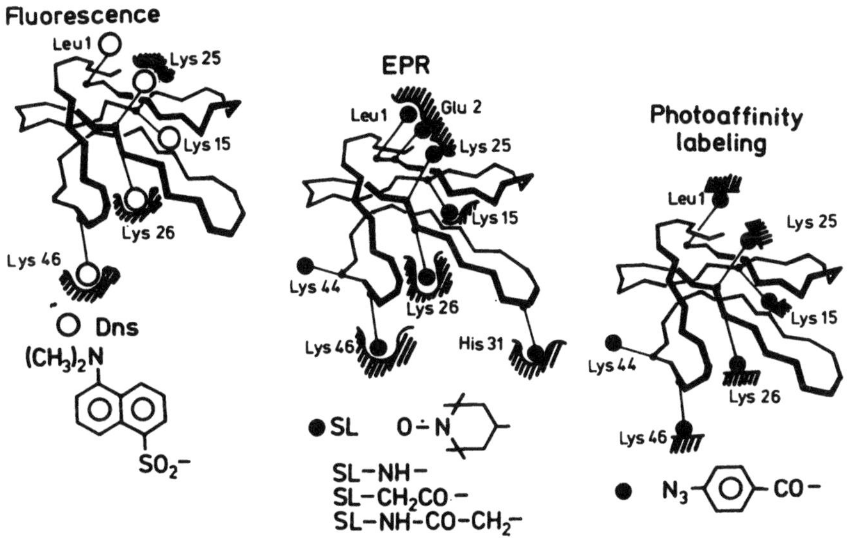

Fig.2 A scheme for AchR-neurotoxin binding as inferred from spectral and chemical data.

seems justifiable to believe that the studies on NT-II derivatives revealed the essential features which can adequately describe the AchR interaction with a whole family of the short-chain neurotoxins. As far as their long-chain counterparts are concerned, the data for several spin-labeled derivatives are indicative of a similar mode of binding to AchR (5,12) . Binding of the long-chain neurotoxins to several AchR subunits (7,9,13), evidences that they occupy a considerable area of AchR.
Owing to high efficiency of photoinduced crosslinking, the NT-II photoactivable derivatives are promising for elucidating the mutual disposition of the bound neurotoxin and AchR subunits.
In combination with the neurotoxins bearing "spectral" labels, these compounds may find application as versatile tools for studies aimed at understanding the spatial organization of the AchR active center.

Acknowledgements

The report is based on experimental work conducted in Moscow and Uppsala (in collaboration with Dr. E.Karlsson).

References

1. Karlsson, E.: Handbook of Experimental Pharmacology (Lee, C.Y., ed.) vol. 52, pp. 159-212, Springer Verlag, Berlin (1979).

2. Difton, M.J., Hider, R.C.: Trends Biochem. Soc. 5, 53-56 (1980).

3. Tsetlin, V.I., Karlsson, E., Arseniev, A.S., Utkin, Yu.N., Surin, A.M., Pashkov, V.S., Pluzhnikov, K.A., Ivanov, V.T., Bystrov, V.F., Ovchinnikov, Yu.A.:FEBS Lett. 106, 47-52 (1979).

4. Ivanov, V.T., Tsetlin, V.I., Karlsson, E., Arseniev, A.S., Utkin, Yu.N.,Pashkov, V.S., Surin, A.M., Pluzhnikov, K.A., Bystrov, V.F.: in Natural Toxins (Eaker,D., Wadström,T., eds.), pp. 523-530, Pergamon Press, Oxford and New York (1980).

5. Tsetlin, V.I., Karlsson, E., Utkin, Yu.N., Pluzhnikov, K. A., Arseniev, A.S., Surin, A.M., Kondakov, V.I., Bystrov, V.F., Ivanov, V.T., Ovchinnikov, Yu.A.: Toxicon 20, 83-93 (1982).

6. Pluzhnikov, K.A., Karelin, A.A., Utkin, Yu.N., Tsetlin, V. I., Ivanov, V.T.: Bioorgan Khim. (USSR), 8,905-913(1982).

7. Witzeman, V., Raftery, M.A.: Biochem. Biophys. Res. Commun. 85, 623-631 (1978).

8. Hucho, F.: FEBS Lett. 103, 27-32 (1979).

9. Nathanson, N., Hall, Z.: J. Biol. Chem. 255, 1698-1703 (1980).

10. Tsetlin, V., Karlsson, E., Utkin, Yu., Arseniev, A., Pluzhnikov, K., Surin, A., Pashkov, V., Bystrov, V., Ivanov, V.: Proceedings of the IIId USSR-FRG Symposium: Voelter, W., Wünsch, E., Orchinnikov, Y., Ivanov, V.(eds), Walter de Gruyter, 389-398 (1981).

11. Staros, V.T., Bayley, H., Standring, D., Knowles, J.R.: Biochem. Biophys. Res. Commun. 80, 568-573 (1978).

12. Ellena, J.F., McNamee, M.G.: FEBS Lett. 110, 301-304 (1980).

13. Oswald, R.E., Changeux, J.-P.: FEBS Lett. 139, 225-229 (1982).

PEPTIDES AS IRON IONOPHORES (SIDEROPHORES)

Günther Winkelmann
Institut für Biologie I, Mikrobiologie I, Universität Tübingen
7400 Tübingen, West-Germany

Introduction

Enzymes containing iron are necessary constituents of all living organisms. The way, however, by which environmental iron is scavenged and transported into the cells is still unresolved and has attracted special attention during the last few years. A substantial body of knowledge has accumulated from the work with microorganisms. Bacteria and fungi produce ferric specific chelating agents which function by removing iron from insoluble ferric hydroxides and transport iron into the cells. Today microbial chelating agents are collectively called siderophores [1]. To qualify as a siderophore the iron chelating agent must be low molecular, repressible as a response to the iron content of the medium, have a large formation constant for ferric iron and serve to solubilize and transport iron into the microbial cell.

1. The diversity of siderophore structures

Although the diversity of siderophores is well documented elsewhere [2,3] a collection of some siderophore structures is presented here with special emphasis on the coordination centre. One part of these siderophores consists of combinations of carboxylic acids (succinic acid, citric acid) with diamines or diaminocarboxylic acids, representing no real peptides, an-

Chemistry of Peptides and Proteins, Vol. 2

other part consists mainly of amino acids and can be regarded as iron-chelating cyclic peptides. Iron ionophores possessing a peptide backbone predominate among eukaryotic microorganisms, like fungi. It appears that the development of a peptide chelate backbone and its corresponding uptake system within the membrane proved to be an advantage during competition for environmental iron. Representatives of the peptidic iron ionophores are the ferrichrome-type siderophores, which possess a cyclohexapeptide moiety containing a tripeptide sequence of N^5-acyl-N^5-hydroxy-L-ornithine and a remaining tripeptide sequence (R', R", R"), which may consist of gly-gly-gly (ferrichrome and malonichrome), gly-ser-gly (ferricrocin), gly-ser-ser (ferrichrysin, ferrichrome A and ferrirubin) or gly-ala-gly (ferrichrome C). The trivalent iron is octahedrally coordinated by three secondary hydroxamate groups, resulting in an uncharged iron (III) chelate. The three N^5-acyl-groups (R) of the hydroxamate ligands around the metal centre may contain acetic acid (ferrichrome, ferricrocin, ferrichrysin), methyl glutaconic acid (ferrichrome A), trans-anhydromevalonic acid (ferrirubin), cis-anhydromevalonic acid (ferrirhodin) or malonic acid (malonichrome).

Attention has been given to the nature of these acyl groups with regard to their iron transport properties in *Ustilago sphaerogena* [4]. Thus, it has been found that the ferrichrome transport system of the fungus is very sensitive to minor structural variations. Uptake rates of ^{59}Fe-labelled semisynthetic derivatives differed depending on the acyl chain length. Branching at the β-position of the hydroxamate acyl function greatly reduced the uptake rates. Similar variations of N-acyl substituents in coprogen led to analogous alterations of the chelate transport rates in Neurospora crassa [5].

2. The peptide backbone of ferrichrome-type siderophores

Compared to the conformational freedom in the metal free ferrichrome ligand, ferrichrome possesses a rigid solution conformation [6], which is in good agreement with its x-ray crystal structure [7]. The six amino acids form a relatively planar β-structure with one or two transannular hydrogen bonds. Furthermore, the solution conformation of the different ferrichrome-type siderophores proved to be largely identical, regardless of the substitution of serine for glycine in the peptide backbone and of the nature of the N-acyl groups around the metal centre. However, it was suggested from the crystal structure of ferrichrysin [8] that glycine at position 1 may be necessary to allow proper conformation of the peptide backbone. Summarizing these results it may be concluded that the peptide backbone represents a constant and characteristc feature of ferrichrome-type siderophores, which allows specific recognition by membrane proteins involved in siderophore transport. As proof of this statement, it was found that a synthetic ferrichrome ligand, containing an opposite chirality of the peptide backbone, such as enantio-ferrichrome [9] possessing D-ornithyl instead of L-ornithyl residues, was not recognized by fungi, which normally transport ferrichrome-type siderophores [10,11]. These studies disprove the idea that siderophores penetrate cell membranes by diffusion only. Although ferrichrome and enantio-ferrichrome have identical dimensions and an identical lipophilicity, their transport properties differed greatly. Concentration-dependent uptake of enantio-ferrichrome revealed a diffusion line, whereas uptake of the natural ferrichrome showed a saturation kinetic which is indicative for a mediated transport process. The inability of fungal siderophore transport systems of recognizing enantio-ferrichrome may, however, not only rely on the opposite chirality of the peptide backbone. Although structure and conformation of peptides have to be considered as the most probable recognition sites of siderophores, the coordination centre may also be of importance.

3. The configuration of the metal centre

During the last few years considerable progress has been made to determine the absolute configuration of siderophores using single crystal x-ray diffraction studies and circular dichroism measurements in solution [12,13,14,15]. Since the hydroxamic acids of siderophores are unsymmetrical and optically active ligands, both optical (Fig. 1) and geometrical (Fig. 2) coordination isomers are possible, yielding theoretically a total of 16 different isomers.

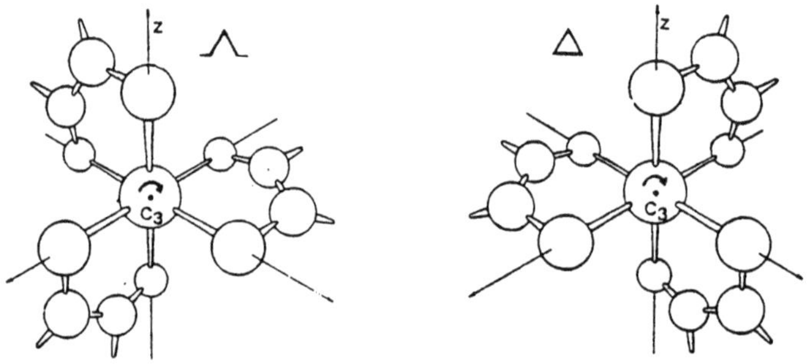

Fig. 1. Optical isomers (Λ and Δ) of a tris(hydroxamato)iron complex.

The question, as to whether different optical isomers are accepted by one siderophore uptake system is still unresolved. One approach to this question is to substitute the kinetically labile ferric ion in siderophores by the kinetically inert chromic ion. Thus, Λ-cis-chromic desferri ferrichrome was taken up by the fungus *Ustilago sphaerogena* as fast as natural ferrichrome indicating that the Λ-cis isomer is one of possible isomers which is recognized by the transport system [14]. However, the possibility that the Δ-cis isomer was excluded from uptake could not be shown, because of the unavailability of this compound for testing. It has been shown that chromic complexes of ferrichrome and ferrichrysin consist exclusively of the Λ-cis isomer [13].

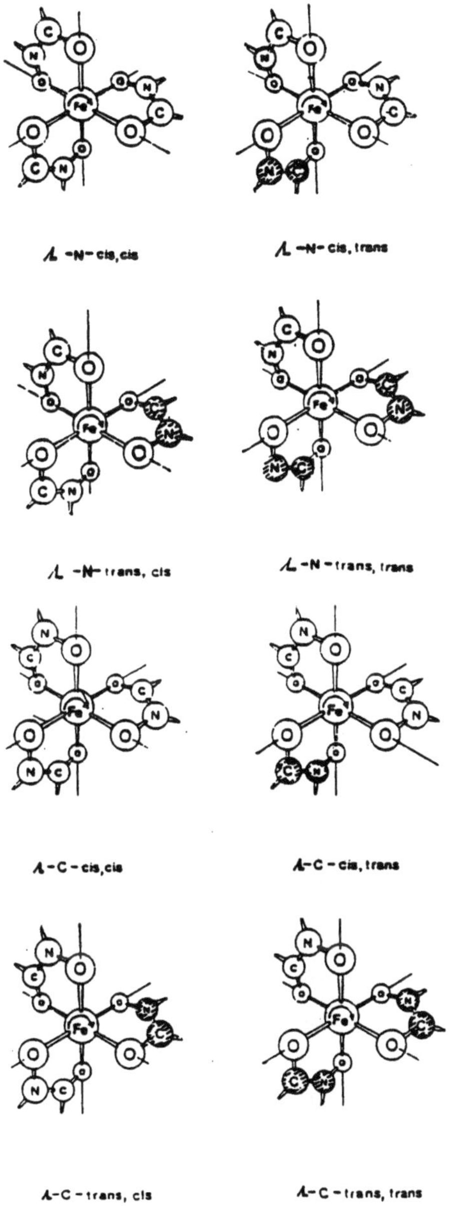

Fig. 2. Eight possible geometrical coordination isomers (Λ-series) of a tris(hydroxamato)iron complex with unsymmetrical bidentate ligands. Nomenclature according to Leong and Raymond [20].

In natural ferric siderophores, however, only the cis, cis-isomer has been found so far. Thus, ferrichrome [7], ferrichrome A [16], ferrichrysin [8] and ferricrocin [15] adopt a Λ-cis configuration, whereas ferric rhodotorulic acid [17] and coprogen [15] prefer the Δ-cis-configuration. In the case of tracetyl fusarinine [18], both, Δ and Λ isomers are formed, depending on the crystallization conditions. In solution, however, the molecule exists predominantly as the Δ-cis isomer.

An approach to the biological activity of Δ-cis isomers is to use synthetic chelators containing an opposite chirality of the peptide backbone. Using the synthetic Δ-cis-enantio-ferrichrome [9], the biological inactivity towards various fungi can be demonstrated [10,11]. However, as discussed above, it remains to be established as to whether the chirality of the peptide backbone or the chirality of the metal center is crucial for stereoselective recognition of siderophores. Unequivocal results may be obtained, when different optical and geometrical isomers of the same ligand are tested. The proportion of isomers may be well balanced as in the case of the ferrioxamines. Ferrioxamine E crystalizes as a racemic mixture of Λ and Δ isomers [19]. Ferrioxamine B exists as a mixture of Λ-cis and Λ-trans isomers in solution, yielding a cis/trans ratio of 65:35 [20]. Contrary to these non-peptidic siderophores, the peptidic siderophores of the ferrichrome-type family possess a backbone, which appears to force the iron centre to adopt predominantly a configuration, which is Λ-cis for the natural ferrichrome-type compounds and Δ-cis for the synthetic enantio-ferrichrome.

Most interestingly, the fungus *Neurospora crassa* is able to take up a Λ isomer (ferrichrocin) as well as a Δ isomer (coprogen) [15]. Whether they are transported by the same transport system, is currently being examined.

1

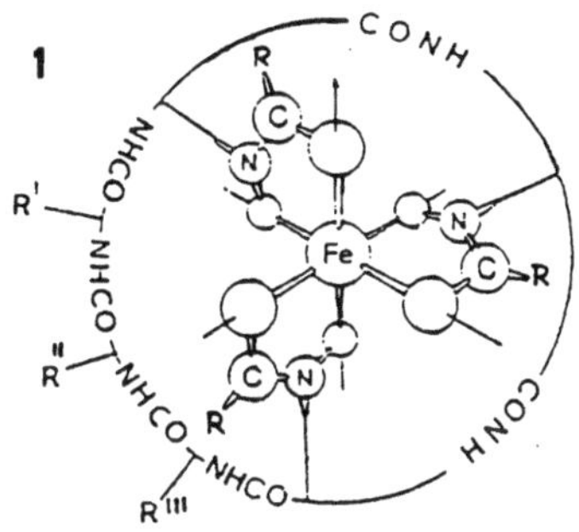

2

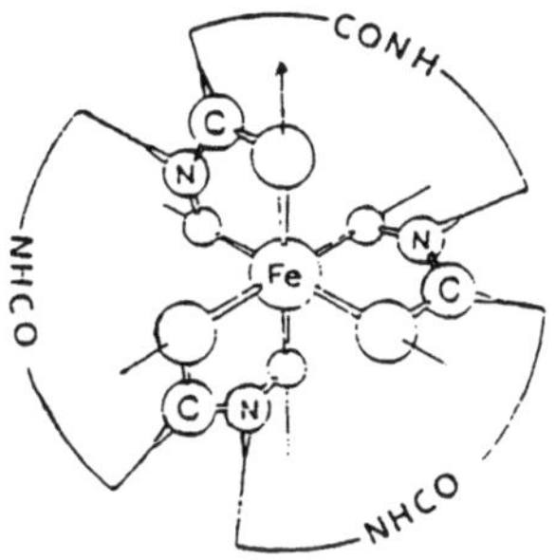

3

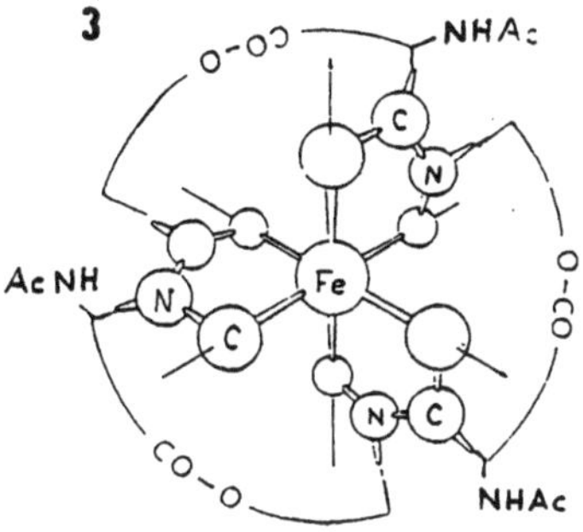

4

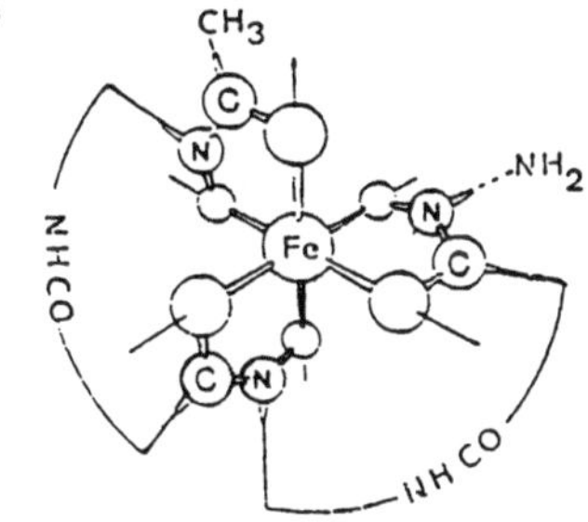

5

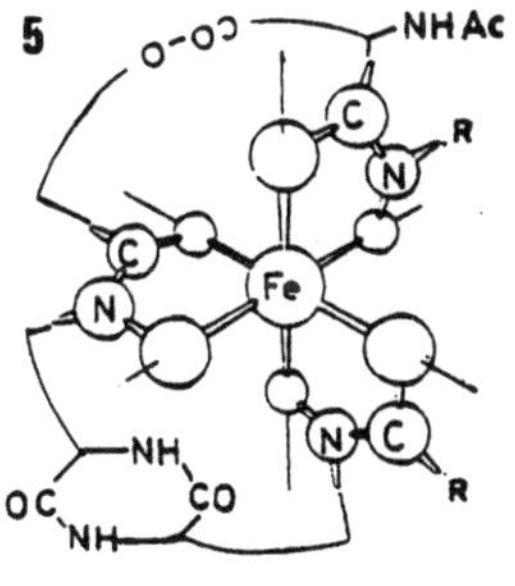

Fig. 3. The absolute configuration of siderophores.
1. Ferrichrome-type siderophores (Λ-cis). 2. Ferrioxamine E (racemic, Λ-cis shown). 3. Triacetylfusarinin (Δ-cis). 4. Ferrioxamine B (Λ-cis/trans, Λ-cis shown). 5. Coprogen (Δ).

To summarize the data of the structure/function relationship of siderophores, it may be concluded that peptides as iron ionophores have achieved the highest degree of conformational and configurational stability among all siderophores which enabled the development of specific high affinity uptake systems for iron (III) chelates and which in an evolutionary sense may have helped to prevent loss of iron by competing organisms.

Acknowledgments

This work was supported by the Deutsche Forschungsgemeinschaft (SFB 76).

References

[1] Lankford, C.E.: Critical Reviews of Microbiology 2, 273-331 (1973).

[2] Neilands, J.B.: Ann. Rev. Biochem. 50, 715-731 (1981).

[3] Winkelmann, G.: In "The Biological Chemistry of Iron" (H.B. Dunford, D. Dolphin, K.N. Raymond and L. Sieker, eds), Reidel Publ. Comp. Dordrecht, Holland 1982, pp. 107-116.

[4] Emery, T. and Emery L.: Biochem. Biophys. Res. Commun. 50, 670-675 (1973).

[5] Ernst, J. and Winkelmann, G.: Arch. Microbiol. 100, 271-282 (1974).

[6] Llinás, M., Klein, M.P. and Neilands, J.B.: J. Mol. Biol. 52, 399-414 (1970).

[7] Van der Helm, D., Baker, J.R., Eng-Wilmot, D.L., Hossain, M.B. and Loghry, R.A.: J. Amer. Chem. Soc. 102, 4224-4231 (1980).

[8] Norrestam, R., Stensland, B., and Bränden, C.I.: J. Mol. Biol. 99, 501-506 (1975).

[9] Naegeli, H.-U. and Keller-Schierlein, W.: Helv. Chim. Acta 61, 2088-2095 (1978).

[10] Winkelmann, G.: FEBS Lett. 97, 43-46 (1979).

[11] Winkelmann, G. and Braun, V.: FEMS. Microbiol. Lett. 11, 237-241 (1980).

[12] Leong, J. and Raymond, K.N.: J. Amer. Chem. Soc. 96, 1757-1762 (1974).

[13] Leong, J. and Raymond, K.N.: J. Amer. Chem. Soc. 96, 6628-6630 (1974).

[14] Leong, J., Neilands, J.B., and Raymond, K.N.: Biochem. Biophys. Res. Commun. 60, 1066-1071 (1974).

[15] Wong, G.B., Kappel, M.J., Raymond, K.N., Matzanke, B., and Winkelmann, G.: J. Amer. Chem. Soc. (in press).

[16] Zalkin, A., Forrester, J.D., Templeton, D.H.: J. Amer. Chem. Soc. 88, 1810-1814 (1966).

[17] Carrano, C.J. and Raymond, K.N.: J. Amer. Chem. Soc. 100, 5371-5374 (1978).

[18] Hossain, M.B., Eng-Wilmot, D.L., Loghry, R.A., van der Helm, D.: J. Amer. Chem. Soc. 102, 5766-5773 (1980).

[19] van der Helm, D. and Poling, M.: J. Amer. Chem. Soc. 98, 82-86 (1975).

[20] Leong, J. and Raymond, K.N.: J. Amer. Chem. Soc. 96, 293-296 (1975).

INTERACTION OF PROTEINASES AND THEIR PROTEIN INHIBITORS AS STUDIED BY THE SPIN-LABEL TECHNIQUE

H.R. Wenzel and H. Tschesche
Universität Bielefeld, Fakultät für Chemie, D-4800 Bielefeld 1

E. von Goldammer and J. Paul
Ruhr-Universität, Institut für Biophysik, D-4630 Bochum 1

Introduction

The current interest in naturally occurring proteinase inhibitors has mainly emerged because of the control functions that these proteins exercise in a variety of biologically important proteolytic processes, e.g. digestion, blood clotting, pressure regulation, fertilization, phagocytosis, complement immune reactions and many others [1]. Moreover, the proteinase-inhibitor complexes involved have often served as model for protein-protein association using various spectroscopic techniques.

We thought that spin-labelling could be advantageously added as a tool to study proteinase-inhibitor interactions, as this method is especially sensitive to structural or conformational changes [2]. The bovine trypsin inhibitor (Kunitz) and its complexes with trypsin or chymotrypsin were chosen as a first test system [3].

Spin-labelling procedure

The single polypeptide chain of the Kunitz-inhibitor consists of 58 amino acid residues, which are cross-linked by three disulfide bridges (fig. 1a). A paramagnetic nitroxyl moiety was to be covalently attached at a single site without affecting the complex formation with proteinases. This could be achieved in two steps:

Chemistry of Peptides and Proteins, Vol. 2

The four lysine residues 15, 26, 41 and 46 were converted to homoarginine residues by O-methylisourea [4]; the α-amino group of arginine-1 remained unaffected and lent itself as the attachment point for a maleimide spin-label reagent (fig. 1b). Amino acid analysis and quantitative ESR measurements indicated the incorporation of 0.85 nitroxyl groups per inhibitor molecule. Neither guanidination nor spin-labelling appreciably changed the activity of the inhibitor against trypsin and chymotrypsin [5].

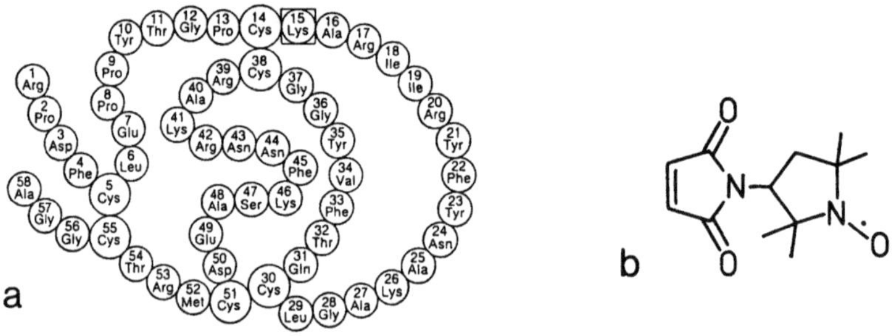

Fig. 1. (a) Amino acid sequence of the bovine trypsin inhibitor (Kunitz).
(b) Structural formula of the spin-label reagent 3-maleimido-2,2,5,5-tetramethylpyrrolidine-1-oxyl.

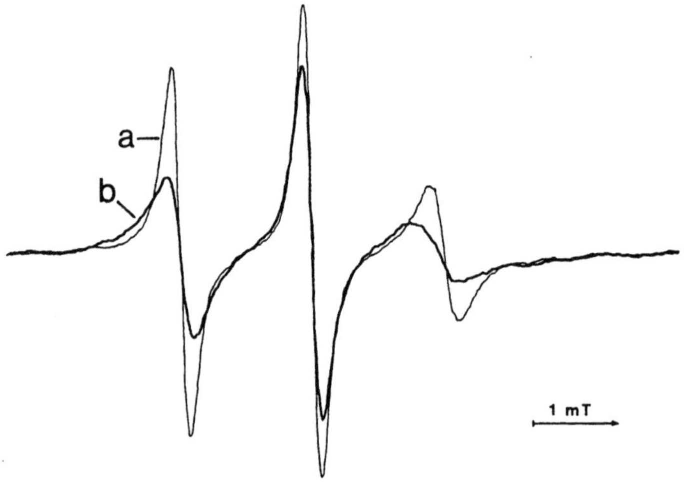

Fig. 2. ESR spectra of (a) the spin-labelled guanidinated inhibitor (50 μM) and (b) its complex with bovine trypsin (inhibitor 50 μM, enzyme 80 μM) in 0.1 M triethanolamine-HCl buffer, pH 7.0.

ESR spectra

The spin-labelled guanidinated inhibitor in solution shows a 'weakly immobilized' ESR spectrum (fig. 2a). Assuming isotropic motion, the rotational correlation time τ of the spin-label can be estimated to be about 2 ns [6], from which a particle diameter d of 2.5 nm can be calculated, when the Debye model of rotation is used: $\tau = \pi \eta d^3/6kT$. As this value is close to that expected for the rotation of the whole inhibitor molecule with its greatest length of 2.9 nm and its greatest diameter of 1.9 nm [7], the nitroxyl group probably has only rather limited mobility relative to the protein.

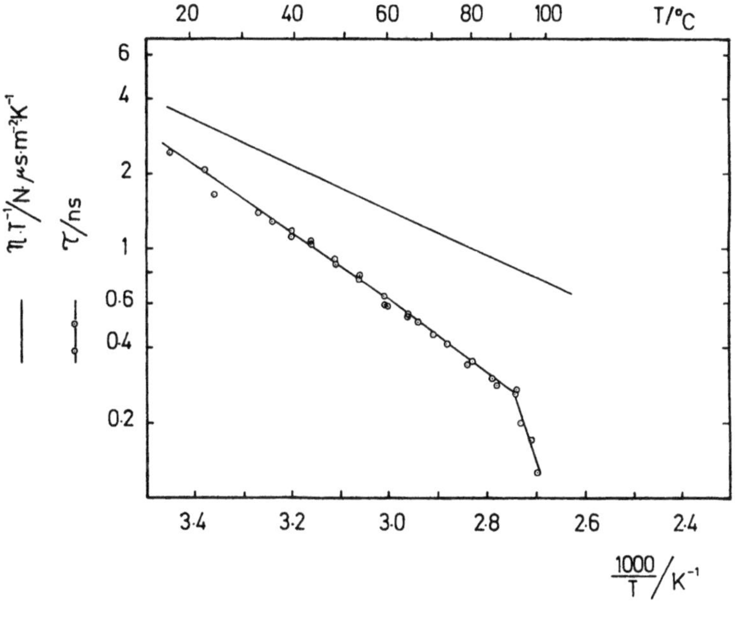

Fig. 3. Arrhenius plot of τ for the spin-labelled guanidinated inhibitor (0.2 mM in 0.1 M triethanolamine-HCl buffer, pH 7.0).

The upper curve of lg η/T vs. 1/T for water is included for comparison.

τ values were determined over the temperature range 15-98°C, the results are presented in fig. 3 as Arrhenius plot. There is a slow and reversible decrease of τ, with a small deviation from linearity around 60°C, when increasing the temperature to 90°C. This behaviour reflects the decrease of viscosity with temperature and is also consistent with a local increase in the motional freedom of the spin-label. A sharp and irreversible break point which is accompanied by loss of signal intensity occurrs at 90°C. It can be related to loosening of the rather rigid structure of the inhibitor and finally its thermal denaturation.

Proteinase-inhibitor complexes

When increasing amounts of trypsin or chymotrypsin are added to the inhibitor solution, the ESR lines become broader and their amplitudes decrease. With complete complex formation a τ of about 5 ns can be determined from the corresponding spectrum (fig. 2b).

The fraction of free inhibitor at each point of a titration can be calculated from the low field line of the ESR spectrum [5]. Fig. 4 shows the resulting ESR titration curves with trypsin and chymotrypsin, the enzymatic activities of diluted incubation mixtures are also included for comparison. Curves with sharp break points are obtained for trypsin, indicative of a 1:1 complex with an extremely small dissociation constant. As expected, enzymatic activity appears when all inhibitor is bound.
The titration curves for chymotrypsin are different: They clearly indicate partial dissociation of the enzyme-inhibitor complex at concentration ratios near unity. This effect is of course more pronounced for the diluted samples used for activity measurements. Quantitative evaluation of the data yields a dissociation constant of about 80 nM for the complex. This value is in the same order of magnitude as the dissociation constant of 11 nM determined by kinetic measurements at pH 8.0 for the complex of chymotrypsin and bovine trypsin inhibitor (Kunitz) specifically guanidinated at lysine-15 [8].

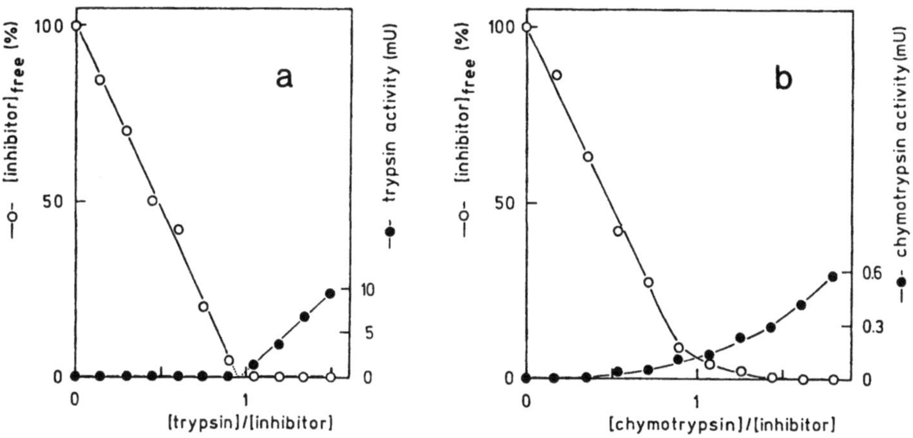

Fig. 4. Titration curves of spin-labelled guanidinated inhibitor with bovine trypsin (a) and bovine chymotrypsin (b).
o: ESR-titrations; inhibitor 50 µM, enzyme concentration varied, 0.1 M triethanolamine-HCl buffer, pH 7.0.
●: Enzymatic activities per ml of the enzyme/inhibitor mixtures diluted 1:100 (a) or 1:50 (b) with 0.2 M triethanolamine-HCl buffer, 20 mM $CaCl_2$, pH 7.8.
Substrates were N-benzoyl-L-arginine-4-nitroanilide for trypsin and N-(3-carboxypropionyl)-L-phenylalanine-4-nitroanilide for chymotrypsin [9].

Discussion

This paper describes spin-labelling in combination with ESR spectroscopy as a promising tool to study the conformation of protein proteinase inhibitors and the interaction with their target enzymes. Bovine trypsin inhibitor (Kunitz) was examined first, because detailed information as to structure and function of this protein is already available. The inhibitor was guanidinated and then specifically spin-labelled at its N-terminus. Labelling at other sites seems possible and is under investigation in our laboratory. The approach can probably be extended to other inhibitors. Moreover, spin-labelling of the other reaction partner, the proteinase, has been achieved [10] and might provide complementary information about enzyme-inhibitor interactions.

Attractive aspects of the spin-label method are its high sensitivity, the possibility to measure in optically opaque solutions and, usually, the absence of interfering signals from the environment.

Acknowledgement

This work has been supported by the Deutsche Forschungsgemeinschaft and the Fonds der Chemischen Industrie.
We thank Bayer AG for supplying us with the bovine trypsin inhibitor (Kunitz) = Trasylol®.

References

1. Holzer, H., Tschesche, H., eds.: Biological Functions of Proteinases, Springer-Verlag, Berlin Heidelberg New York 1979.
2. Berliner, L.J., ed.: Spin Labeling Theory and Applications, Academic Press, New York 1976.
3. Wenzel, H.R., Tschesche, H., v.Goldammer, E., Netzelmann, U.: Biophys. Struct. Mechan. 7, 285 (1981).
4. Chauvet, J., Acher, R.: Biochem. Biophys. Res. Commun. 27, 230 (1967).
5. Wenzel, H.R., Tschesche, H., v.Goldammer, E., Netzelmann, U.: FEBS Lett. 140, 53 (1982).
6. Matheson Jr., R.R., Dugas, H., Scheraga, H.A.: Biochem. Biophys. Res. Commun. 74, 869 (1977).
7. Huber, R., Kukla, D., Rühlmann, A., Epp, O., Formanek, H.: Naturwissenschaften 57, 389 (1970).
8. Vincent, J.-P., Schweitz, H., Lazdunski, M.: Eur. J. Biochem. 42, 505 (1974).
9. Fritz, H., Trautschold, I., Werle, E.: in Methoden der enzymatischen Analyse (Bergmeyer, H.U., ed.) 3rd ed., vol. I, 1104, Verlag Chemie, Weinheim 1974.
10. Kosman, D.J.: J. Mol. Biol. 67, 247 (1972).

AN INHIBITOR OF ELASTASE FROM ANEMONIA SULCATA

H. Tschesche and H. Kolkenbrock
Lehrstuhl Biochemie, Fakultät für Chemie, Universität Bielefeld
4800 Bielefeld, Germany

Elastase inhibitors, especially those inhibiting human leukocyte elastase, have gained particular interest as potential therapeutical agents that could be used to prevent leukocyte tissue damage in inflammatory processes. Such processes are the subject of current investigations in numerous laboratories, e.g. the processes of lung emphysema and chronic inflammatory polyarthritis. The undesired pathophysiological and destructive processes, e.g. in lung emphysema are due to tissue degradation by elastase and in rheumatoid arthritis are thought to be due to cartilage degradation by both leukocyte elastase and collagenase [1,2]. The leukocyte enzymes can be released from the granula of invading leukocytes by several events such as cell death, frustrated phagocytosis, exocytosis and other processes. Under normal physiological conditions the deleterious effect of the elastase released is counteracted mainly by the antielastolytic effect of the serum and plasma α_1-proteinase inhibitor (formerly α_1-antitrypsin).

Only a very few elastase inhibitors have so far been discovered in nature, e.g. in egg white [3], in legumes seed [4] and in mammalian plasma the α_1-proteinase inhibitor [5].

The sea anemone, Anemonia sulcata, has previously been shown to contain an interesting number of low molecular weight trypsin inhibitors of the Kunitz type family [6] and also various peptide toxins [7]. Our investigations on Anemonia sulcata extracts revealed the presence of low amounts of an elastase inhibitor.

The inhibitor was isolated from the animals after homogenisation by extraction with water:ethanol (1:1) and was processed further as outlined in the scheme.

Chemistry of Peptides and Proteins, Vol. 2

Isolation of a Elastase-Inhibitor from the Sea Anemone Anemonia sulcata

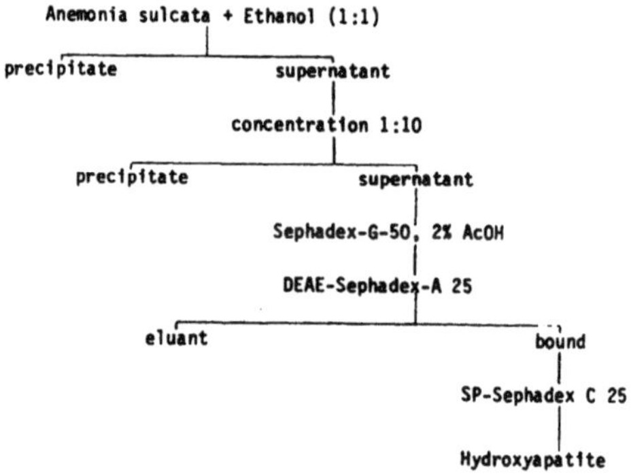

Purification was achieved by the following steps: (a) removal of precipitated contaminations after concentration 1:10 by rotary evaporation followed by centrifugation; (b) gel filtration on Sephadex G-50 in 2 % acetic acid; (c) ion equilibrium chromatography on DEAE-Sephadex A-25 in 0.1 molar Tris/HCl pH 8.0 using NaCl-gradient elution, Fig. 1; (d) ion equilibrium chromatography on SP-Sephadex C-25 in 0.05 molar sodium acetate pH 4.0 with NaCl-gradient elution, Fig. 2; (e) and two final steps of hydroxyapatite chromatography using 5 mmolar phosphate buffer pH 7.2 for elution. Fig. 3.

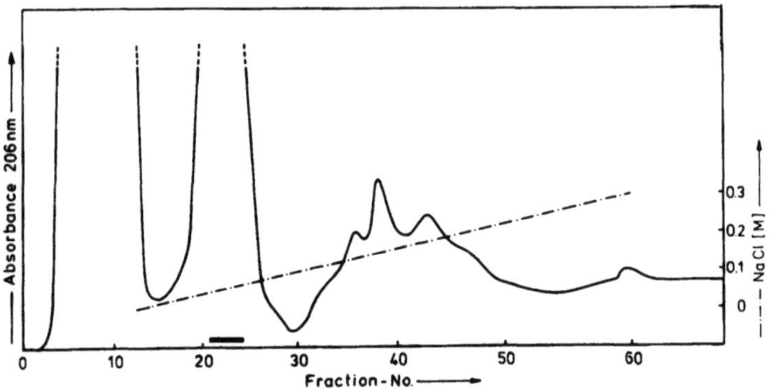

Fig.1. Elution profile of the elastase-inhibitor on a hydroxyapatite column Sephadex C-25 column (35x60 cm),equilibrated with 0.1 M Tris/HCl buffer pH 8.0 and eluted with the same buffer on a 0-0.35 molar NaCl gradient. Fractions were collected in 25 ml aliquots.

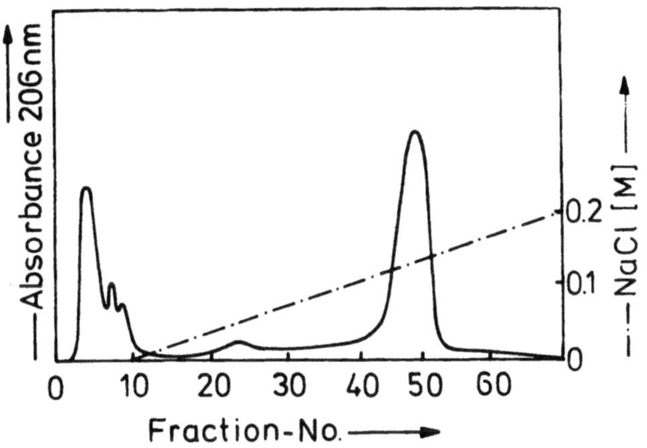

Fig.2. Elution profile of the elastase-inhibitor on a SP-Sephadex C-25 column (2x30 cm), equilibrated with 0.05 molar sodium acetate buffer pH 4.0 and eluted with the same buffer on a 0-02 molar NaCl gradient. Fractions were collected in 10 ml aliquots.

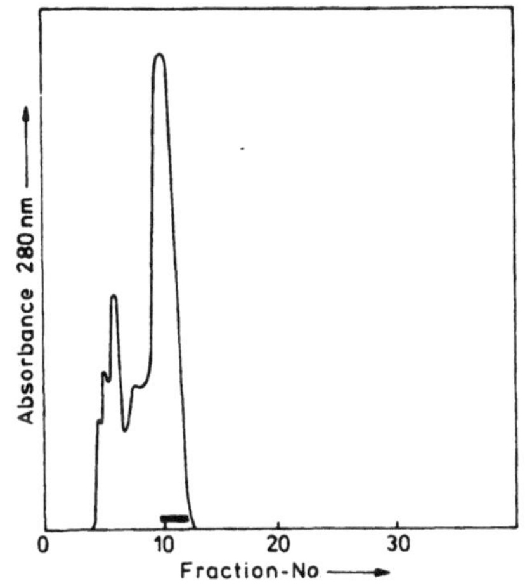

Fig.3. Elution profile of the elastase-inhibitor on a hydroxyapatite column (0.7x60 cm),equilibrated with 5 mmolar potassium phosphate buffer pH 7.2.Fractions were collected in 2 ml aliquots.

An inhibitor preparation obtained showed apparent homogeneity in the isoelectric focussing on a Servalyt-Precote ampholyte. The isoelectric point was determined to be 6.2.

The amino acid composition is given in Table 1. The inhibitor contains six half-cystine residues forming three disulfide bridges. It contains no carbohydrates.

Amino Acid	16 h	32 h	70 h	
	moles amino acid pro mole protein			
Asx	4,874	4,871	4,755	5
Thr	2,965	2,753	2,610	3
Ser	2,911	2,928	2,724	3
Glx	2,213	2,146	2,114	2
Pro	4,099	4,022	3,297	4
Gly	4,871	4,778	4,532	5
Ala	2,062	2,141	2,201	2
Cys	5,347	5,288	4,696	6
Val	1,900	1,990	1,986	2
Met	1,781	1,959	1,846	2
Ile	3,045	2,969	2,907	3
Leu	4,097	4,103	3,951	4
Tyr	2,055	1,889	1,702	2
Lys	1,996	1,985	1,942	2
His	0,947	0,974	1,020	1
Arg	2,157	2,051	2,052	2
				48

Table 1. Amino Acid Analysis of the Elastase-Inhibitor from Anemonia sulcata.

The inhibition of the pancreatic elastase with the Anemonia sulcata inhibitor is rather strong, the inhibition of human leukocyte elastase is somewhat weaker ($K_{ass} \sim 10^7$ l/M), Fig. 4 and 5. The reaction of the inhibitor with the elastase seems to follow a 1:1 stoichiometry as was estimated on the weight basis of 80 % pure enzyme and a MW of 5117 for the inhibitor.

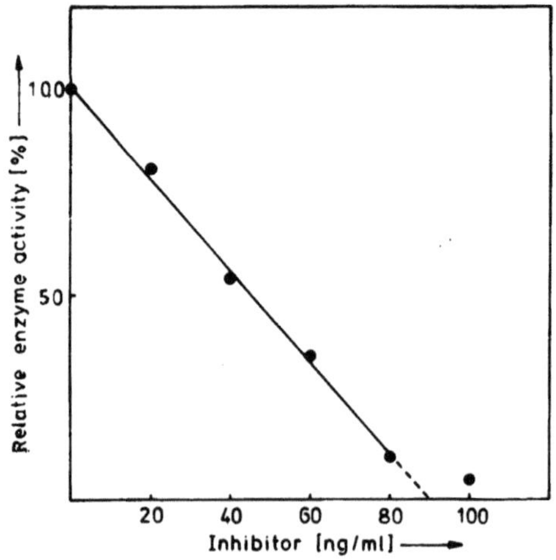

Fig. 4. Titration of porcine pancreatic elastase with the elastase inhibitor. The decrease in elastase-catalyzed hydrolysis of Suc [Ala]3 Nan was assayed at 25 °C in 0.2 molar triethanolamine buffer, pH 7.8, measured at 405 nm.

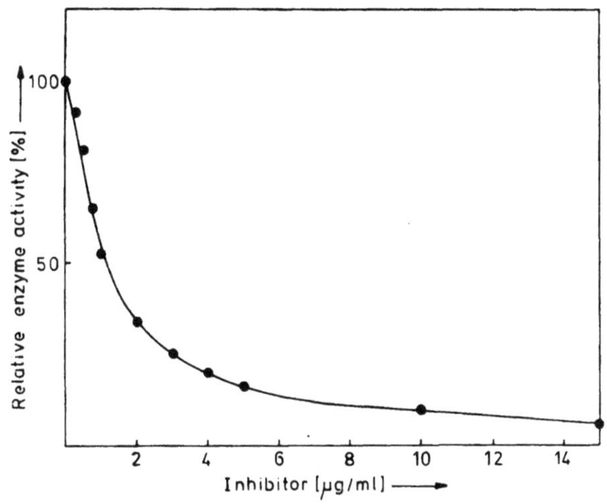

Fig. 5. Titration of human leucocytic elastase with the elastase inhibitor. The decrease in elastase-catalyzed hydrolysis of Suc $[Ala]_2$ ValNan was assayed at 25 °C in 0.2 molar triethanolamine buffer pH 8.0 containing 1 molar NaCl, at 405 nm.

The amino acid sequence was determined by sequencing three peptides obtained after cleavage of the entire polypeptide chain by cyanogen bromide cleavage of the reduced and carboxymethylated inhibitor. The overlap of the peptides was determined by a single sequence determination of the first thirty amino terminal residues of performic acid oxidized inhibitor.

The sequence of the residues in the elastase inhibitor from anemonia sulcata is given in Table 2.

```
Lys-Pro- Asp -Cys-Pro-Leu-Ile-Cys-Thr-Met-Gln- Tyr -Asp-Pro-Val-
                    5                  10                  15
Cys-Gly-(Ser)-Asp-Gly-Ile-Thr-Tyr-Gly-Asn-Ala-(Cys)-Met-Leu-Leu-
                   20                  25                  30
Gly-Ala- Ser -Cys-Arg-Ser-Asp-Thr-Pro-Ile-Glu- Leu -Val-His-Lys-
                   35                  40                  45
Gly-Arg-Cys.
```

Table 2. Amino acid sequence of the elastase-inhibitor from anemonia sulcata. Amino acids in brackets need further confirmation.

From the sequence data obtained it became obvious on comparison with the known families of proteinase inhibitors that the elastase inhibitor from anemonia sulcata is not a member of the family of Kunitz-inhibitors [8] as are the trypsin inhibitors already sequenced. Surprisingly, the elastase inhibitor belongs to the family of Kazal inhibitors [8,9]. In this family important antagonists are found, such as the pancreatic secretory trypsin inhibitors of mammalian species, Fig. 7., or the seminal plasma inhibitors [9]. A detailed comparison of the sequences will be given in a separate paper.

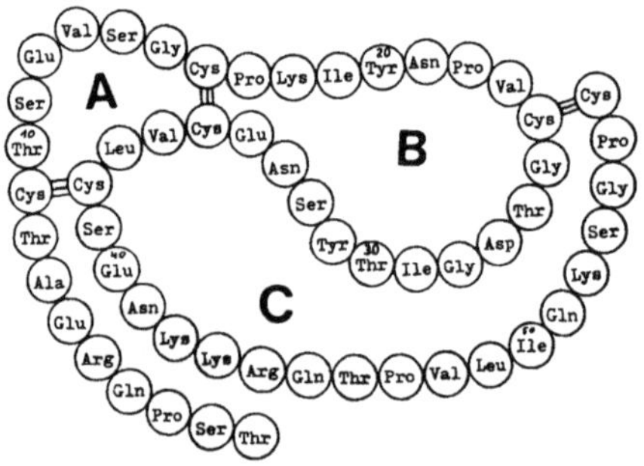

Fig. 6. The complete covalent structure of the porcine pancreatic secretory trypsin inhibitor I.

Thus, it became obvious that the species of the Anemonia sulcata contains both families of inhibitors, Kunitz and Kazal type. This is even more striking since the Anemonia sulcata is a member of the coelenterata a rather old and primitive organism in nature.

On the basis of the sequence homology one could predict the same disulfide-bridged covalent structure and the same folding pattern determined for the pancreatic and seminal Kazal inhibitors.

It now becomes obvious that the A-loop is subject to ring size variation, while the B- and C-loops with 20 and 33 amino acid residues, respectively, are strongly preserved. Thus, the Kazal type family of proteinase inhibitors also comes from a very old gene as has been demonstrated for the Kunitz family of proteinase inhibitors. It is surprising that the half-cystine residues IV and V are 6 residues apart, while in all other members of the family these residues are separated by only two amino acid residues [9].

The physiological function of this inhibitor is not yet known. Obviously it possesses the same reactive site methionine residue for the interaction with leukocyte elastase as the human plasma α_1-proteinase inhibitor, for which it has been established that the inhibitory activity against human leukocyte elastase is abolished upon oxidation to methionine sulfone by various oxidizing agents including cigarette smoke.

References

1. Starkey, P.M. (1977): Elastase and Cathepsin G, the Serine Proteinases of Human Neutrophil Leucocytes and Spleen, in: Proteinases in Mammalian Cells and Tissues (Barrett, A.J., ed.), pp. 57 - 89, North-Holland Publ. Comp., Amsterdam - New York - Oxford.
2. Woolley, D.E., Tetlow, L.C., and Evanson, J.M. (1980): Collagenase Immunolocalization Studies of Rheumatoid and Malignant Tissues in: Collagenase in Normal and Pathological Connective Tissues (Woolley, D.E. and Evanson, J.M., eds) pp. 105 - 125, John Wiley & Sons, Cichester - New York - Brisbane - Toronto.
3. Gertler, A. and Feinstein, G. (1971): Inhibition of Porcine Elastase by Turkey Ovomucoid and Chicken Ovoinhibitor, Eur. J. Biochem. 20, 547 - 552.
4. Wilson, K.A. and Laskowski, M. (1975): Partial Amino Acid Sequence of Trypsin Inhibitor II from Garden Bean, Phaseolus Vulgaris, with Location of Trypsin and Elastase Reactive Sites. J. Biol. Chem. 250, 4261 - 4267.
5. Heimburger, N. and Haupt, H. (1966): Zur Spezifizität der Antiproteinasen des Humanplasma für Elastase. Klin. Wschr. 44, 1196 - 1198.
6. Wunderer, G., Bêress, L., Machleidt, W. and Fritz, H.: Broad Specifity Inhibitors from Sea Anemones. Methods in Enzymology 45, 881 - 888.
7. Bêress, L., Bêrres, R., and Wunderer, G.: Isolation and Characterisation of Three Polypeptides with Neurotoxic Activity from Anemonia sulcata. FEBS-Letters 50, 311 - 314.
8. Tschesche, H. (1974): Biochemistry of Natural Proteinase Inhibitors. Angew. Chem. Int. Ed. Engl. 13, 14 - 28.
9. Tschesche, H., Kupfer, S., Klauser, R., Fink, E., and Fritz., H. (1976): Structure, Biochemistry and Comparative Aspects of Mammalian Seminal Plasma Acrosin Inhibitors. Protides of the Biological Fluids - 23rd Colloquium (Peeters, H. ed.) Oxford and New York.

PHOSPHORYLATED MEMBRANE PROTEINS PP 105 AND PP 135 AS MARKERS OF METASTASIZING CELLS?

F. Alfred Anderer, Joachim Pfeifle

Friedrich-Miescher-Laboratorium der Max-Planck-Gesellschaft, D-7400 Tübingen, Germany

Introduction

Molecular mechanisms relevant for the development of metastasizing cells in cancer are still unknown. In search for experimental systems which, at least in part, simulate the release of metastasizing cells from primary tumors, we established several tumor cell lines able to undergo growth transition from solid tumors in vivo to in vitro adherent growth in monolayers and growth in suspension cultures, followed by reimplantation into animals developing solid tumors again. Distinct alterations of the cell surface pattern during these growth transitions might represent a potential molecular basis of cell-to-cell communication and one of the most potent mechanisms might reside in the endogenous protein phosphorylation systems on the cell surface (1).

Results

Endogenous phosphorylation on the cell surface of intact tumor cells was studied with STU 51A-232B and STU-D1756 mouse fibroblasts, K42-A2 hamster fibroblasts and HeLa cells (2). All four tumor cell lines could undergo growth transition from in vitro adherent growth to growth in suspension cultures and

Chemistry of Peptides and Proteins, Vol. 2

vice versa. During logarithmic growth in suspension two major phosphoproteins were accessible to phosphorylation with exogenous ATP or GTP. The reaction was not dependent on cyclic nucleotides cAMP, cGMP or cCMP. The apparent molecular weights of the two major phosphoproteins were 105 000 and 135 000 (105 000 and 128 000 in human cells) and ^{32}P incorporation into these two phosphoproteins comprised 60% of total phosphorylation. Evidence that pp105 and pp135 were located on the cell surface could be obtained from treatment of phosphorylated cells with exogenous trypsin. Both proteins were completely degraded within 5-10 min and full regeneration on the cell surface required 1-2 hours. Kinetic studies of endogenous phosphorylation indicated that the time course of the reaction was dependent on the nucleotide triphosphate precursor and on the cell population density. The initial rapid increase of pp135 phosphorylation at the high cell population density indicated a peculiar role of the cell number in the reaction scheme. Therefore, we studied endogenous phosphorylation with increasing cell population density. ^{32}P-incorporation into membrane pp135 showed an exponential increase with increasing cell number. This exponential increase was not altered when phosphorylation was performed in the presence of exogenous casein, i.e. casein phosphorylation did not compete with pp135 phosphorylation. These findings indicated a possible phosphorylation mechanism of pp135 dependent on the number of cell collisions since these increase exponentially with increasing cell population density. Phosphorylation of pp105 did not follow these exponential criteria.

On adherent cells grown in monolayers pp105 and pp135 were not accessible to endogenous phophorylation. To exclude the possibility that lack of phosphorylation was due to lack of cell collisions, adherent cells were phosphorylated by shaking them with the same number of ^{3}H-fucose prelabeled suspension cells. No specific phosphorylation of pp105 and pp135 on adherent cells was observed since minor amounts of ^{32}P incorporation were strictly associated with ^{3}H-fucose labeling thus indicating that ^{32}P incorporation was due to contamination with suspension cells. After removal of adherent cells from the substrate with EGTA, endogenous phosphorylation of the retained adhesive plaques yielded predominant phosphorylation of pp105 and pp135 (3). A similar phosphorylation pattern was obtained with the substrate-attached material of suspension cells derived from cell collisions with the wall of the culture vessel. Passive adsorption of secreted components could be ruled out. These findings indicate a potential role of pp 105 and pp135 in cell adhesion.

The distribution of pp105 and pp135 in logarithmically growing tumor cells was determined after subcellular fractionation by endogenous phosphorylation. In the cytosol of suspension cells both phosphoproteins were present in higher amounts whereas in the nuclei only traces could be found. Both phosphoproteins were also present in the cytosolic fraction of adherent cells. The identity of cytosolic and membrane pp 105 and pp135 could be demonstrated by the identity of the phosphopeptide patterns after V8-protease treatment. The patterns also showed that pp 105 and pp135 were not distinct states of processing of one and the same phosphoprotein. It is noteworthy, that solid tumors developing from tumor cells reimplanted into mice did not express any pp 105 and pp 135 but re-expression of these phosphoproteins was again observed after growing the cells of the solid tumor in monolayer and suspension cultures.

The distribution of pp 105 and pp 135 in normal mouse tissue was investigated with tissue homogenates phosphorylated endogenously or with an exogenous kinase. pp 105 was predominantly present in proliferating tissues, such as the foetus and the juvenile thymus, whereas the distribution criteria for pp 135 were less distinct.

Isolation of pp 105 and pp 135 was started with the cytosolic

Table 1. Purification of cytosolic pp 105 and pp 135 from STU 51A-232B mouse fibroblasts grown in suspension (data refer to cpm ^{32}P/mg protein).

step	purification factor pp 105 / pp 135	total recovery % pp 105 / pp 135
cytosol	1 / 1	100 / 100
40-70% ammonium sulfate	2.5 / 3.7	74 / 92
DEAF-sepharose	143 / 35	49 / 36
hydroxylapatite + phenylsepharose	384 / 1535*	35 / 23

* purity (Coomassie Blue-staining): pp 105 >95%; pp 135 >60%

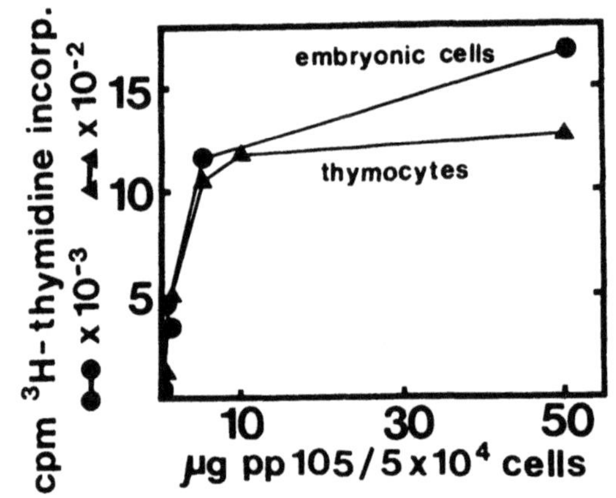

Figure 1. Incorporation of ^{3}H-thymidine into 5 x 10^{4} resting embryonic cells or thymocytes derived from STU mice, in the presence of increasing amounts of pp 105 isolated from STU 51A-232B tumor cells.

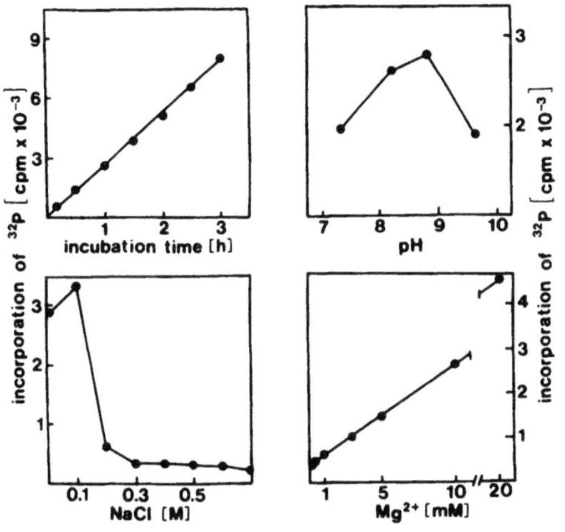

Figure 2. Autophosphorylation of pp 105 in the presence of γ[32P]ATP: a) time course, b) dependence on pH, c) dependence on NaCl concentration, d) dependence on Mg^{++} concentration.

fraction of tumor cells grown in suspension. pp 105 could be purified to >95% and pp 135 to >60% (Table 1). In a screening for potential functions of the two phosphoproteins we found that pp 105 could stimulate mitosis in resting embryonic cells or juvenile thymocytes, an activity not observed with the pp 135 sample. In Fig.1 the dependence of ^{3}H-thymidine incorporation into the stimulated cells is given for increasing concentrations of pp 105. Furthermore, it could be demonstrated that pp 105 is capable of autophosphorylation in the presence of ATP or GTP. In Fig.2 the time course of autophosphorylation was linear for at least 3 hours with an optimum between pH 8.5 and 8.8 and at physiological salt conditions. Autophosphorylation proceeded even in the presence of EDTA but Mg^{++} stimulated the reaction with increasing concentrations. In further screening studies we did not succeed in specifying any function for pp 135.

Highly purified pp 105 was used to prepare a monospecific antiserum which allowed the localization of pp 105 on the cell surface as well as in the cell by means of indirect immunofluorescence. Only suspension cells showed specific binding of anti pp 105 antibodies on the cell surface. In the cytoplasma of suspension cells pp 105 was accumulated in small granular structures. In adherently grown tumor cells pp 105 can be localized in cytoplasmatic structures near the nucleus whereas in adherently grown embryonic cells pp 105 appeared to be strictly associated with structures of the microfilament bundles.

The additional location of pp 105 in proliferating normal tissue and its mitogenic activity possibly indicates a key role in regulation of proliferation. Neoplastic transformation of a cell may partially influence this regulation which is reflected by the transitions of pp 105-associated highly ordered cytosolic structures in embryonic cells to less well-organized structures in adherent tumor cells. These structures were apparently completely disorganized in suspsension cells associated with the expression of pp 105 on the cell surface. In metastasizing tumor cells the potential function of pp 105 may be based on its mitogenic activity possibly aiding in implantation of these tumor cells in tissues other than the original one.

References

1. Greengard, P.: Science 199, 146-152 (1978).
2. Pfeifle, J., Hagmann, W., Anderer, F.A.: Biochim.Biophys. Acta 670, 274-284 (1981).
3. Hagmann, W., Anderer, F.A.: Biochem.Biophys.Res.Commun. 104, 1585-1591 (1982).

THE ACROSOMAL MEMBRANE SYSTEM AND ITS ROLE IN MAMMALIAN FERTILIZATION

Edda Töpfer-Petersen, Adriana Hinrichsen-Kohane, Christian Schmoeckel, Wolf-Bernhard Schill
Department of Dermatology, University of Munich, Munich,Germany

Introduction

For mammalian spermatozoa the fertilization process is associated with a variety of morphological and biochemical changes occurring after ejaculation during their residence in the female reproductive tract(1). This process, termed capacitation, is a poorly understood phenomenon which consists in a series of endogenous physiological alterations of the sperm cell, particularly of the spermatozoan membranes. These alterations lead to the acrosome reaction, by means of which spermatozoa gain progressively the ability to penetrate the outer investments of the ovum, cumulus oophorus and corona radiata and attach and bind to the zona pellucida, the site of species-specific gamete recognition (2). After penetration of this acellular layer, the spermatozoon finally fuses with the oocyte vitellus . Removal of epididymal and seminal plasma proteins coating the sperm surface, reduction of the negative net charge by loss of sialic acid as well as a well-defined sequence of biochemical alterations of the surface proteins (3) seem to be prerequisites for the subsequent rearrangement of intrinsic membrane proteins leading to protein-poor areas of high fluidity. These areas are presumably the sites where membrane fusion between the plasma membrane and the underlying outer acrosomal membrane occurs(4). This vesiculation process marks the start of the acrosome reaction allowing the release of acrosomal enzymes, such as hyaluronidase, which presumably facilitates the passage through the cumulus mass. The complete loss of the acrosomal cap occurs when the spermatozoon reaches the surface or has already penetrated the zona pellucida. At this stage another enzyme, acrosin, starts to play a central role. This trypsin-like enzyme is partially found associ-

Chemistry of Peptides and Proteins, Vol. 2

ated with the inner acrosomal membrane. The main physiological role attributed to acrosin is that of favouring the penetration through the zona pellucida which allows the spermatozoon to reach and to fuse with the oocyte vitelline membrane (5).
The characterization of structure and composition of spermatozoan membranes is essential for a better understanding of the complex mechanism of fertilization. The aim of these investigations was therefore to gain more insight into the architecture and biochemistry of the acrosomal membrane sytem of mammalian spermatozoa.

Morphological studies

The distribution of the externally disposed carbohydrate side-chains of the sperm surface has been extensively studied. For this purpose, plant lectins have been proven to be useful in studying such membrane associated problems concerning localization and function of carbohydrates and their glycoconjugates (6). The most widely used lectin is concanavalin A (Con A) which binds preferentially to mannose-rich carbohydrate moieties, whereas Ricinus communis agglutinin-120 (RCA-120) is a highly specific lectin interacting only with the terminal galactose of the exposed membrane glycan chains.
Using boar spermatozoa as an experimental model, the carbohydrate structure of the sperm surface was analyzed by fluorescence microscopical methods with FITC-conjugated Con A and FITC-RCA-120. The Con A binding sites as well as the RCA-120 binding sites were distributed over the whole sperm surface with predominance of the acrosomal cap (7).
Fluorescence microscopy provides mainly informations concerning the cell surface. Therefore the ultrastructural distribution of membrane saccharides of the inner cellular membranes was investigated by transmission electron microscopy using the lectin-peroxidase technique (8).
Incubation with Con A results in a remarkable staining of the enveloping plasma membrane (Fig.1a). However, the plasma membrane overlying the acrosome is very sensitive to cell manipulation . Hence, a certain percentage of the spermatozoa lose the plasma membrane showing the typical rolling formation (Fig.1b). The outer acrosomal membrane is now exposed and stains with Con A

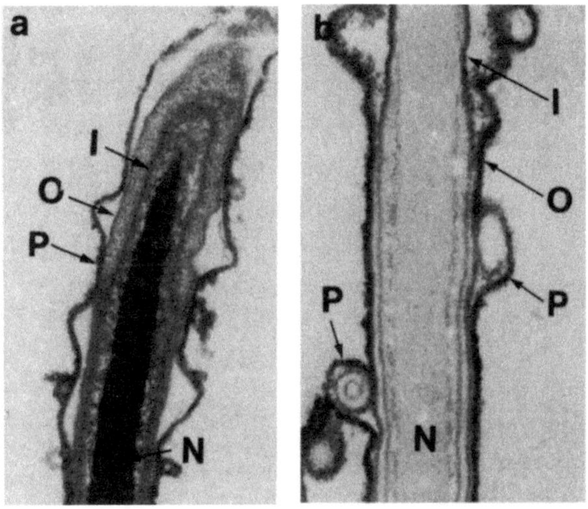

Fig.1 . Ultrastructural localization of Con A receptors after incubation with Con A.
a. A positive staining of the intact plasma membrane (P). Uranyl and lead citrate staining. x 48 000.
b. The plasma membrane (P) showing the typical rolling formation in the region of the equatorial segment, both acrosomal membranes, outer acrosomal membrane (O) and inner acrosomal membrane (I) are seen to be stained with Con A. Nucleus (N); Lead citrate staining; x 55 000.

demonstrating the presence of mannose containing structures. The heterogenity observed in the acrosomal matrix, indicates damages of the acrosome. The lectin penetrating the acrosome is now distributed along the inner acrosomal membrane (8). Similar results could be obtained with RCA-120, both, indicating the complexity of the carbohydrate structures of the acrosomal membrane components.

The carbohydrate content of many species is well-documented (6) either by lectin-based techniques or radio-labelling procedures. However, the carbohydrate content of the acrosomal membranes is poorly described. But there seems to exist species-specific differences in the carbohydrate content and lectin binding ability of the acrosomal membranes, when comparing the data available from diverse species.

The possibility of isolating the outer acrosomal membrane (OAM) of boar spermatozoa, in contrast to the inner acrosomal membrane, tempted us to undertake the characterization of this membrane. Homogenization and centrifugation through modified colloidal silica (9) rendered a largely homogenous membrane fraction with low contamination of plasma membrane vesicles and cytoplasmatic droplets as revealed by electron microscopy.

For evaluation of purity and further characterization of the isolated membrane fraction, an antiserum was raised in rabbits. The anti-OAM-IgG fraction was isolated by ammonium sulfate precipitation and ion-exchange chromatography. Using the sandwich technique with a second antibody (anti-IgG) conjugated to ferritin,a distinct antibody-precipitation ,visualized by the electron dense ferritin, could be observed on the outer acrosomal membrane by electron microscpoy (Fig.2a).The slight staining of the plasma membrane may be due to plasma membrane contaminations within the isolated outer acrosomal membrane fraction (Fig.2b). It cannot be excluded, that due to complexity of membrane structures some antigenic components are constituents of both spermatozoan membranes, the plasma membrane and the outer acrosomal membrane.

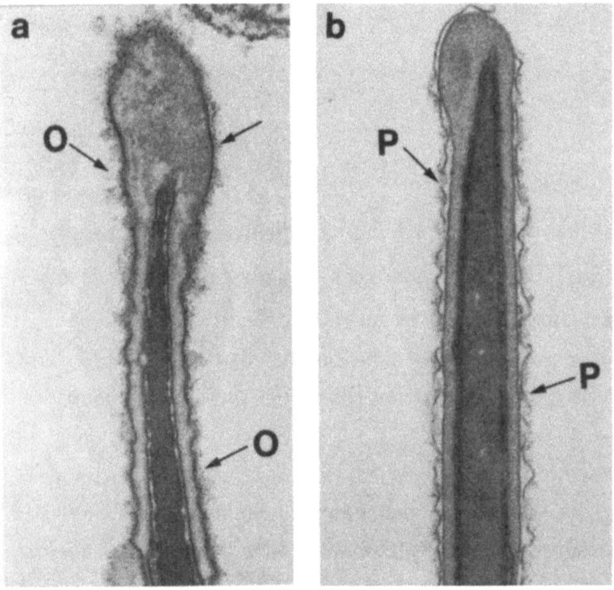

Fig.2 .Boar spermatozoa treated with anti-OAM-IgG and GAR-ferritin.
a. The plasma membrane (P) is disrupted, a distinct staining of the underlying outer acrosomal membrane (O) could be observed. Uranyl and lead citrate staining; x33 000.
b. Slight staining of the intact plasma membrane (P). Uranyl and lead citrate staining ; x 34 000; (unpublished data).

Biochemical studies

Glycoproteins, constituents of the acrosomal membrane system, seem to play an essential role during the process of capacitation and acrosome reaction as was demonstrated recently by lectin binding studies (3). In order to characterize the protein components of the outer acrosomal membrane which are responsible for the lectin-membrane interaction , affinity chromatography on immobilized lectins and electrophoretic methods were employed. The isolated membrane fraction contained a mixture of at least ten protein components (10) which could be discriminated by SDS-polyacrylamide gel electrophoresis (SDS-PAGE) ranging in molecular weight between 450 and 15 kdal. To determine which of these proteins are responsible for the binding of RCA-120 to the membrane, the terminal galactose of the externally disposed glycan chains of the isolated membrane fraction was radio-labeled by galactose oxidase mediated ^{3}H-sodium borhydride reduction. Subsequent to the detergent extraction with 1% sodium cholate, the receptor proteins were isolated by means of affinity chromatography on agarcse conjugated RCA-120. Most of the membrane-bound activity could be released from the column by washing with 0.2 M galactose. Analysing the RCA-120 receptor fraction by high resolution gel filtration on a TSK-3000 column it was found that the activity was exclusively eluted near the void volume indicating the high molecular weight of the RCA-120 receptor/s.SDS-PAGE of the ^{3}H-labeled membrane revealed only one galactoprotein with the molecular weight of about 450 kdal,which predominantly contributes to the RCA-120 properties of the outer acrosomal membrane documented by electron microscopical studies.

Screening of the Con A (or other lectins) binding properties of the solubilized membrane components partially isolated by chromatographic methods was performed by means of an enzyme-linked-lectin-assay (10). This sensitive method allows a suitable control of high pressure liquid chromatography eluates for their different lectin binding abilities.

In the case of the solubilized membrane fraction chromatographed on the TSK 3000 column by HPLC, the enzyme-linked-lectin-assay revealed that more than one glycoprotein shows binding capacity to Con A. Con A binding could be detected together with the ^{3}H-activity near the void volume of the column and predominantly in the range of 100-40 kdal (Fig.3).

Electrophoretic analysis including lectin staining procedures demonstrated that the high molecular weight peak represents a mixture of glycoprotein aggregates, of the RCA-120 receptor protein with the molecular weight of 450 kdal and a Con A binding glycoprotein with the molecular weight of 340 kdal. Four to five further Con A receptor proteins could be identified, the main proteins corresponding to the molecular weights of 120, 110, 88 and 66 kdal. The two bands with molecular weights of 110 and 88 stained only weakly with Coomassie blue (unpublished data).

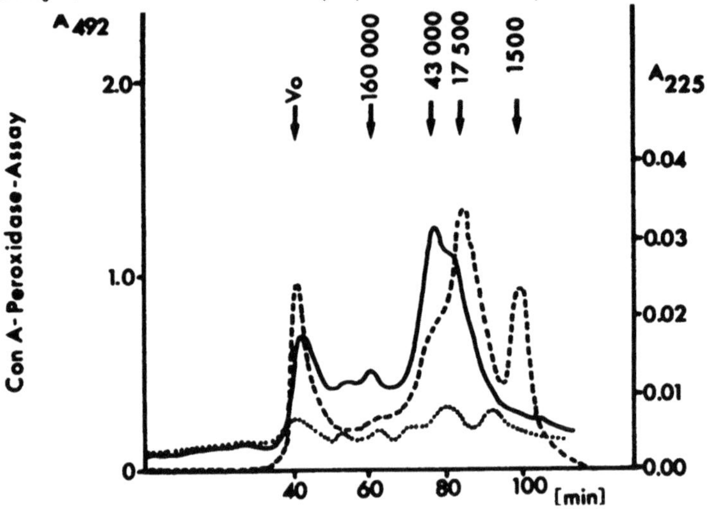

Fig.3. High resolution gel filtration on a TSK 3000 column (7.5x600 mm); 0.1 M sodium phosphate, pH 7.4, 0.1 M sodium sulfate, 1% sodium cholate, ---- absorbance at 225 nm —— enzyme-linked-lectin-assay (ELLA) with Con A-peroxidase, measured at 492 nm ···· ELLA with Con A-peroxidase in the presence of 0.2 M α-methyl mannoside. Sample: 200 μl solubilized outer acrosomal membrane fraction.

We have tentatively identified by different chromatographic and electrophoretic methods the receptor sites for RCA-120 and Con A of the outer acrosomal membrane of boar spermatozoa. Much work remains to be carried out regarding the biochemical characterization of each of these components and, above all, the elucidation of the biological role displayed by these factors during the mechanism of fertilization.

References

1. Johnson M.H.: J.Reprod. Fert. 44, 167-184 (1975).
2. Peterson R.N., Russell L., Bundman D., Frien M.: Science 207, 73-74 (1979).
3. Schwarz M.A., Koehler J.K.: Biol.of Reprod. 21, 1295-1307 (1979).
4. O`Rand M.G.: The spermatozoon. Ed. D.W. Fawcett and J.H. Bedford. Urban and Schwarzenberg, Munich and Baltimore. pp 195-204 (1979).

5. McRorie R.A., Williams W.L.: Annual Rev. of Biochem. 43, 777-803 (1974).
6. Koehler J.K.: Arch Androl. 6, 197-217 (1981).
7. Töpfer-Petersen E., Schmoeckel C., Schill W.-B.: J. Reprod. Fert.,subm.
8. Töpfer-Petersen E., Schmoeckel C., Schill W.-B.: Andrologia,in press.
9. Töpfer-Petersen E., Schill W.-B.: Andrologia, 13, 174-175 (1981).
10. Töpfer-Petersen E., Schill W.-B.: Int. J. Androl., subm.

Acknowledgement

The excellent technical assistance of Miss A. Weiß is gratefully acknowledged. The ultrastructural investigations were made possible through the highly skilled technical assistance of Mrs. E. Januschke.

This research was supported by Deutsche Forschungsgemeinschaft Schi 86/7-2.

PENICILLIN-LIKE ACTION OF A BACTERICIDAL PROTEIN (COLICIN M) OF <u>ESCHERICHIA COLI</u>

Volkmar Braun und Ralf Dreher

Microbiology II, University of Tübingen, D-7400 Tübingen, Germany

Introduction

Bacteriocins are proteins synthesized by certain bacterial strains and which kill other bacteria usually of the same group as the producing organisms. The bacteriocins of Escherichia coli are called colicins. They are of interest with regard to their specific toxic effects, their synthesis and secretion and their uptake by sensitive cells. Their modes of action are very diverse. Some are enzymes having ribonuclease or deoxyribonuclease activity, while another group of colicins form ion-permeable channels in the cytoplasmic membrane (1, 2). Colicin M is unique among the colicins in that it causes lysis of sensitive cells (3).

Uptake of colicin M by sensitive strains of E. coli. Reversible translocation across the outer membrane

E. coli cells are surrounded by an outer membrane and an inner (cytoplasmic) membrane. Both membranes are impermeable to proteins unless a specific translocation system for a protein exists. Uptake of colicin M requires a receptor protein at the cell surface, encoded by the fhuA (tonA) gene, the tonB gene product whose function is unknown and the tolM gene product (4, 5, 6). Further uptake from the initial binding site at the cell surface protein requires cellular energy (6) and an active tonB gene. Colicin M stays at the receptor protein in energy-depleted cells and in tonB mutants. Under these conditions colicin M prevents adsorption of the bacterial virus T5, that binds to the same

Chemistry of Peptides and Proteins, Vol. 2

receptor protein as colicin M. In energized tonB$^+$ cells colicin M is taken up within 4 min after addition to a cell culture. After that time it is no longer accessible to the colicin-inactivating agents trypsin, sodium dodecyl sulfate (SDS) or colicin-specific antibodies (7) which do not penetrate the outer membrane.

Colicin M does not induce a cellular reaction sequence that finally leads to cell lysis. It has to act continuously until the onset of lysis. The action of colicin M requires calcium ions. Addition of the calcium complexing compound EGTA shortly before lysis sets in (after 20 min) prevents lysis. Interestingly, upon treatment with EGTA colicin M again becomes accessible to the inactivating agents trypsin, SDS and antibodies. Replacement of EGTA by calcium again leads to lysis. The reversible translocation across the outer membrane suggested by this experiment is corroborated by the use of a temperature sensitive tolerant mutant. This mutant is insensitive to colicin M at 42°C but sensitive at 30°C (8). When cells treated with colicin M at 30°C are shifted to 42°C before lysis commences they do not lyse. Colicin M remains cell-bound and upon lowering of the incubation temperature to 30°C cells start to lyse. As in the case of treatment with EGTA colicin M is resistant to trypsin, SDS, and antibodies at 30°C but sensitive at 42°C. Ca^{2+} and the function specified by the tolM gene apparently fix the colicin at the target site inside the cell. After removal of Ca^{2+}, or inactivation of the mutated TolM function at 42°C, colicin M is released from the target site, moves toward the cell surface, stays there and is accessible to the inactivating compounds.

Mode of action of colicin M

The primary toxic effect of colicin M is directly related to cell lysis since the decrease of surviving cells and lysis occur simultaneously (3). In an osmotically stabilizing medium (16 % sucrose) spheroplasts are formed. The rod-like shape of normal E. coli cells is converted to large round cells. An example is given of cells treated for 50 min with colicin M under osmotic protection (Fig. 1). Eventually the spheroplasts also lyse. This penicillin-like action suggests murein (peptidoglycan) as target of colicin M action. Similar to the β-lactam antibiotics, inhibition of murein biosynthesis by colicin M would lead to lysis of cells due to the high osmotic pressure of the cytoplasm which is no longer counterbalanced by the defective rigid murein layer. Therefore growing cells of strain W7 dap lys were pulse-labeled with radioactive diaminopimelate, the murein-specific amino acid. Four minutes after addition of colicin M the rate of diaminopimelate incorporation was reduced and after 16 min had slowed to only 10 % of the control rate without colicin M. Lysis started 14 min after administration of colicin M (9). Lysis was accompanied by the release of murein degradation products. It is important to note that inhibition of murein synthesis preceded murein degradation. These results were corroborated by data obtained with a biosynthetically active membrane fraction which showed that the incorporation of the precursors of murein, UDP-N-acetylmuramyl pentapeptide and UDP-N-acetyl glucosamine, into insoluble material (murein) was also inhibited by colicin M (9). It is concluded that colicin M like many β-lactam antibiotics triggers autolysis by inhibition of murein biosynthesis,

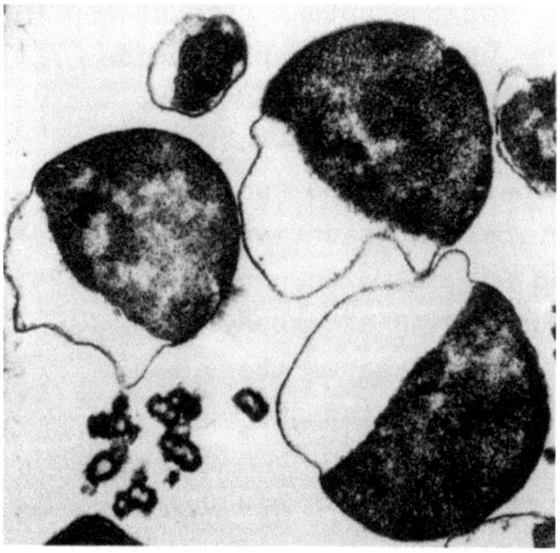

Fig. 1. Spheroplasts of E. coli induced by colicin M.

or that it shifts the balance between the activity of synthetic and lytic enzymes in favor of the latter.

Functional domains of colicin M

Colicin M consists of a single polypeptide with a molecular weight of 27.000 (7). At least two regions on the polypeptide chain appear to be important for its toxicity. One forms the recognition site which binds to the FhuA receptor protein at the cell surface and is required for the uptake through the outer membrane to the target site which presumably resides in the cytoplasmic membrane. The other region forms the active center that inhibits murein biosynthesis. We used limited proteolysis to see whether portions of the polypeptide chain could be assigned to these functions. Degradation of colicin M with carboxypeptidase B released 0.5 mol of arginine and 0.25 mol of lysine per mol protein suggesting a carboxy terminal amino acid sequence -lys-arg. Upon release of these basic charged amino acids the protein precipitated and became inactive. These two terminal amino acids apparently are very important for keeping the colicin soluble in an aqueous solvent.

Colicin M was then degraded with various endoproteasesin order to obtain products which were still active in one or both the functions mentioned above. Proteinase K yielded a limited proteolysis. Fig. 2 illustrates such a degradation.

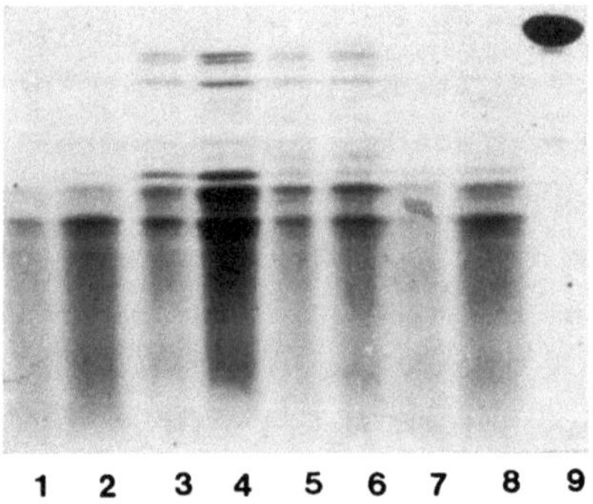

Fig. 2. Degradation of colicin M with proteinase K. Lanes 1 and 2, 5 min incubation at 37°C with an enzyme: colicin ratio of 1:120. Lanes 3-8, enzyme: colicin 1:600, lanes 3,4 for 10 min, lanes 5,6 for 20 min, lanes 7,8 for 30 min. Lane 9, undegraded colicin M, concentration as used in the experiments 1-8.

The samples 1-8 shown in Fig. 2 still inhibited growth of E. coli cells at a dilution of $1:10^4$. The untreated colicin M was active at a dilution up to 2×10^5. The comparison of lanes 1-8 with lane 9 indicated that no colicin was left after proteolysis. Therefore, degradation products were apparently active toward whole cells where binding to the receptor, uptake through the outer membrane and action at the target had still to take place. The result that one or several of the degradation products still killed whole cells was surprising. To confirm that no colicin M had remained in the sample which could account for the toxicity, the products were separated by high pressure liquid chromatography. The column clearly separated chymotrypsin A (mol. wt. 25 000), cytochrome c (mol. wt. 12 800), and insulin A und B chains (mol. wt. 2 500 and 3 500 respectively) (not shown).

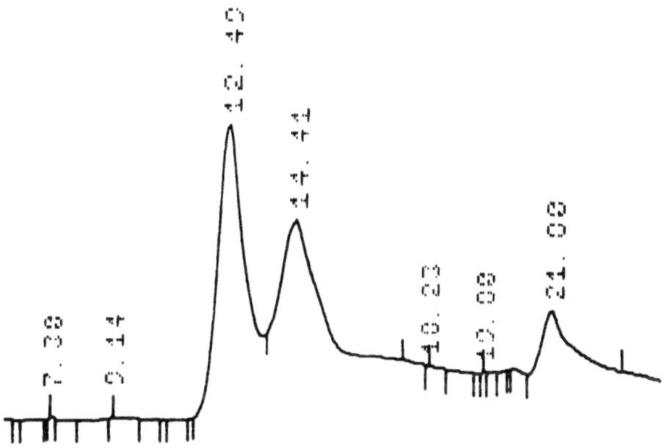

Fig. 3. Separation of colicin M from its degradation products obtained with proteinase K at an enzyme:colicin ratio of 1:200 (15 min incubation at 4°C). HPLC, TSK gel G3000 SW; elution with 50 mM sodium phosphate, pH7, 0.05 % SDS, 1 ml/min; pressure 25 bar.

Colicin M was eluted after 12.40 min. The main fraction of the degradation products was eluted after 14.41 min which corresponded to a molecular weight of approximately 13 000. The colicin M peak fraction was still active at a dilution of 10^3, the peak fraction of the degradation products at a dilution of 10^4. This result proved that the total activity of colicin M toward whole cells, uptake and lysis, resides in a fragment of the molecule. Half of the molecule is either dispensable for the activity or two fragments cooperate without being covalently linked. We are currently isolating the degradation products on a preparative scale and sequencing the products to the point where they can be aligned along the polypeptide chain to assign the functional domains.

Acknowledgment

This work was supported by the Deutsche Forschungsgemeinschaft (SFB 76) and the Fonds der Chemischen Industrie. We thank B. Johnson for reading the manuscript.

References

1. Braun, V., Hantke, K.: Organization of Prokaryotic Cell Membranes, (B.K. Ghosh, Ed.) Vol. II p. 1-73 CRC Press, Boca Raton. 1981.

2. Konisky, J.: The Bacteria (L.N. Ornston, J.R. Sokatch, Ed.) Vol. 6, p. 71-136, Academic Press, New York, 1978.

3. Braun, V., Schaller, K., Wabl, M.R.: Antimicrob. Agents Chemother. 5, 520-523 (1974).

4. Braun, V., Schaller, K., Wolff, H.: Biochim. Biophys. Acta 328, 87-97 (1973).

5. Kadner, R.J., Heller, K., Coulton, J.W., Braun, V.: J. Bacteriol. 143:256-264 (1980).

6. Braun, V., Frenz, J., Hantke, K., Schaller, K.: J. Bacteriol. 142, 162-168 (1980).

7. Schaller, K., Dreher, R., Braun, V.: Bacteriol. 146, 54-63 (1981).

8. Schaller, K., Krauel, A., Braun, V.: J. Bacteriol. 147 135-139 (1981).

9. Schaller, K., Höltje, J.-V., Braun, V.: J. Bacteriol., submitted.

STRUCTURE, FUNCTION AND BIOSYNTHESIS OF THE SPERM PROTEINASE ACROSIN

Werner Müller-Esterl, Hans Fritz

Abteilung für Klinische Chemie und Klinische Biochemie in der Chirurgischen Klinik Innenstadt der Universität München
D-8000 München, Germany

Roselinde Fock-Nüzel, Friedrich Lottspeich and Agnes Henschen

Max-Planck-Institut für Biochemie
D-8033 Martinsried, Germany

Introduction

Mammalian fertilization comprises a complex series of morphological and biochemical events which are eventually terminated by the fusion of the haploid male and female gamete to form the diploid zygote. At the present, the molecular basis of sperm-egg fusion is only poorly understood. Acrosin, a major proteinase of spermatozoa, has been implicated in the mediation of sperm binding to the zona pellucida of the ovum (1) and the sperm penetration through the zona pellucida (2). Two lines of evidence suggest a crucial role of acrosin in the lysis of the zona pellucida glycoprotein layer: first, isolated zona pellucida is effectively degraded by acrosin <u>in vitro</u> (3), and second, sperm penetration through the zona is completely prevented <u>in vivo</u> in the presence of synthetic acrosin inhibitors (4).

We have isolated acrosin from boar spermatozoa and characterized biochemical, immunological, and biological features of the sperm proteinase.

Chemistry of Peptides and Proteins, Vol. 2

Results and Discussion

Acrosin (EC 3.4.21.10) was extracted from boar spermatozoa at pH 2.6 and purified to apparent homogeneity by a five-step isolation procedure resulting in a 160-fold purification and a total yield of 41 % (5).

Antibodies against purified boar acrosin were raised in rabbits. Titers were checked by the double antibody ELISA technique with acrosin immobilized on ELISA plates and anti-rabbit IgG conjugated to alkaline phosphates (from goat) as the second antibody. Acrosin-directed antibodies were immunoselected by affinity chromatography on acrosin-Sepharose. Double immunodiffusion test according to Ouchterlony indicated a single precipitin line in protein staining with Coomassie Brilliant Blue and in active enzyme staining using N^{α}-benzoyl-arginine-ß-naphthylamide as the substrate and Fast Garnet Blue as the coupling dye (6). (Fig. 1).

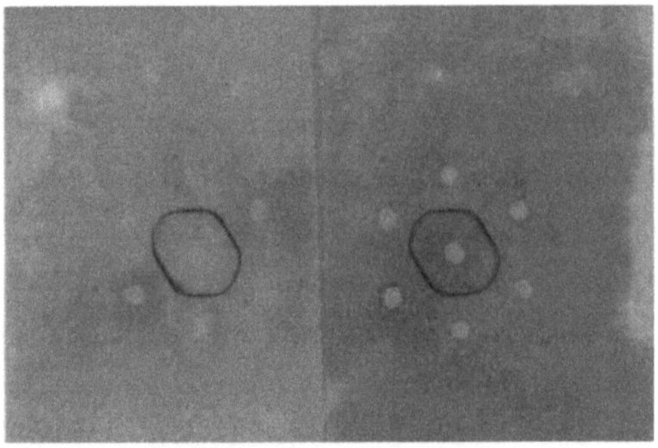

Fig. 1. Double immunodiffusion test according to Ouchterlony. Center well: boar acrosin; peripheral wells: anti-boar acrosin-IgG. Left: protein staining. Right: active enzyme staining.

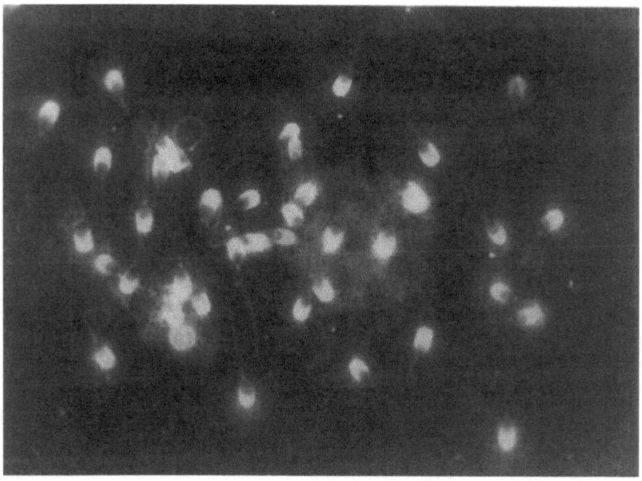

Fig. 2. Immunofluorescent staining of acrosin in boar spermatozoa.

Immunofluorescent localization of acrosin using specific antibodies against acrosin and FITC-coupled anti-rabbit-IgG antisera (from goat) as the second antibody revealed that acrosin resides in the acrosome, a lysosome-like structure on the apical part of the sperm head (7) (Fig. 2).

Purified boar acrosin has been extensively characterized (6) (cf. Table 1). The molecular weight of the proteinase is 40 000 as judged by SDS electrophoresis under non-reducing conditions. A pI of 10.5 was determined by analytical isoelectric focusing. A carbohydrate content of 4 % (w/w) has been calculated. Amino acid analysis revealed that the protein backbone comprises at least 328 amino acid residues corresponding to a calculated molecular weight of 35 700 for the protein moiety (Table 2). The high molar percentage of proline (10.9 %) and arginine (7.6 %) in boar acrosin is striking. On the basis of our amino acid composition, we calculated a polarity index of 41.2 % according to Calpaldi. A molar extinction coefficient ($E^{1\%}_{280\ nm}$) of 16.8 was determined.

Properties		Comments
Molecular weight	40 000	by SDS electrophoresis
	42 000	by gel filtration
pI	10.5	
carbohydrate content	4 % (w/w)	5 mol glucosamine/mol protein
		2 mol mannose/mol protein
		1 mol fucose/mol protein
		1 mol galactose/mol protein
N-terminal amino acid	Val/Arg	heavy chain/light chain
Specific activity	37 U/mg	using N^{α}-benzoyl-L-arginine-4-nitro-anilide as the substrate (6)
$E^{1\%}_{280\ nm}$	16.8	

Tab. 1. Properties of boar acrosin.

Amino acid	Boar acrosin	Human acrosin	Ram acrosin
Asx	23	31	24
Thr	20	19	19
Ser	18	25-26	25
Glx	27	30-32	29
Pro	40	31-34	36
Gly	34	38	31
Ala	19	27	25
Cys	8	8	12
Val	24	22	26
Met	6	4	4
Ile	17	21	18
Leu	21	22	21
Tyr	10	10	10
Phe	11	10-11	10
His	6	6-7	9
Lys	16	18	16
Arg	25	19	25
Trp	$\geq$3	—	3
Total	328	341-349	343

Tab. 2. Amino acid compositions of boar, human, and ram acrosin (8).

Acrosin belongs to the class of trypsin-like enzymes in that it preferentially splits Arg-X and Lys-X bonds (9). Common chromogenic substrates for acrosin activity assays are N^{α}-benzoyl-L-arginine-4-nitroanilide, N^{α}-benzoyl-L-arginine-ethyl-ester, and N^{α}-benzoyl-D,L-arginine-ß-naphthylamide. A variety of natural and synthetic trypsin inhibitors are potent inactivators of acrosin and vice versa; rare exemptions from the general rule are the human seminal plasma inhibitors HUSI-I, which exclusively inhibits trypsin (10), and the boar sperm acrosin inhibitor, which preferentially inhibits acrosin (11). The most potent synthetic acrosin inhibitors known so far were recently synthetized by Borin et al. (12). K_i ranging from 4.7 x 10^{-7} M to 9.7 x 10^{-9} M were found for tripeptide aldehydes derived from microbial leupeptins.

Sequence analysis of boar acrosin (13) indicates that the proteinase comprises two chains which are most probably disulfide-bridged (Fig. 3). The major chain contains the active site of the proteinase and corresponds to a typical serine proteinase in its activated state. A minor chain of molecular weight close to 3 500 represents a remainder of the propart of the enzyme and confers the carbohydrate moiety of the enzyme (Fig. 4). The homology to the propart of thrombin (A-chain) and several other serine proteinases is obvious.

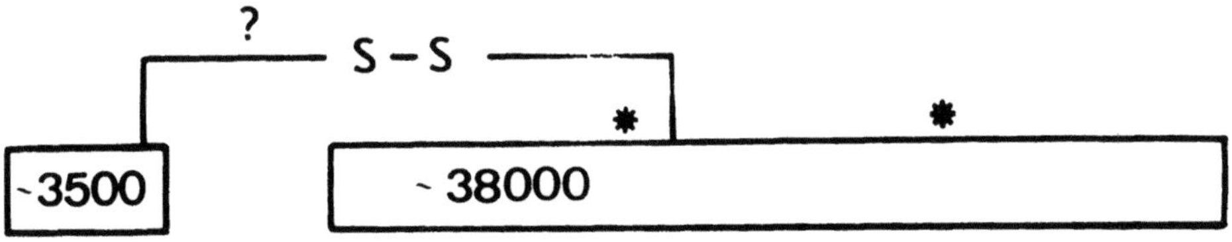

Fig. 3. Gross structure of boar acrosin. The relative position of the active site amino acid histidine (position 47) and serine (position 195) are indicated by an asterisk.

```
                    CHO
      1              |
BA  H2NArg-Asp-Asn-Ala-Thr-Cys-Asp-Gly-Pro-Cys-Gly-Leu-Arg-Phe-
       287
BT     -Lys-Thr-Phe-Gly-Ala-Gly-Glu-Ala-Asp-Cys-Gly-Leu-Arg-Pro-
                                              1
BC                                        H2N-Cys-Gly-Val-Pro-Ala-
        3
HU     -Ser-Pro-Pro-Glu-Glu-Leu-Lys-Phe-Gln-Cys-Gly-Gln-Lys-Thr-

BA      Tyr-Ser-Lys-COOH

BT      Leu-Phe-Gly-Lys-

BC      Ile-Gln-Pro-Val-

HU      Leu-Arg-Pro-Arg-COOH
```

Fig. 4. Primary structure of the boar acrosin light chain (BA), bovine prothrombin A-chain (BT) (14), bovine chymotrypsinogen (BC) (15), and human urokinase (HU) (16). Identical residues are underlined.

Fig. 5 and 6 summarize the sequence data so far evaluated for boar acrosin. An extensive homology to other members of the serine proteinase superfamily is evident. Particularly, the region of the so-called His-loop of serine proteinases (acrosin position 32 - 48) comprising the active site amino acid histidine (acrosin position 47) is highly conserved. Likewise, the active site segment of acrosin including the amino acid serine (chymotrypsin position 195), which is part of the charge-relay system of serine proteinases, has been largely preserved during divergent evolution of the proteinases. The amino acid Asp (chymotrypsin position 189) determining the primary specificity of the proteinase is identical for acrosin, trypsin, and plasmin. Similarly, the amino acid Asp in position 194 is conserved which probably forms a ion-pair with the N-terminal amino acid of the heavy chain (valine) as in chymotrypsin.

1 5 10

BA Val-Val-Gly-Gly-Met-Ser-Ala-Glu-Pro-Gly-Ala-Trp-Pro-Trp-

HP Val-Val-Gly-Gly-Cys-Val-Ala-His-Pro-His-Ser-Trp-Pro-Trp-

BC Ile-Val-Asn-Gly-Glu-Glu-Ala-Val-Pro-Gly-Ser-Trp-Pro-Trp-

BT Ile-Val-Gly-Gly-Tyr-Thr-Cys-Gly-Ala-Asn-Thr-Val-Pro-Tyr-

15 20 25

BA Met-Val-Ser-Leu-Gln-Ile-Phe-Met-Tyr-His-Asn-Asn-Arg-Arg-

HP Gln-Val-Ser-Leu-Arg-Thr-Arg-Phe-Gly- – – – – – – – – – – –

BC Gln-Val-Ser-Leu-Gln-Asp-Lys-Thr-Gly- – – – – – – – – – – –

BT Gln-Val-Ser-Leu-Asn- – – – -Ser-Gly- – – – – – – – – – – –

30 40

BA Tyr-His-Thr-Cys-Gly-Gly-Ile-Leu-Leu-Asn-Ser-His-Trp-Val-

HP Met-His-Phe-Cys-Gly-Gly-Thr-Leu-Ile-Ser-Pro-Glu-Trp-Val-

BC Phe-His-Phe-Cys-Gly-Gly-Ser-Leu-Ile-Asn-Glu-Asn-Trp-Val-

BT Tyr-His-Phe-Cys-Gly-Gly-Ser-Leu-Ile-Asn-Ser-Gly-Trp-Val-

45 50

BA Leu-Thr-Ala-Ala-His-Cys-Phe-Lys-Asn-Lys-

HP Leu-Thr-Ala-Ala-His-Cys-Leu-Glu-Lys-Ser-

BC Val-Thr-Ala-Ala-His-Cys-Gly-Val-Thr-Thr-

BT Val-Ser-Ala-Ala-His-Cys-Tyr-Lys-Ser-Gly-

Fig. 5. Primary structure of the N-terminal part of the boar acrosin heavy chain (BA) (13), human plasmin (HP) (17), bovine chymotrypsin (BC) (15), and bovine trypsin (BT) (18). Identical residues are underlined, – denote deletions/insertions.

BA -Ser-Thr-Asn-Val-Cys-Ala-Gly-Tyr-Pro-Arg-Gly-Lys-Ile-Asp-

HP -Ser-Thr-Glu-Leu-Cys-Ala-Gly-His-Leu-Ala-Gly-Gly-Thr-Asp-

180
BC -Asp-Ala-Met-Ile-Cys-Ala-Gly- – -Ala-Ser-Gly-Val- – -Ser-

BT -Ser-Asn-Met-Phe-Cys-Ala-Gly-Tyr-Leu-Glu-Gly-Gly-Lys-Asp-

BA -Thr-Cys-Gln-Gly-Asp-Ser-Gly-Gly-Pro-Leu-Met-Cys-Arg-Tyr-

HP -Ser-Cys-Gln-Gly-Asp-Ser-Gly-Gly-Pro-Leu-Val-Cys-Phe-Glu-

190 200
BC -Ser-Cys-Met-Gly-Asp-Ser-Gly-Gly-Pro-Leu-Val-Cys-Lys-Lys-

BT -Ser-Cys-Gln-Gly-Asp-Ser-Gly-Gly-Pro-Val-Val-Cys-Ser-Gly-

BA -Arg-Ala-Glu-Asn-Gly-Phe-Val-Ser-

HP -Lys-Asp-Lys-Tyr-Ile-Leu-Gln-Gly-

210
BC -Asn-Gly-Ala-Trp-Thr-Leu-Val-Gly

BT -Lys- – – – – – – – -Leu-Gln-Gly-

Fig. 6. Primary structure of the active site segment of boar acrosin heavy chain (BA), human plasmin (HP), bovine chymotrypsin (BC), and bovine trypsin (BT). Identical residues are underlined, – denote deletions/insertions.

The overall homology of boar acrosin and human plasmin amounts to 53 % of the amino acid residues in identical positions; in general, 81 % of the conserved amino acid and 100 % of the invariant amino acids of the serine proteinase superfamily are found in identical positions. Obviously, the cellular proteinase acrosin derives from an ancestor common to the secretory and plasmatic proteinases, as it is endowed with the main structural features typical for these classes of proteolytic enzymes. Yet, there is a difference: the peptide chain of activated acrosin has a molecular weight of close to 38 000. Thus, it is considerably longer than in other well-characterized serine proteinases with an average chain length of 25 000 Dalton. We suspect that these insertions and/or extensions are primary candidates for the hydrophobic binding sites postulated for boar acrosin (19), which might have been specifically acquired by a cellular-type of proteinase during evolution.

The biosynthesis of acrosin is confined to the testicular compartment. From our isolation procedure, we have calculated that an average of 2.4×10^{-18} mol acrosin/spermatozoon is synthesized (6). Assuming a volume of 1.5×10^{-15} liter for the acrosomal compartment, a local concentration of 1.6 mM results for acrosin. Hence, a powerful proteolytic equipment is located on the sperm tip requiring an effective regulatory system that controls the release of the proteolytic potential. Present evidence suggests acrosin is present in native spermatozoa as an inactive precursor (proacrosin) which is converted to the active enzyme at the time of the acrosome reaction in close vicinity to the ovum (20, 21). Future efforts are directed towards the isolation and molecular characterization of the zymogen of acrosin.

References

1. Saling, P. M.: Proc. Natl. Acad. Sci. U. S. A. 78, 6231-6235 (1981).
2. McRorie, R. A., Williams, W. L.: Ann. Rev. Biochem. 43, 777-803 (1974).
3. Stambaugh, R., Buckley, J.: Science 161, 585-586 (1969).
4. Zaneveld, L. J. D., Robertson, R. T., Williams, W. L.: FEBS Lett. 11, 345-347 (1970).
5. Müller-Esterl, W., Kupfer, S., Fritz, H.: Hoppe-Seyler's Z. Physiol. Chem. 361, 1811-1821 (1980).
6. Müller-Esterl, W., Fritz, H.: Methods Enzymol. 80, 621-632 (1981).
7. Schill, W.-B., Schleuning, W.-D., Fritz, H., Wendt, V., Heimburger, N.: Naturwissenschaften 62, 540 (1975).
8. Brown, C. R., Hartree, E. F.: Biochem. J. 175, 227-238 (1978).
9. Schiessler, H., Schleuning, W.-D., Fritz, H.: Hoppe-Seyler's Z. Physiol. Chem. 356, 1931-1936 (1975).

10. Schiessler, H., Arnhold, M., Ohlsson, K., Fritz, H.: Hoppe-Seyler's Z. Physiol. Chem. 357, 1251-1260 (1976).

11. Tschesche, H., Wittig, B., Decker, G., Müller-Esterl, W., Fritz, H.: Eur. J. Biochem., in press (1982).

12. Borin, G., Chessa, G., Cavaggion, G., Marchiori, F., Müller-Esterl, W.: Hoppe-Seyler's Z. Physiol. Chem. 362, 1435-1445 (1981).

13. Fock-Nüzel, R., Lottspeich, F., Henschen, A., Müller-Esterl, W., Fritz, H.: Hoppe-Seyler's Z. Physiol. Chem. 361, 1823-1828 (1980).

14. Magnusson, S., Petersen, T. E., Sottrup-Jensen, L., Claeys, H.: Proteases and Biological Control, Reich, E., Rifkin, D., Shaw, E., eds., Cold Spring Harbor Laboratories, Cold Spring Harbor, p. 123-149 (1975).

15. Braun, J. R., Hartley, B. S.: Biochem. J. 101, 214-228 (1966).

16. Günzler, W. A., Steffens, G. J., Ötting, F., Buse, G., Flohe, L.: Hoppe-Seyler's Z. Physiol. Chem. 363, 133-141 (1982).

17. Wiman, B.: Eur. J. Biochem. 76, 129-137 (1977).

18. Mikes, O., Holeysovsky, V., Tomasek, V., Sorm, F.: Biochem. Biophys. Res. Comm. 24, 346-352.

19. Müller-Esterl, W., Fritz, H.: Hoppe-Seyler's Z. Physiol. Chem. 361, 1673-1682 (1980).

20. Polakoski, K. L., Zaneveld, L. J. D.: Methods Enzymol. 45, 325-329 (1979).

21. Polakoski, K. L., Parrish, R. F.: J. Biol. Chem. 252, 1888-1894 (1977).

STRUCTURE-FUNCTIONAL STUDIES OF DNA-DEPENDENT RNA-POLYMERASE

V.M.Lipkin; O.Y.Chertov, I.A.Makarova, G.S.Monastyrskaya, E.D.Sverdlov, Y.A.Ovchinnikov

Shemyakin Institute of Bioorganic Chemistry, USSR Academy of Science, Moscow

Introduction

Elucidation of the transcription mechanism requires detailed knowledge of the organization of RNA polymerase active centres. This can be achieved only after determining the primary and spatial enzyme structures.

In studying the primary structures of large β- and β'-subunits of RNA polymerase (molecular weights 150 000 and 155 000, respectively) we used parallel sequencing of proteins and corresponding structural genes *rpoB* and *rpoC*. As a result the complete amino acid structures of the proteins which consist of 1342 and 1407 amino acid residues, respectively, were established (1,2).
The combined study of the amino acid sequence of the protein and the nucleotide sequence of its structural gene made it possible to identify positions of peptide fragments in the protein polypeptide chain and to provide their overlapping. Thus, the most complicated difficulty of the protein structural analysis was overcome.
On the other hand the known amino acid peptide sequence simplified the analysis of the nucleotide sequence. Knowledge of the peptide sequences would permit determining the template chain and translation frame in DNA subfragments and aligning some subfragments into an uninterrupted chains. Such an approach to the study of the DNA and protein sequences allowed us

Chemistry of Peptides and Proteins, Vol. 2

to avoid errors which appeared in the course of investigation of some regions of nucleotide and polypeptide chains.

Primary structures of some fragments of rpoBC operon were studied in a number of laboratories (3,4). The nucleotide sequence of one of these fragments, corresponding to the region from 60 to 100 amino acids of β-subunit of RNA polymerase, is shown in Fig.1. Comparison of these structures made it possible to discover a number of divergences between them. One should pay attention to a large number of deletions and insertions in structures "b" and "c" comparing with the structure "a" found by us. As a result the amino acid sequence from "b" in its major part differs from the amino acid sequence determined by us, the correctness of which is beyond doubt as it is completely confirmed by the analysis of corresponding peptides of the β-subunits. As in the Dennis structure "c" the absence of three nucleotides did not result in the shift of the translation frame, in this case the structures of only two amino acid residues appeared to be wrongly established (Fig.1). However, the absence of one more nucleotide in the other region (not shown in Fig.) did not allow Dennis to carry out the correct deduction of the amino acid sequence.

Recently Squires et al. (5) determined the primary structure of the fragment of rpoBC operon, coding the C-terminal sequence of β'-subunit, which also differs from the C-terminal sequence determined in our laboratory. The main difference is absence of adenine at position 7584 and insertion of thymidine between cytidine 7717 and adenine 7718. That resulted in the complete divergence of the Squires sequence with 45 amino acid residues from the corresponding sequence of β'-subunit (1133-1177) determined by us. The substantial part of our structure in this region is confirmed by the analysis of the corresponding peptides and thus its correctness is undoubtless.

These examples clearly show that only the combined analysis of primary structures of DNA and the corresponding protein gives reliable results.

Investigation of enzyme interaction with its inhibitors, in particular with the antibiotics of rifamycin series, is an

```
a)  CAG AGC TAC AGC GGT AAT TCC GAG CTG CAA TAC GTC AGC TAC CGC CTT GGC
     60                                 70
    Gln-Ser-Tyr-Ser-Gly-Asn-Ser-Glu-Leu-Gln-Tyr-Val-Ser-Tyr-Arg-Leu-Gly-

b)  CAG oGCT ACA GCG GTA ATT CCG AGC TGC AATAAC GTC AoCT AC oG CCT TGG
         ↑                                 ↑          ↑      ↑
    Gln- Ala-Thr-Ala-Val-Ile-Pro-Ser-Cys-AsnAsn-Val-T hr-Th  r-Pro-Trp-

a)  GAA CCG GTG TTT GAC GTC CAG GAA TGT CAA ATC CGT GGC GTG ACC TAT TCC
                80                                 90
    Glu-Pro-Val-Phe-Asp-Val-Gln-Glu-Cys-Gln-Ile-Arg-Gly-Val-Thr-Tyr-Ser-

b) CGA ACCGGCTTTTTT GAC GTC CAG GAA ooo ooo oo CCG TGG CGT GAC CTA TTC
         ↑  ↑↑                      ↑↑↑ ↑↑↑ ↑↑
   Arg-ThrGlyPhePhe-Asp-Val-Gln-Glu-            Pro-Trp-Arg-Asp-Leu-Phe-

a)  GCA CCG CTG CGC GTT AAA CTG CGT CTG GTG ATC TAT GAG CGC GAA GCG CCG
                            100                                     110
    Ala-Pro-Leu-Arg-Val-Lys-Leu-Arg-Leu-Val-Ile-Tyr-Glu-Arg-Glu-Ala-Pro-

b) CGC ACC GCoG CGC GTT AAA CTG CGT CTG GTG ATC TAT GAG CGC GAA GCG CCG
             ↑
   Arg-Thr-Al a-Arg-Val-Lys-Leu-Arg-Leu-Val-Ile-Tyr-Glu-Arg-Glu-Ala-Pro-

c)  GCA CCG CTG CGC GTT AAA CTG CGT CTG GTG ATC TAT Goo oGC GAA GCG CCG
                                                     ↑↑  ↑
    Ala-Pro-Leu-Arg-Val-Lys-Leu-Arg-Leu-Val-Ile-Tyr-G    ly-Glu-Ala-Pro-
```

Fig.1 The nucleotide sequence of the operon *rpoBC* fragment determined a) in our laboratory (1), b) by Gurevich A.I. et al. (3), c) by Dennis P.P. et.al. (4). Arrows mark the differences in sequences "b" and "c" comparing with "a": missed nucleotides are marked by "o" and inserted ones by italics. Amino acid sequences of the studied peptides are underlined and the sequences wrongly derived from "b" and "c" are italicized.

important step in understanding the mode of action and structure of RNA polymerase active site. Along with study of RNA polymerase from wild strain, sensitive to rifamycin action, investigation of rifamycin-resistant RNA polymerase isolated

from rpoB255 mutant strain was also carried out. In the β-subunit of this RNA polymerase aspartic acid resudue 516 is substituted for valine residue (6).
To establish if the mutant enzyme preserves its ability to bind antibiotic we used rifamycin property of quenching Trp residue fluorescence upon its binding to RNA polymerase. The data on fluorimetric titration of RNA polymerase of the wild and

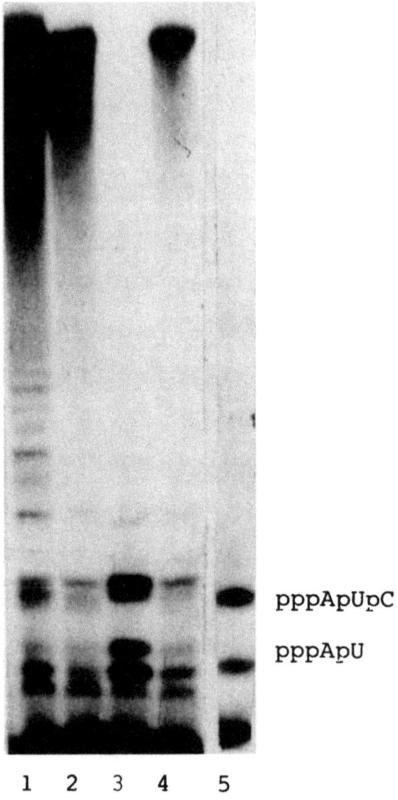

Fig.2.Electropherograms of products of the "cyclic" synthesis in 25% polyacrylamide gel in urea 1) Rif^s RNA polymerase; 2) Rif^r RNA polymerase; 3) Rif^s RNA polymerase+rifampicin; 4) Rif^r RNA polymerase+rifampicin; 5) standards for RNA oligonucleotides.

mutant strains with rifamycin SV and rifampicin allowed us to determine that in case of rifamycin-resistant (Rif^r) RNA polymerase the formation of the antibiotic-enzyme complex (1:1) occurs, however, dissociation constants of such complex (10^{-7}M) are considerably higher then those for rifamycin-sensitive (Rif^s) RNA polymerase.

It was established that rifamycin binding to RNA polymerase does not inhibit at least the formation of the first phosphodiester bond and its inhibiting action is manifested at the early stages of elongation. In particular McClure et al. (7) supposed that antibiotic prevents the transfer of nascent RNA inside the transcription complex. However, the date on maximum length of RNA, which is synthesized by Rif^s RNA polymerase in the presence of rifamycin, are contradictory. Therefore we investigated the "cyclic" synthesis of RNA oligonucleotides. As a template we used the Bsp fragment of phage T7 DNA, containing promoters A_o, A_1, A_2 and A_3. Analysis of the formed oligonucleotides by means of electrophoresis in polyacrylamide gel (Fig.2) showed that under the conditions of cyclic synthesis on T7 DNA in the presence of rifampicin Rif^s RNA polymerase synthesizes mainly di- and trinucleotides. These results are in accordance with the data of Gralla obtained for lacUV5 promoter (8).

The fact, rifampicin does not inhibit trinucleotide synthesis but inhibits the synthesis of longer oligonucleotides, can be explained if one assumes that the antibiotic bound to RNA polymerase occupies the region of RNA product binding site. In this case one can expect some analogy in spatial structures of rifamycin and RNA. To reveal this analogy we built the corresponding Dreiding model using the date on X-ray analysis of rifamycin B (9) and hybrid poly(rA)poly(dT) (10). Comparison of the models allowed us to find out the considerable similarity of spatial structures of rifamycin and dinucleotide fragment of RNA (Fig.3).

If to superpose the plane of aromatic rings of rifamycin with the heterocycle plane of purine nucleotide as well as posi-

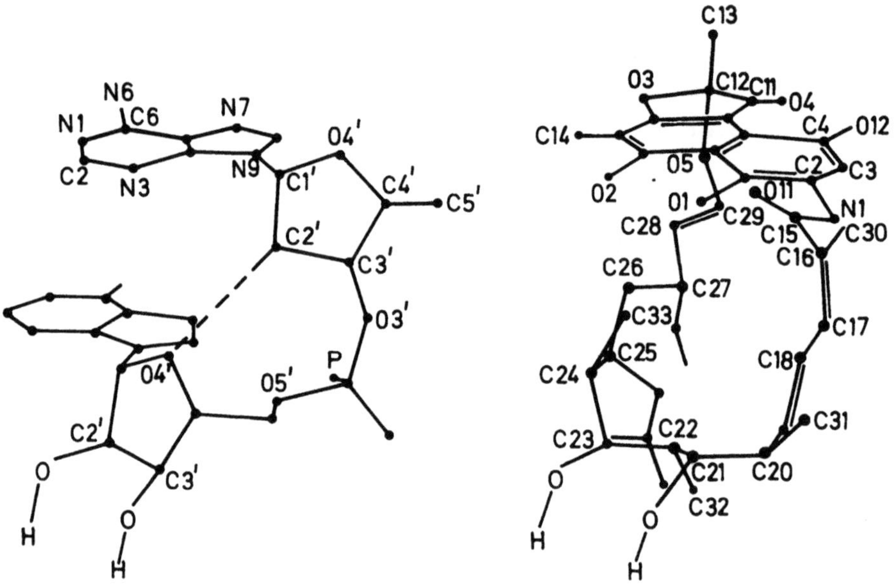

Fig.3 Spatial models of rifamycin B and RNA dinucleotide fragment.

tions of atoms in bond C3-C4 of the antibiotic with atoms N9-C1 of glycoside bond, respectively, then direction of the ansa chain is analogous to that for sugar phosphate skeleton of dinucleotide, and position of functionally important hydroxyl groups at C21 and C23 of the antibiotics coincide, respectively, with 3'- and 2'-groups groups of the second ribose residue.

All the data obtained were used to built the model of interaction of rifamycin with RNA polymerase. According to this model the antibiotic bound to the enzyme occupies the place of the first and the second nucleotides of the synthesized pentaribonucleotide and sterically prevents the normal synthesis of tetra- and more long oligonucleotides. In the formation of the binding site of antibiotic the aspartic acid residue 516 of β-subunit forming the hydrogen bond with hydroxyl group at C23 of

rifamycin takes part. The hydrogen bond is absent in the rifamycin-resistant enzyme in which the aspartic acid residue 516 is substituted for the valine residue. This results in the sharp decrease of the stability of the formed complex.
If the proposed model is correct and the spatial arrangement of rifamycin in complex with RNA polymerase is analogous to the spatial arrangement of the dinucleotide fragment of the synthesized RNA oligonucleotide, the aspartic acid residue 516 of the β-subunit should form the hydrogen bond also with 2'-hydroxyl group of the nascent RNA. As such hydrogen bond cannot be formed in case of rpoB255 mutant, then Rif^s and Rif^r RNA polymerases should differ in the strength of binding of the synthesized RNA oligonucleotides. The presented investigations (Fig.2) confirmed that the cyclic synthesis of oligonucleotides with the length of the chain from 4 to 10 residues for Rif^r RNA polymerase, which is capable for the normal synthesis of di- and trinucleotides as well as long polymer chains of RNA, is supressed in contrast to Rif^s enzyme.

References

1. Ovchinnikov, Yu.A., Monastyrskaya, G.S., Gubanov, V.V., Guryev, S.O., Chertov, O.Yu., Modyanov, N.N., Grinkevich, V.A., Makarova, I.A., Marchenko, T.V., Polovnikova, I.N., Lipkin, V.M., Sverdlov, E.D: Eur. J. Biochem., 116, 621-629 (1981).

2. Ovchinnikov, Yu.A., Monastyrskaya, G.S., Gubanov, V.V., Gyryev, S.O., Salomatina, I.S., Shuvaeva, T.M., Lipkin, V.M., Sverdlov, E.D.: Dokl. Akad. Nauk SSSR, 261, 763-768 (1981).

3. Gurevich, A.I., Avakov, A.E., Kolosov, M.N.: Bioorg. Khim., 5, 1735-1739 (1979).

4. Delcuve, G., Downing, W., Lewis, H., Dennis, P.P.: Gene, 11, 367-373 (1980).

5. Squires, C., Krainer, A., Barry, G., Shen, W.-F., Squires, C.L.: Nucl. Acid. Res., 9, 6827-6840 (1981).

6. Ovchinnikov, Yu.A., Monastyrskaya, G.S., Gubanov, V.V., Lipkin, V.M., Sverdlov, E.D., Kiver, I.F., Bass, I.A., Mindlin, S.Z., Danilevskaya, O.N., Khesin, R.B.: Mol.

Gen. Genet., 184, 536-538 (1981).

7. McClure, W.R., Cech, C.L.: J. Biol. Chem., 253, 8949-8956 (1978).

8. Carpousis, A.J., Gralla, J.D.: Biochemistry, 19, 3245-3253 (1980).

9. Brufani, M., Fedeli, W., Giacomello, G., Vaciago, A.: Experientia, 20, 339-342 (1964).

10. Zimmerman, S.B., Pheiffer, B.H.: Proc. Natl. Acad. Sci. USA, 78, 78-82 (1981).

AFFINITY LABELING OF THE BINDING SITE FOR INITIATION SUBSTRATE IN E.coli DNA-DEPENDENT RNA POLYMERASE BY ADENOSINE-5'-TRI-METAPHOSPHATE

Yu.V.Smirnov, V.M.Lipkin, Yu.A.Ovchinnikov
M.M.Shemyakin Institute of Bioorganic Chemistry, Academy of Sciences of the USSR, Moscow, USSR

M.A.Grachev, A.A.Mustaev
Novosibirsk Institute of Organic Chemistry, Siberian Branch of the Academy of Sciences of the USSR, Novosibirsk, USSR

Introduction

DNA dependent RNA-polymerase is the key enzyme of nucleic acid template biosynthesis which provides complementary copying of the gene information. The enzyme is a complex multisubunit protein interacting with templates, substrates and various regulatory factors.At present the primary structure for all the subunits of E.coli RNA-polymerase have been established(1,2,3, 4). However,the spatial structures of the enzyme catalytic sites are still obscure. In our opinion, the spatial structure of substrate binding sites can be studied by means of RNA polymerase affinity modification with a set of nucleoside-5'-triphosphate analogs having incorporating reactive groups in different parts of NTP molecule. Isolation of peptides with attached analogs from RNA-polymerase labeled subunits and determination of their structure must be mostly useful in elucidation of the polypeptide chain regions forming the enzyme substrate binding sites.

The RNA polymerase-DNA complex is supposed to have two substrate binding sites (5). One of them, an initiation site, binds nucleoside-5'-triphosphate included in the first position of

Chemistry of Peptides and Proteins, Vol. 2

synthesized RNA. The other, an elongation or polymerization site, binds a substrate inserted in the second and subsequent positions. The affinity of substrates to the second site is more pronounced than to the first one. The initiation site prefers prefers a purine nucleoside and it is not specific to the substrate triphosphates moiety (6). In contrast, the elongation site is very specific to the triphospate moiety.
In order to purify peptides with the attached substrate, the following criterions should be used in searching for the NTP analogs : a) reactive group must be as small as possible, the analog differs slightly from the native NTP; b) the specificity must be limited to provide minimal heterogeneity upon isolation of modified peptides; c) the chemical reaction yield must be quite high at labeling. Investigation on affinity modification of RNA-polymerase must be carried out in complete system with template.

Results

Holoenzyme RNA-polymerase preparation containing 80% of the subunit as judged from Coomassie G-250-stained SDS polyacrylamide gel electrophoresis was used in our experiments. The conditions of experiments, except those specially described, were the same as in (8,9).
Adenosine-5'-trimetaphosphate (ATMP) was chosen as an analog of NTP with a reactive function in the triphosphate moiety. According to the data of Knorre et al (7) nucleoside-5'-trimetaphosphates are known to be active phosphorylating agents (Fig.1). They are formed as intermediate compounds by activation of the nucleoside-5'-triphosphate group with different condensing reagents, e.g. dicyclohexylcarbodiimide, and reacts with different types of nucleophiles - amines, alcoholes and others. They are rapidly hydrolysed in the presence of water. Knorre et al. (7) used nucleoside-5'-trimetaphosphates to incorporate different

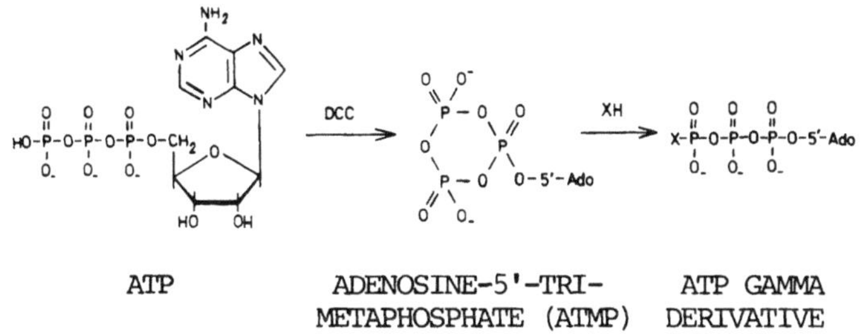

Fig.1. Formation of adenosine-5'-trimetaphosphate upon activation of the ATP triphosphate group with dicyclohexylcarbodiimide and formation of γ-derivatives of ATP.

substitutes into triphosphate regions thus constructing NTP analogs with substitutes at γ-phosphate. Now we used adenosine-5'-trimetaphosphate as an affinity reagent for modification of E.coli RNA polymerase initiation site.

Testing inhibition of RNA-polymerase complex with different templates, it was found out that at concentration more than 10^{-5}M ATMP it starts inhibiting RNA synthesis. Studying kinetics of RNA polymerase inactivation under the ATMP action in the system where promoter-contained Bsp-I DNA fragment of phage T7 was used as template, it was found out that the degree and rate of inactivation increased with enhancing concentration of ATMP (Fig.2, curves 2 and 3). ATP protects the enzyme from inactivation (Fig.2, curve 1). In the presence of UTP a considerable increase of rate and degree of inactivation was observed (Fig.2, curve 4).

It was established that ^{32}P-labeled ATMP modified the RNA-polymerase-template complex. Radioactivity was mainly located in β-subunit (after denaturation and separation of the modified complex with SDS-polyacrylamide gel electrophoresis) though β'-, σ - and apparently α- subunits were also labeled. It can be explained by nonspecificity of labeling because of high chemical reactivity of ATMP, which can modify side chains of lysine, histidine, threonine, serine and other amino acid residues with a mobile hydrogen atom.

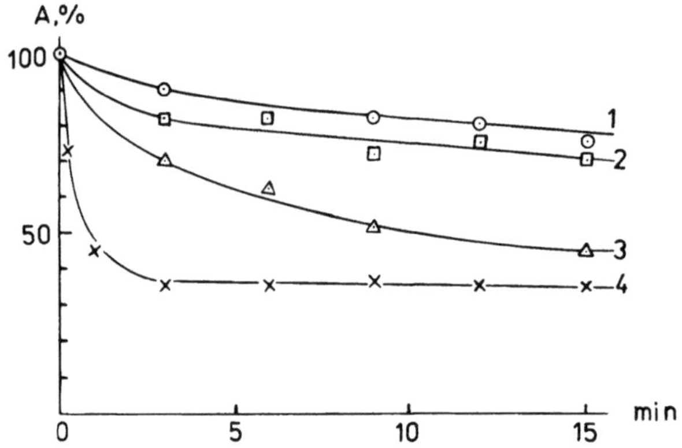

Fig.2. Kinetics of inactivation of RNA polymerase - promoter-containing fragment of phage T7 DNA complex with ATMP: 1-5 mM ATP + 0.5 mM ATMP, 2-0.1 mM ATMP, 3-0.5 mM ATMP, 4-0.1 mM UTP + 0.1 mM ATMP. Ordinate - activity of RNA polymerase (in % relative to the initial one).

To avoid nonspecificity of labeling, use was made of the technique of incorporation of radioactive label with the second substrate which was successfully applied by Sverdlov et al (8,9). The RNA polymerase - DNA complex premodified with ATMP, in which This analog, covalently bound to the initiation site, was expected to be able to form the phosphodiester bond with another nucleoside-5'-triphosphate entered at the enzyme elongation site (Fig.3).

The system consisting of RNA polymerase and template (Bsp-I DNA fragment of T7 phage, poly d(A-T) or calf thymus DNA) was incubated for 5 min at 37° to form a binary complex, then was cooled to room temperature and ATMP was added to concentration $5x10^{-4}$M. After complete hydrolysis (30 min) of ATMP, which did not react with the protein, the second substrate, α-^{32}P UTP, was added at concentration $8x10^{-6}$M. If the DNA fragment of phage T7 was used as template, radioactive UTP was added together with ATMP since the structure of promoters makes possible obtaining the expected synthesis product (pppApU*) (Fig.4). Then the system was denaturated with SDS and mercaptoethanol and subjected to SDS-polyacrylamide gel electrophoresis. One of the obtainned X-ray patterns

RNA POLYMERASE - POLY /d(A-T)/ COMPLEX

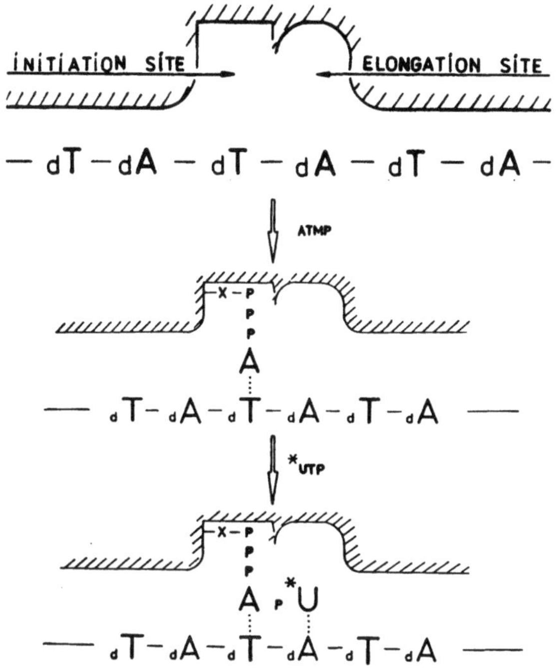

Fig.3. Formation of phosphodiester bond between the analog covalently bound to the initiation site and radioactive UTP positioned in the enzyme elongation site.

A_0	G - U - U - G - G - C - U - U - U - A - G - G - A -
A_1	A - U - C - G - A - G - A - G - G - G - A - C - A -
A_2	G - C - U - A - G - G - U - A - A - C - A - C - U -
A_3	A - U - G - A - A - A - C - G - A - C - A - G - U -

Fig.4. Structures of RNA-chains, which can be synthesized on promoter sites of the BSP-I - fragment of phage T7 DNA.

is presented in Fig.5. There is only one radioactive band (with the exception of start) located in the β-subunit. The same results were obtained using other templates. The β- and β'-subunits are resolved unsufficiently in the given system and the-

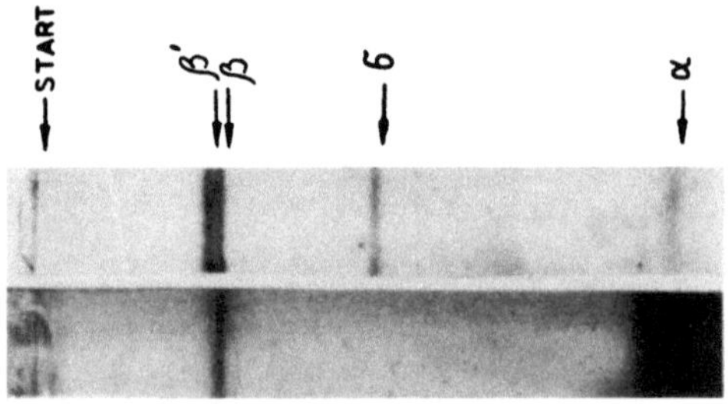

Fig.5. Radioautograph of SDS-polyacrylamide gel electrophoretic analysis of incorporation of - ^{32}P UTP into RNA polymerase - poly/d(A-T) complex premodified with ATMP. Arrows indicate to subunit positions according to the Coomassie G-250 staining.

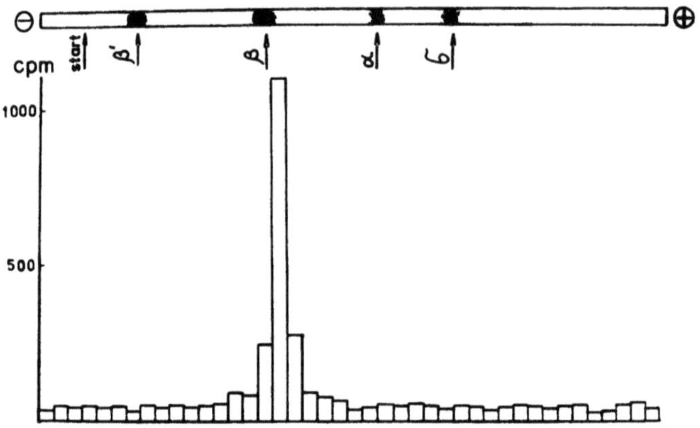

Fig.6. Cellulose acetate electrophoretic analysis of incorporation of ^{3}H UTP into RNA polymerase - calf thymus DNA complex premodified with ATMP. Arrows indicate to subunit positions according to the amido black 10B staining.

re is a large background in position of α-subunit. For unequivocal identification of the modified subunit of RNA polymerase such experiment was repeated, and after modification the RNA-polymerase complex was separated from low molecular components of the mixture by Bio-Gel A-5m gel filtration. Further, the desalted and lyophilized preparation was separated by cellulose acetate electrophoresis in denaturing ammonium-borate buffer with 8M urea according to Zillig (10). The result is given in Fig.6. Only one band corresponded to the β-subunit and no radioactivity was observed in β'- and α-subunit positions. The radioactivity peak shifted to more acidic components. We suppose that it took place due to the additional charge given the β-subunit by attached pppApU*. Electrophoretic separation in cellulose acetate is known to depend only on the charges because of its large pores.

Thus, the use of radioactive UTP in combination with reactive adenosine-5'-trimetaphosphate enabled us to label specifically the initiation binding centre in the RNA-polymerase complex, since nonspecifically bound ATMP molecules cannot form internucleotide bonds with the second substrate. The forming short transcript pppApU* (for the template of the DNA fragment of phage T7) covalently bound to the β-subunit is apparently not capable of translocating, that is the reason of the inhibiting action of ATMP. The analog can be used for isolation the modified peptide of the β-subunit.

References

1. Ovchinnikov Yu.A., Monastyurskaya G.S., Gubanov V.V., Guryev S.O., Salomatina I.S., Shuvaeva T.M., Lipkin V.M., Sverdlov E.D.: Dokl.Akad.Nauk SSSR, 261, 763-768, (1981).
2. Ovchinnikov Yu.A., Monastyrskaya G.S., Gubanov V.V., Guryev S.O., Chertov O.Yu., Modyanov N.N., Grinkevich V.A., Makarova I.A., Marchenko T.V., Polovnikova I.N., Lipkin V.M., Sverdlov E.D.: Eur.J.Biochem. 116, 621-629, (1981).
3. Ovchinnikov Yu.A., Lipkin V.M., Modyanov N.N., Chertov O.Yu. Smirnov Yu.V.:FEBS Letters, 76, 108-111, (1977).
4. Burton Z., Burgess R.R., Lin J., Moore D., Holder S., Gross C.A.: Nucleic Acid Res. 9, 2889-2903, (1981).
5. Chamberlin M.J.: RNA Polymerase ,eds. Losick R., Chamberlin M., Cold Spring Harbor Laboratory, 17-67, (1976).
6. Malygin A.G., Shemyakin M.F.:FEBS Letters, 102, 51-54,(1979).

7. Knorre D.G., Kurbatov V.A., Samukov V.V. : FEBS Letters, 70, 105-108, (1976).

8. Sverdlov E.D., Tsarev S.A., Modyanov N.N., Lipkin V.M., Grachev M.A., Zaichikov E.F., Pletnev A.G. Bioorgan.Chim. 4, 1278-1280, (1978).

9. Sverdlov E.D., Tsarev S.A.: Bioorgan.Chim. 6, 1110-1113,(1978)

10. Fujiki H., Zurek G.:FEBS Letters, 55, 242-244, (1975).

LOCALIZATION OF MUTATIONS RESULTING IN RIFAMICIN-RESISTANCY OF *E.coli* RNA POLYMERASE

Sergei O.Guryev, Galina S.Monastyrskaya, Valentine V.Gubanov, Nadezhda F.Kalinina, Valeri M.Lipkin, Eugene D.Sverdlov

Shemyakin Institute of Bioorganic Chemistry, USSR Academy of Sciences, 117988 Moscow, USSR

Alexander I.Gragerov, Vadim G.Nikiforov, Irina F.Kiver, Irina A.Bass, Olga N.Danilevskaya, Sophia Z.Mindlin

Institute of Molecular Genetics, USSR Academy of Sciences, 123182 Moscow,USSR

Introduction

To understand the mechanism underlying the operation of RNA-polymerase and the structure of its active centre, it is useful to investigate the interaction between the enzyme and its inhibitors, e.g. antibiotics of the rifamicin series.

Of great help here are mutations that entail the enzyme inability to bind the inhibitor. The mutations of rifamicin resistance, both, recessive (rif^r) and dominant (rif^d) are known to be localized in the rpoB gene encoding the β-subunit of *E.coli* RNA-polymerase (1-4).

However, the exact location of these mutations within the gene is unknown. For an accurate localization of the mutations one has to determine the nucleotide sequence of the mutant and wild-type genes. We have previousely determined the primary structure of the rpoB gene cloned from the rpoB 255 mutant chromosome /8/. Here we shall present the data on determining the sequence of the wild-type rpoB gene and of four independently isolated rif^r mutants.

Materials and Methods

The method for obtaining the rif^r mutations in the cloned fragment of the rpoB gene and the properties of the obtained mutations will be published later. Selection of the mutations was based on the ability of bacterial DNA fragments to be incorporated into a chromosome from a plasmid as a result of the homologeous recA-dependent recombination. The study of the pIBI9 plasmid which carries the EcoRI-fragment of the mutant rpoB gene (rpoB255 mutation), incorporated in the pBR322 vector, showed that the portion of cells capable of growing in the presence of rifampicin is 10^{-3} - 10^{-4}, whereas the spontaneous frequency of appearance of the rif^r mutants is 10^{-7} - 10^{-8}.

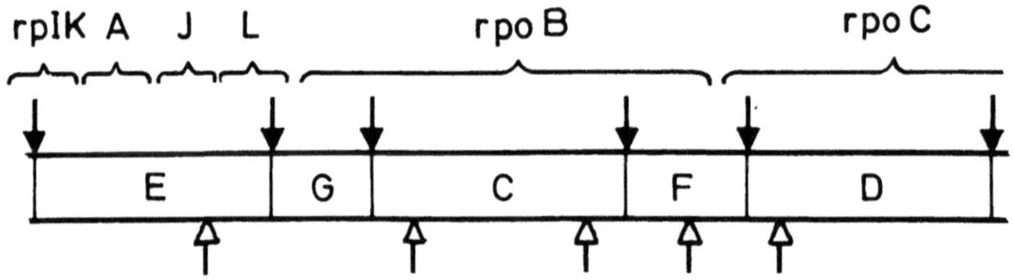

Fig.1. Restriction map of *E.coli* DNA within the rpoB gene. The sites for restriction endonucleases EcoRI (↓) and PstI(↑) are shown.

DNA of the pIBI830 plasmid, containing the EcoRI-fragment of the wild-type rpoB gene interested in the pMB9 vector (Fig.1), was treated with IM hydroxylamine at pH5-6 (69°C, 30-60 min.). Hydroxylamine was removed by dialysis against TES.*Ecoli* C600 cells were transformed using the treated preparations (transformation efficiency is 10-20% of the control one). The colonies grown in Petri dishes were suspended in a physiological solu-

tion and seeded on agar with rifampicin. From the resistant clones the plasmid DNA was isolated and its ability to confer the rifampicin-resistance to sensitive cells was examined. On the whole 15 clones which contain rifampicin-resistance mutations in the pIBI830 plasmid were selected in such a way.
As the rpoB255 mutation is situated in the central PstI-fragment of the rpoB gene (Fig.1), other dominant rifampicin-resistant mutations were assumed to be localized within the same region. Therefore we clone the corresponding PstI-fragment from the phage $\lambda rif^{d}18$. A mixture of $\lambda rif^{d}18$ DNA-pBR322 DNA was treated with restrictase PstI and ligated. *E.coli* C600 (Rec^{+}) cells were transformed with the obtained mixture and the colonies which carry the hybrid plasmids (Tc^{r}, Ap^{s}) were selected. The transformants, which give the rifampicin-resistant progeny cells with high frequency, were picked out among them. In such a manner the plasmid pBK2645 was selected. The enzymes used, the procedure for obtaining the restriction endonucleases and the methods for analyzing the DNA primary structure have been earlier described /5/.

Results and Discussion

According to the studies of our laboratory and of other investigators /6,7/ mutations to dominant rifampicin resistance, rpoB3 and rpoB255 , are localized in the fragment of the central part of the rpoB gene. To localize these mutations the plasmids were constructed containig Pst-fragments isolated from the wild-type rpoB gene /8/ and $\lambda rif^{d}18$ phages carrying the rpoBC operon with the corresponding mutation.
Primary structures of the cloned fragments were determined by the modified /9/ Maxam-Gilbert method /10/ and compared with the structure of the corresponding fragment of the rpoB gene containing the rpoB255 mutation /8/. This comparison revealed that the rpoB255 mutation is in position 1547 (Fig.2), and the

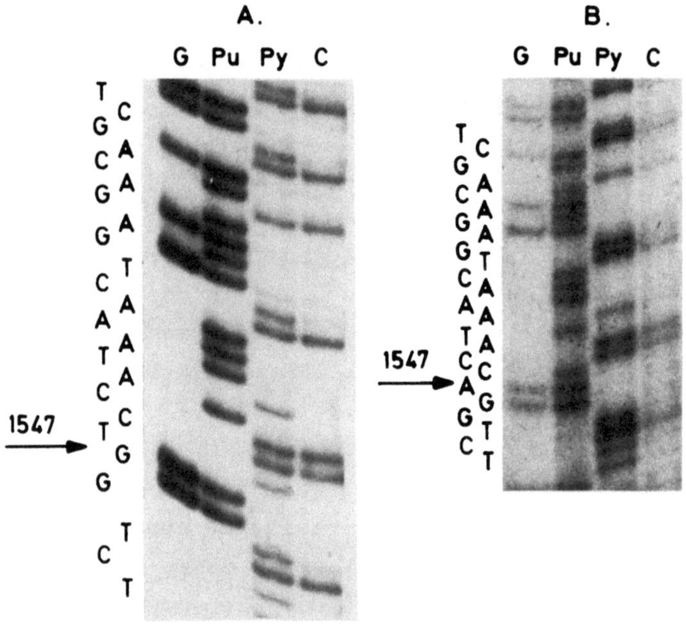

Fig.2. Autoradiograms of polyacrylamide gels used for analyzing the DNA sequences of fragments of wild (a) and mutant $rif^{d}21$ (b) rpoB genes. The region of the rif^{d} mutation is represented. Arrows show the site where the substitution takes place. A number of nucleotides from the rpoB genestarting point is given.

mutation rpoB3 is in position 1592 (Fig.3).

The first of these mutations is the A.T.→T.A. transversion which results in the substitution of an aspartic acid residue (516) in the wild-type β-subunit by a valine residue. The second mutation is the C.G.→T.A. transition leading to the replacement of serine (531) by a phenylalanine. Three other rif^{r} mutations were obtained by treating the plasmids, containing the EcoRI-C (Fig.1), with hydroxylamine.

Analysis of primary structures of Pst-fragments, isolated from these plasmids, showed that all three mutations are C.G.→T.A. transitions, one of them being coincident with the rpoB3 mutation. Two other mutations turned out identical and are the substitution of CG (1691) by AT entailing the presence of Leu in the mutant fragment instead of Pro (564). Disposition of mutations is show in fig.4.

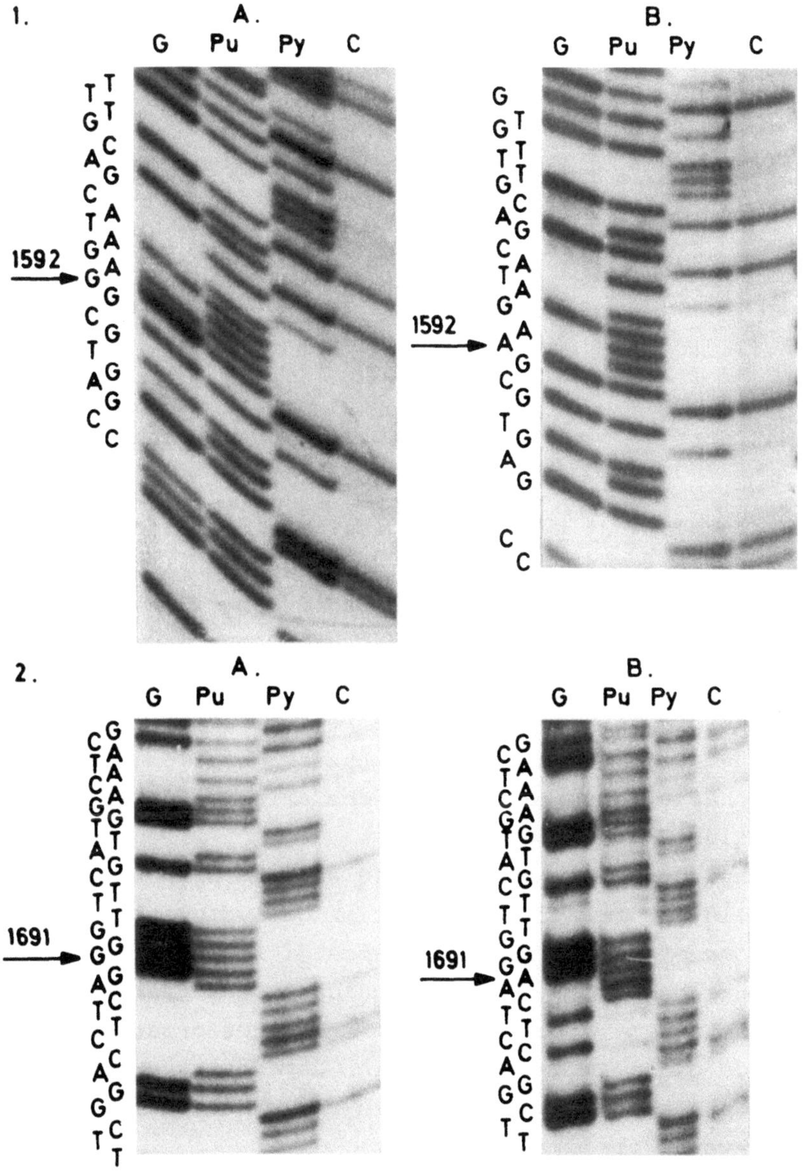

Fig.3. Autoradiograms of the polyacrylamide gels used for analyzing the DNA sequences of the fragments of wild (a) and mutant (b) rpoB genes; I - for rpoB 3 mutation, II - for two rif^r mutations.

```
1540   T                                                          T
       ↑                                                          ↑
TTT ATG GAC CAG AAC AAC CCG CTG TCT GAG ATT ACG CAC AAA CGT CGT ATC TCC
Phe-Met-Asp-Gln-Asn-Asn-Pro-Leu-Ser-Glu-Ile-Thr-His-Lys-Arg-Arg-Ile-Ser-
        ↓                                                           ↓
        Val                                                         Phe
     (rpoB255)                                            (rpoB3, rifr5)

GCA CTC GGC CCA GGC GGT CTG ACC CGT GAA CGT GCA GGC TTC GAA GTT CGA GAC
Ala-Leu-Gly-Pro-Gly-Gly-Leu-Thr-Arg-Glu-Arg-Ala-Gly-Phe-Glu-Val-Arg-Asp-

                                                      T           1701
                                                      ↑
GTA CAC CCG ACT CAC TAC GGT CGC GTA TGT CCA ATC GAA ACC CCT GAA GGT CCG
Val-His-Pro-Thr-His-Tyr-Gly-Arg-Val-Cys-Pro-Ile-Glu-Thr-Pro-Glu-Gly-Pro-
                                                       ↓
                                                      Leu
                                                 (rifr3,rifr4)
```

Fig.4 Disposition of mutant substitutions in the rpoB gene of RNA polymerase.

It was supposed that the dominant rifamicin-resistant mutations are the double substitutions /11/. If so, the rif^r5 mutation, whose position coincides with the position of the dominant one, must not be dominant and vica versa. The investigation of the mutations found in our work is in progress. The studied mutations indicate probably to the region essential for rifamicin binding. Some of them replace the amino acids with functional side chains (Asp and Ser) for the amino acids with hydrophobic residues (Val and Phe).

The carboxyl group of aspartic acid and the hydroxy group of serine could participate in formation of hydrogenic bonds with oxygen atoms of rifamicin. Appearance of rifamicin resistance due to the substitution Pro→Leu indicates that for the interaction of the enzyme with rif the definite conformation of the polypeptide chain is necessary which depends on a proline residue. When styding the rifamicin resistance one should bear in mind several circumstances.

The first one is evident: as these mutations do not influence the main functions of RNA polymerase, the mutant groups are not taken into account when building the models of the enzyme active centre.
Secondly, since the alkylating analogs of rifamicin modify β- and δ-subunits /12/ it is quite possible that rifamicin binding to the β-subunit takes place in the vicinity of the contact site of these two subunits. Probably the mutations of rifamycin resistance indicate the position of this site. It is noteworthy that Traverse observed the difference in the binding of the δ-subunit to the mutant core-enzyme compared to wild-type RNA polymerase /13/. The rif^r mutations are known to suppress the mutant phenotype of rpoD mutations/14/. This observations may support the idea that this region of rifampicin and δ-subunit binding are located near each other. One of the localized mutations of rpoB255 is the substitution of the aspartic acid residue for valine. The δ-subunit is a very acidic protein /15/ and,therefore,if this mutation is near the binding site of the δ-subunit, the elimination of the acidic group, taking place as a result of the mutation, can lead to the increase of the association constant of the δ-subunit with the mutant enzyme compared to the wild-type one. Thirdly, according to our experiments /16,17/, which were confirmed by other investigators /18/, 5'- and 3'-terminal groups of the growing RNA in the transcribing complex contact with the β- and δ-subunits at the first stages of the synthesis. It means that the product binding site of the RNA-polymerase is also near the region of the contact of these two subunits. Besides, Krakow /19/ supposed that the δ-subunit is bound to the product-site and complexes for it with the product because the δ-subunit dissociation from the transcribing complex after RNA length reaches appoximately 10 residues. Krakow /19/ has already supposed that the δ-subunit is bound to the product binding site and competes for it with the product. According to his hypothesis the product pushes out the δ-subunit from the enzyme. Consequently, both, rifamicin and

product of the binding to RNA-polymerase are near the boundary which separates β- and δ-subunits. It may be that the regions of binding of rifamicin and the product even overlap in such a way that rifamicin either sterically blocks the elongation according to McClure and Chech /20/ or weakens the product binding to the transcribing complex according to Zillig et al /21/. From all the data given above one can arrive to the conclusion that the binding sites for rifamicin, product and the δ-subunit, are localized near each other, and, consequently, mutations of resistance to the antibiotic can serve as a marker in determining all these regions.

References

1. Neil, A., Zillig, W.: FEBS Lett. 11. 165-168 (1978).
2. Babinet, C.: Biochemie, 53, 507-516 (1971).
3. Mindlin, S.Z., Ilyina, T.S., Voeykova, T.A., Velkov, V.V.: Mol. Gen. Genet. 115, 115-121 (1972).
4. Ivakura, I., Ishihama, A., Jura, T.: Mol. Gen. Genet. 121 181-196 (1973).
5. Ovchinnikov, Yu.A., Monastyrskaya, G.S., Gubanov, V.V., Guryev, S.O., Chertov, O.Yu., Modyanov, N.N., Grinkevich, V.A., Makarova, I.A., Marchenko, T.V., Polovnikova, I.N., Lipkin, V.M., Sverdlov, E.D.:Eur. J. Biochem. 116, 621-629 (1981).
6. Kirscbaum, J.B., Konrad, E.B.: J. Bacterial. 116, 517-526 (1973).
7. Mindlin, S.Z., Ilyina, T.S., Gorlenko, J.M., Chachikyan, N.A., Kovalev, Yu.N.: Genetica 12 (12) 116-130 (1976).
8. Ovchinnikov, Yu.A., Monastyrskaya, G.S., Gubanov, V.V., Lipkin, V.M., Sverdlov, E.D., Kiver, I.F., Bass, I.A., Mindlin, S.Z., Danilevskaya, O.N., Khesin, R.B.:Mol. Gen. Genet, 84, 536-538 (1981).
9. Ovchinnikov Yu.A., Guryev, S.O., Krayev, A.S., Monastyrskaya, G.S., Skryabin, K.S., Sverdlov, E.D., Zakharyev, V.M., Bayev, A.A.: Gene, 6, 235-249 (1979).
10. Maxam, A., Gilbert, W.: Proc. Nat. Acad. Sci. USA, 74, 560-564 (1977).

11. Newman, A., Heyward, R.S.: Mol. Gen. Genet. 177, 527-533 (1984).

12. Stender, W., Stutz, A.A., Scheit, K.H.: Eur. J. Biochem. 56, 129-133,(1975).

13. Travers, A.A.: FEBS Lett. 10, 71-75 (1976).

14. Yura, T., Ishihawa, A.: Ann. Rev. Genet. 13, 57-59 (1979).

15. Burton, Z., Burgess, R.R., Lin, J., Moore, D., Holder, S., Gross, C.A.: Nucleic Acid Res. 9, 2889-2898 (1981).

16. Sverdlov, E.D., Tsarev, C.A.: Bioorgan Chem. 6(7), 1110-1113 (1980).

17. Sverdlov, E.D., Tsarev, S.A., Begar, V.A.: FEBS Lett. 114, 111-114 (1980).

18. De Riemer, L.H., Meazes, C.: Biochemistry 20, 1612-1617 (1981).

19. Krakow, J.S., Phodes, G., Jovin, T.M.: in RNA Polymerase, Edit. by Losic, R., Chanberlin, M., 127-158 (1976).

20. McClure, W.R., Jovin, T.M.: J. Biol. Chem. 250, 4073-4076 (1975).

21. Zillig, W., Zechel, K., Rabussay, D., Schachner, M., Seth, V.S., Palm, P., Heil, A., Seifert, W.: Cold Spring Harbor Symp. Quant. Biol. 35, 47-54 (1971).

ROLE OF CYCLIC AMP IN THE REGULATION OF GENOME EXPRESSION OF EUKARYOTS

Evgenii S.Severin
Institute of Molecular Biology, Academy of Sciences of USSR, Moscow 117984 USSR

Protein Kinases and Phosphorylation Processes

One of the main mechanisms of cell activity regulation is the enzymatic modification of proteins, in particular their phosphorylation. Such phosphorylation leads either to a change in the enzyme activity or to other modifications of proteins. In order to elucidate the role of phosphorylation processes in the regulation of cell activity, highly purified preparations of protein kinases from various sources were obtained and studied. In our experiments, we mainly used the enzyme from a highly differentiated brain tissue and protein kinase from proliferating culture Physarum polycephalum. The comparative analysis of some parameters of the two above enzymes is shown in Table 1.

It follows from Table 1 that the isolated homogeneous cAMP-independent protein kinase from the synchronous culture Physarum polycephalum consists of four identical subunits with MW 68,000. The principal protein substrates in vitro are casein, phosvitin and the C-terminal fragment of histone H1 (1). The enzyme has a unique substrate specificity: it phosphorylates both acidic proteins (casein, phosvitin) and the basic protein, histone H1.

Another enzyme that was closely studied in our laboratory was cAMP-dependent protein kinase which we isolated from pig brain (2).

Chemistry of Peptides and Proteins, Vol. 2

TABLE 1. Comparative characteristics of protein kinases from pig brain and Physarum polycephalum.

Parameter	Pig brain enzyme	Enzyme from Physarum polycephalum
Molecular weight	180,000	260,000
Subunit structure	R_2C_2 (regulatory subunit (R), 53,000; catalytic subunit (C), 38,000)	C_4 (catalytic subunit (C), 68,000)
Effector regulation	dependent on cAMP	independent of cyclic nucleotides, Ca, dsRNA
Substrate specificity		
a) nucleotide substrate	ATP ($1.2 \cdot 10^{-5}$M)	ATP ($1.5 \cdot 10^{-5}$M) GTP
b) protein substrate	lysine-rich histones (N-terminus)	casein, phosvitin, histone H1 (C-terminus)
c) phosphorylated amino acid residues	H1 - Ser-37 H2a -Ser-19 H2b -Ser-14, Ser-36	Thr
Effect on cell cycle of Physarum polycephalum	1.5-hr delay of mitosis	1.5-hr advancement of mitosis

cAMP-Dependent Protein Kinases

The whole enzyme of cAMP-dependent protein kinase is an active complex which dissociates under the action of cAMP into two different subunits, regulatory (R) and catalytic (C); the latter catalyzing the phosphotransferase reaction (see Eqns(1) and (2).

$$RC + cAMP \rightleftharpoons R{\cdot}cAMP + C \qquad (1)$$

$$ATP + H1 \xrightleftharpoons{C} ADP + H1 - P \qquad (2)$$

Using the quenched-flow method, the main constants of the phosphotransferase reaction (K_s, k_2, k_3 and k_4) which are shown in the scheme below (Eqn 3) were estimated (3):

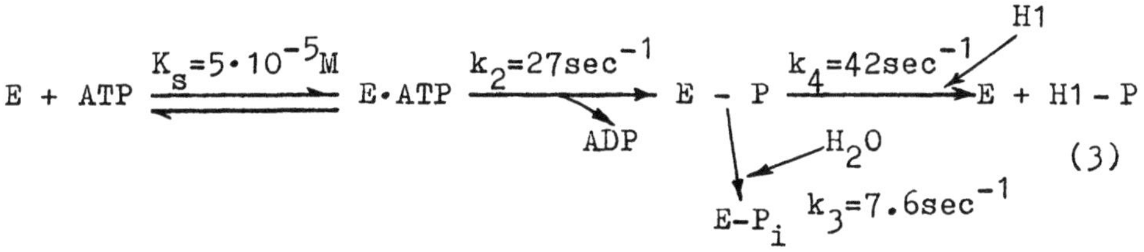

(3)

As follows from the scheme, the constant of phosphoryl enzyme hydrolysis (k_3=7.6sec^{-1}) is lower than the constant of formation of enzyme phosphoform (k_2=27sec^{-1}). Such a ratio of constants in the absence of the second substrate should lead to accumulation of the phosphoryl enzyme, and this enabled us to obtain it in the individual state. The ratio k_4/k_3 indicates that formation of the enzyme phosphoform can be detected only in the absence of the second substrate.

The functional topography of the active site of cAMP-dependent protein kinase was investigated using original synthetic ATP analogs; some of them appeared irreversible inhibitors of the enzyme. In the case of adenosine 5'-chloromethylpyrophosphonate, incorporation of 1.2 equivalent of the compound per 1 mol enzyme was observed. pK of the modified group was about 6.5. A suggestion is possible that the functional group blocked due to alkylation of the compound by the $ClCH_2$-group and the catalytic group of the enzyme is the same histidine residue. Another irreversible inhibitor of protein kinase was adenosine 5'-(n-fluorosulphonylphenylphosphate). In this case pK of the modified group was around 4.7 that might indicate the formation of an anhydride bond with the ω-carboxyl group of glutamic or aspartic acids in the active site (4). Finally, treatment of the catalytic subunit with "oxo-ATP", followed by sodium borohydride reduction, brought about irreversible inhibition of the enzyme activity. This reaction proceeded strictly in the ATP-site of the active center (protection from inhibition with excess ATP, incorporation of one equivalent of the ^{14}C-labeled inhibitor, isolation of the individual ^{14}C-labeled peptide after tryptic digestion)(5). These studies have shown that the lysine residue of the active site is subjected to selective modification. Besides, the following amino acid sequence for the isolated ^{14}C-labeled peptide was established: Val Gly Pro L$\overset{*}{y}$s Arg. One can assume that this peptide is a part of the cationic locus of the active site which is involved in the binding of the negatively charged polyphosphate chain of ATP (Fig. 1).

Thus, the ATP-binding site of the active center of protein kinase consists of at least three loci:

1) the hydrophobic cleft of the adenine moiety binding;

2) the cationic locus that binds the polyphosphate chain of ATP (Lys, Arg);

3) the catalytic site (histidine residue).

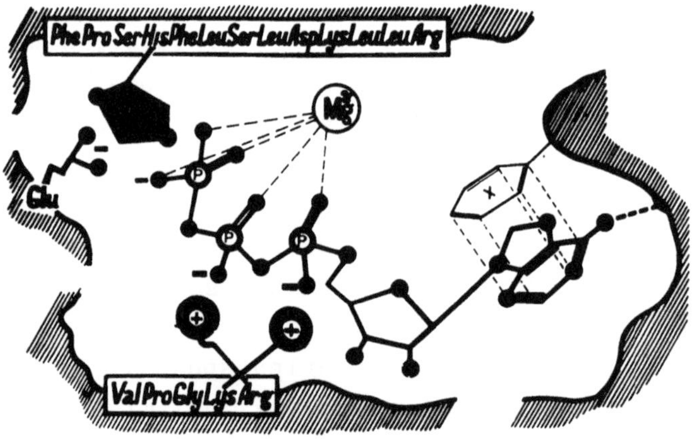

Fig. 1. A scheme of the ATP-binding site of catalytic subunit of cAMP-dependent protein kinase .

Nuclear Translocation and Biological Action of Regulatory Subunit of cAMP-Dependent Protein Kinase

While the function of the catalytic subunit of cAMP-dependent protein kinases as a factor modifying protein substrates has been closely studied, the biological role of the regulatory subunit still remains obscure. The question whether the regulatory component has its own independent functions, other than inhibition of the phosphotransferase activity of the catalytic subunit, requires a detailed investigation. The regulatory subunit was assumed to affect the transcription level by directly interacting with the template, likewise a cAMP-binding protein of bacteria. To check this suggestion, we analysed the changes in the template activity of chromatin after its treatment with the regulatory subunit in vitro. It appeared that the addition of the homogeneous regulatory subunit of pig brain protein kinase into the RNA polymerase reaction mixture caused an increase in the template activity of chromatin (6)(Fig. 2).

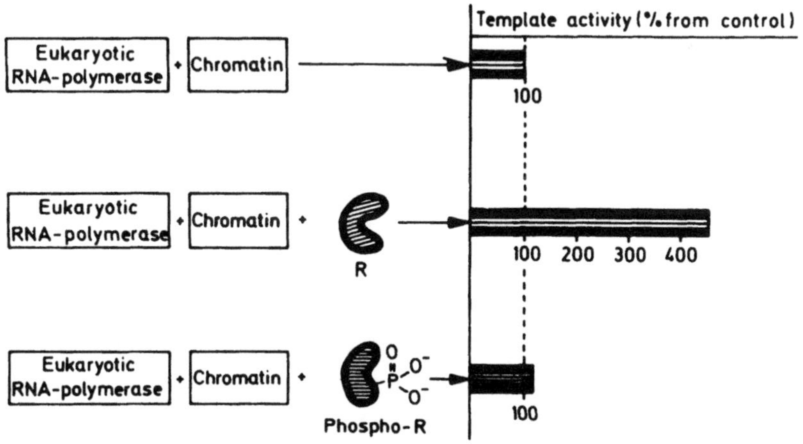

Fig. 2 . Effect of regulatory subunit of cAMP-dependent protein kinase on template activity of chromatin.

Besides the estimated binding constant ($10^{10}M^{-1}$), high specificity of action of the regulatory subunit on chromatin is also evidenced by the experiments on chromatin treatment with the phosphoform of the regulatory subunit. In this case, no effect on the template activity of chromatin was observed (7).

In the light of the above findings, a particular interest acquires investigation of the mechanisms of nuclear translocation and acception of protein kinase by structural elements of genome in normal and tumor cells. In normal cells like Chinese hamster fibroblasts (CHO) or mouse fibroblasts (3T3) we observed a pronounced translocation of exogenous protein kinase and its subunits separately into interphase nuclei and acception of these proteins by metaphase chromosomes. Whereas nuclear translocation of the ^3H-labeled regulatory subunit and acception by chromosomes of the SV40-transformed 3T3 cells were not detected. Addition into the system of the regulatory subunit-cAMP complex led to essential recovery of nuclear translocation of the labeled protein and its acception by metaphase chromosomes. In the cell culture of human spontaneous cancer, KB cells, the regulatory subunit penetrates the nucleus but is not accepted by chromosomes. Addition of cAMP does not produce any effect in this case. The results of our study indicate that transformation of 3T3 cells by virus SV-40 does not cause any modifications in the specific sites for binding the regulatory subunit to chromatin but affects the system of cAMP biosynthesis decreasing the intracellular level of the latter that leads to disruption of transport of cAMP-binding proteins through the nuclear membrane. In the case of neoplastic transformation of another type (KB cells), it is the mechanism of interaction of the regulatory subunit with the genetic material of cancer cells that is affected.

Thus, besides inhibition of the phosphotransferase activity, the regulatory subunit of cAMP-dependent protein kinase plays an important role in the processes related to the genome expression. A change in the template activity of chromatin if the latter interacts with the regulatory subunit can result in the synthesis of new protein products. We made an attempt to identify the product formed when the cells were treated with the regulatory subunit. For this purpose, into the culture medium of 3T3 cells the ^{14}C-labeled amino acids were added and in a certain time the newly-synthesized proteins of the cytosol were separated by a two-dimensional polyacrylamide gel electrophoresis by O'Farrell. 4 hr after treatment of the cells with the regulatory subunit among the synthesized proteins one protein which was not observed in the control experiments was detected. This protein designated by us as P-15 had a MW of 15,000 and the isoelectric point 6.3 (8). Any modifications of the regulatory subunit (phosphorylation, proteolysis, denaturation) led to the disappearance of protein P-15 that was indicative of the specificity of the above process. In the virus-transformed 3T3 cells in the presence of cAMP, synthesis of this protein took place even in the absence of the regulatory subunit. At the same time, the intensity of the detected product considerably increased after treatment of the cells with the regulatory subunit. The experimental results suggest that the changes in the cells transformed by virus SV-40 are to some extent similar to those observed in the normal cells after treatment with the regulatory subunit. The nature and biological role of protein P-15 are the subject of our future investigations.

References

1. Solomoniya, R. O., Nesterova, M. V., Kostyuk, D. A., Severin, E. S.: Biochem. Int. 3, 147-155 (1981).
2. Nesterova, M. V., Sashchenko, L. P., Vasiliev, V. Yu., Severin, E. S.: Biochim. Biophys. Acta 377, 271-286 (1975).
3. Gabibov, A. G., Kochetkov, S. N., Smirnov, I.V., Sashchenko, L. P., Severin, E. S.: Eur. J. Biochem. 115, 297-301 (1981).
4. Gulyaev, N. N., Tunitskaya, V. L., Baranova, L. A., Nesterova, M. V., Murtuzaev, I. M., Severin, E. S.: Biokhimiya

(Rus) 41, 1241-1249 (1976).

5. Kochetkov, S. N., Bulargina, T. V., Sashchenko, L. P., Severin, E. S.: Eur. J. Biochem. 81, 111-118 (1977).

6. Nesterova, M. V., Barbashov, S. F., Aripdzhanov, A. A., Abdukarimov, A., Severin, E. S.: Biokhimiya (Rus) 45, 979-991 (1980).

7. Nesterova, M. V., Ulmasov, Kh. A., Abdukarimov, A., Aripdzhanov, A. A., Severin, E. S.: Exp. Cell Res. 132, 367-373 (1981).

8. Nesterova, M. V., Glukhov, A. I., Severin, E. S.: Mol. Cell. Biochem. (1982)(in press).

CLONING AND EXPRESSION OF THE HUMAN LEUCOCYTE INTERFERON GENE

Eugene D.Sverdlov, Sergey A.Tsarev, Elena M.Khodkova, Galina S.Monastyrskaya, Vladimir A.Efimov, Oksana G.Chakhmakhcheva

Shemyakin Institute of Bioorganic Chemistry, USSR Academy of Sciences, 117988, Moscow, USSR

Valentin D.Solovyev, Vladimir P.Kusnetsov

Gamaleya Institute of Epidemiology and Microbiology USSR Academy of Medical Sciences, Moscow, USSR

Introduction

Interferons (IFNs) are a family of secreted proteins or glycoproteins which cause viral resistance while acting on their target cells (1,2). They possess also immunomodulatory and antiproliferative properties. IFNs are presently being evaluated for activity to cure different human diseases. However, until recently IFNs have not been available for clinical trials. At present the genetic engineering open the way of obtaining the IFNs by means of microbiological synthesis. Different kinds of human interferon genes have been recently cloned in E.coli and their structures have been established (3,4). Several groups have already reported efficient synthesis of IFN β_1 (5), αIFNs A, B and D (6,7) in Escherichia coli and αIFN D in yeast (8). We describe here the expression of another human IFN gene, IFN-F, in E.coli directed by different bacterial promoters.

Discussion

The way of the cloning of the corresponding c-DNA have been described in our recent paper (9).It includes isolation of polyadenylated m-RNA from human leucocyte cells stimulated for

Chemistry of Peptides and Proteins, Vol. 2

interferon production, obtaining of its c-DNA using the synthetic deoxyhexadecanucleotide as a primer for reverse transcriptase, insertion of the c-DNA connected with Hind III linkers into Hind III site of the pBR 322 vector plasmid, transformation E.coli K-12 strain 802 and screening of the clones grown on nitrocellulose filters by hybridization with ^{32}P-labelled synthetic hexadecanucleotides. The complete nucleotide sequence of the Hind III insert of the recombinant plasmid isolated from one of the positive clones was determined by the Maxam-Gilbert procedure. The comparison of the structure with the known structures of the IFNs genes (4) showed that insertion includes a complete coding sequence for the mature interferon F and a part coding for five acids of the precursor signal peptide. The restriction map of insertion is presented in Fig.1.

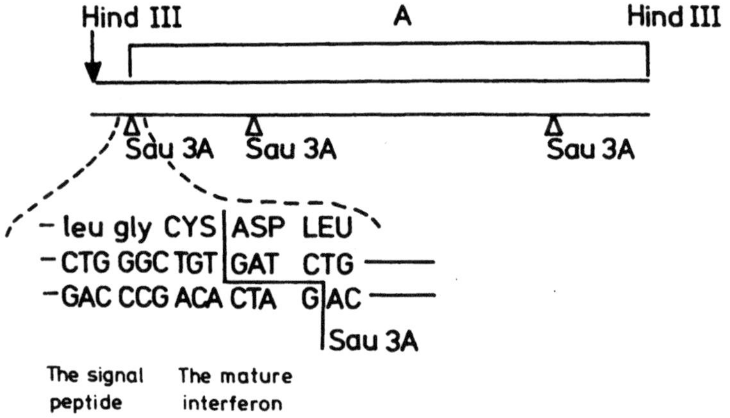

Fig.1. Sau 3A restriction map of the cloned IFN-F gene and the sequence surrounding the first site.

Antiviral activity of preinterferons is known to be rather low (6). Since the bacteria have no machinery for correct processing of eucaryotic proteins it is necessary to make a program of the mature protein synthesis in bacteria by means of in vitro operation.

The way to express the interferon F gene directly in E.coli was similar to that used for other genes (6,7). One of the three Sau 3A restriction endonuclease sites in insertion is located near the borderline separating the signal and mature parts of the gene, between codons 1 and 2 of the mature IFN gene. Using incomplete splitting of the gene with the Sau 3A, we were able to isolate fragment of the gene containing almost complete, but the first cysteine codon TGT, mature gene.
This fragment was ligated to a short synthetic DNA fragment containing the EcoRI and Sau 3A sticky ends as well as the duplex incorporating codon for cysteine and ATG translational initiation codon. The gene created after ligation contains the ATG initiation codon in addition to the sequence coding for mature IFN.
Therefore, the gene could be expected to direct the synthesis of fMet-IFN-F. However, as the most bacterial proteins do not contain N-terminal methionine residues, one can conclude that bacteria have a machinery for splitting the N-fMet. Because of that the final result of the IFN gene expression could be hoped to be the mature IFN.
For the expression of the gene, it is necessary to insert it into a bacterial plasmid downstream from a regulatory region containing the suitable bacterial promoters and ribosome binding sites. We have compared three kinds of the regulatory units listed in Table 1.
Two of the regulatory regions, a 285 base pair EcoRI fragment containing two 95 base pair lac UV-5 promoter fragments, separated by a 95 base pair heterologous spacer (10) and a part of trp operon, containing trp promoter, operator and trp leader ribosome binding site without ATG codon of the leader peptide programmed the direct expression of interferon. The third regulatory region was a 203 base pair EcoRI fragment containing the lac UV-5 promoter, operator, ribosome binding site and the sequence coding for the first 7 amino acids of the E.coli

ß-galactosidase structural gene. In this case the expression plasmid directed the synthesis of the fusion protein containing 7 amino acids of the ß-galactosidase, a proline residue, the codon of which is generated by the fusion of the promoter fragment to the gene, a methionine residue and the IFN-F mature protein.

Interferon activity was determined in extracts of E.coli cells containing the recombinant plasmids. The results of determination are presented in Table 1. Antiviral activity is observed in the clones containing IFN expression plasmids with promoters oriented so that transcription proceeds to the IFN gene but not in the extracts of clones containing the promoters in reverse orientation.

Table.1 Antiviral activity of the extracts from E.coli cells carrying the IFN plasmids.

plasmid	promoter	promoter orientation	interferon activity (U per 1 culture)
pBR 322	–	–	10
pBR 322-trp	–	–	10
pTMK-IFN-F-7	–	–	10
pTMK-IFN-F-71	lac^2	correct	10^6
pTMK-IFN-F-74	lac^2	opposite	10
pTMK-IFN-F-76	lac 203	correct	10^5
pTMK-IFN-F-78	lac 203	opposite	10
pTMKS-IFN-F-71	trp	correct	10^7-10^8
pTMKS-IFN-F-72	trp	correct	10^7-10^8

Extracts were prepared for IFN assay as described (6). IFN activity in the supernatants was determined by cytopathic effect inhibition assay (13) using vesicular stomatitis virus. pBR 322-trp: plasmid with trp promoter fragment but without IFN gene. pTMK-IFN-F-7 plasmid contains IFN-F gene without promoter. lac^2- tandem lac UV-5 promoter (10). lac 203 - lac UV-5 promoter with the part of the ß-galactosidase gene. trp - promoter of the trp operon of E.coli.

The IFN produced by E.coli behaves like human leucocyte interferon. It is stable to treatment at pH 2 and is neutralized by the antibodies to human leucocyte IFN but does not react with antihuman fibroblast or immune IFN antibodies.
The activity of the extracts increases from lac 203 UV-5 promoter to trp promoter. Low activity in the first case may be caused by synthesis of the fusion protein, but the sharp difference between double lac UV-5 and trp promoters neads some comments. The difference is much more pronounced than that in case of the expression of the IFN-ß gene under control of the same promoters (11). It is noteworthy, that the 5'end of m-RNA, the synthesis of which is directed by the lac UV-5 promoter connected to the IFN-F gene, may fold in specific secondary structure (Fig.2) involving an initiating AUG codon and ribosome binding site. This structure may be responsible for the low level of the gene expression at the translation step (12).

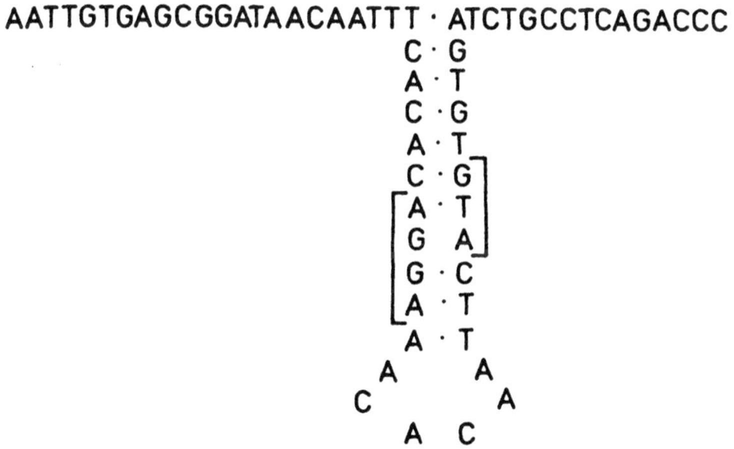

Fig.2. Proposed secondary structure of the 5'terminal segment of the IFN m-RNA corresponding to the recombinant plasmid with lac UV-5 promoter (10) fused to the IFN gene. The brackets indicate the ribosome interaction sequences.

The best producer clones give a high expression of the human leucocyte interferon. Assuming a specific activity of the IFN $4 \cdot 10^8$ $U \cdot mg^{-1}$, the yield of interferon per liter bacterial culture is approximately 250 µg. This is much more than one could isolate from human blood. Investigations of the biological properties of the bacterial IFN-F are in progress.

References

1. Isaacs, A., Lindenmann, J.: Proc. R. Soc., B147, 258-267 (1957).
2. Stewart, W.E.II.: The Interferon System (Springer, New York 1979).
3. Mantei, N., Schwarzstein, M., Streuli, M., Panem, S., Nagata S., Weissman, C.: Gene, 10, 1-10.
4. Goeddel, D.V., Leung, D.W., Dull, T.J., Gross, M., Lawn, R.M., McCandliss R., Seeburg, P.H., Ullrich, A., Yelverton, E., Gray, P.W.: Nature, 290, 20-26, (1981).
5. Nagata, S., Taira, H., Hall, A., Johnsrud, L., Streuli, M., Ecsodi, J., Boll, W., Weissmann, C.: Nature, 284, 316-320 (1980).
6. Goeddel, D.V., Yelverton, E., Ullrich, A., Heyneker, H.L., Miozzari, G., Holines, W., Seeburg, P.H., Dull, T., May, L., Stebbing, N., Crea, R., Maeda, S., McCandliss, R., Sloma, A., Tabor, J.M., Gross, M., Familletti, P.C., Pestka, S.: Nature, 287, 411-416 (1980).
7. Yelverton, E., Leung, D., Weck, P., Gray, P.W., Goeddel, D.V.: Nucl. Acid. Res., 9, 731-741 (1981).
8. Hitzeman, R.A., Hagie, F.E., Levine, H.L., Goeddel, D.V., Ammerer, G., Hall, B.D.: Nature, 293, 717-722 (1981).
9. Ovchinnikov, Yu.A., Sverdlov, E.D., Tsarev, S.A., Khodkova, E.M., Monastyrskaya, G.S., Efimov, V.A., Chachmachtcheva, O.G., Solovyev, V.D., Kusnetsov, V.P., Kavsan, V.M.: Doklady Academii Nauk USSR, 262, 725-728 (1982).
10. Johnsrud, L.: Proc. Nat. Acad. Sci. USA, 75, 5314-5318, (1978).
11. Goeddel, D.V., Shepard, H.M., Yelverton, E., Leung, D., Crea, R., Sloma, A., Pestka, S.: Nucl. Acid. Res., 8, 4057-4074 (1980).
12. Iserentant, D., Fiers, W.: Gene, 9, 1-12 (1980).
13. Nosick, D.N., Aspetov R.D., Novochatsky A.S.: Antibiotics (USSR), 1, 49 (1979).

List of Contributors

Abdurekhmanova, J. 319
Adylbekov, M. 319
Alakhov, Y.B. 187
Ancans, J. 261
Anderer, F.A. 357
Aschauer, H. 179

Bali, J.-P. 255
Bass, I.A. 403
Bektas, S. 179
Beni, C. 103
Berga, D. 261
Berndt, H. 97
Bessler, W.G. 87
Bieber, F. 179
Bode, K. 127
Bodenmüller, H. 165
Bogachuk, A.P. 195
Boheim, G. 215, 281
Borin, G. 255
Bosch, R. 215
Brandenburg, D. 11
Braun, V. 371
Braunitzer, G. 179
Breslav, M. 41
Brückner, H. 215
Buć, P. 71
Büllesbach, E. 17, 23

Calker, D. van 267
Casaretto, M. 35
Chakhmakhcheva, O.G. 421
Chertov, O.Y. 387
Chipens, G. 41, 261

Danho, W. 17, 35
Danilevskaya, O.N. 403
Deigin, V.I. 47, 145
Dovgas, N.V. 187
Dreher, R. 371

Efimov, V.A. 421

Faasch, H. 55
Fähnle, M. 115, 121
Fock-Nüzel, R. 377
Franke, M. 273
Fritz, H. 377

Gattner, H.-G. 17
Geiger, R. 273
Gespach, C. 137
Göhring, W. 29, 137
Goldammer, E. von 343
Grachev, M.A. 395
Gragerov, A.I. 403
Grigoryeva, V. 41
Grishin, E.V. 301, 309
Grotjahn, L. 155
Gubanov, V.V. 195, 403
Guryev, S.O. 195, 403

Hagenmaier, H. 155
Hamprecht, B. 267
Hanke, W. 281
Harper, G.P. 171
Hartter, P. 215
Hass, W. 55
Hegemann, P. 227
Heinzel, W. 103
Hensen, A. 377
Hesch, R.-D. 35
Hinrichsen-Kohane, A. 363

Ivanov, V.T. 47, 79, 291, 327

Johnson, R.B. 87
Jung, G. 215, 281

Kahl, J.-U. 63
Kalbacher, H. 103
Kalinina, N.F. 403
Katz, E. 215
Kessler, H. 233
Khodkova, E.M. 421
Kiver, I.F. 403
Kiyatkin, N.I. 301
Klein, B. 127
Kleinschmidt, T. 179
Klotz, G. 11
König, W.A. 55, 215
Kolkenbrock, H. 349
Korneyev, A. 319
Kostina, M.B. 203
Kozhich, A.T. 79
Krikis, A. 261

Kusnetsov, V.P. 421
Kuzin, A. 209

Lemke, H.D. 249
Lipkin, V.M. 195, 387, 395, 403
Lobbia, A. 255
Löffler, F. 267
Lottspeich, F. 377

Mästle, W. 121
Makarova, I.A. 387
Maser, F. 127
Mikhaleva, I.I. 47, 79
Mindlin, S.Z. 403
Mishlyakova, N. 261
Miura, T. 63
Monastyrskaya, G.S. 195, 387, 403, 421
Moroder, L. 29, 255
Motuz, L.P. 187
Müller, J. 103
Müller-Esterl, W. 377
Mustaev, A.A. 395
Mutter, M. 127
Mutule, I. 41
Mutulis, F. 41

Nassal, M. 71
Nikiforov, V.G. 403

Obermeier, R. 3
Oesterhelt, D. 227, 249
Ovchinnikov, Y.A. 241, 387, 395

Paul, J. 343
Pfeifle, J. 357
Pletnev, V. 209
Pluzhnikov, K.A. 327
Prosolova, T.K. 301

Reiser, G. 267
Rohrschneider, S. 273
Romani, S. 29
Rothe, M. 115, 121

Sadykov, A. 319
Salikhov, S. 319
Salomatina, I.S. 195
Sanguansermsri, T. 179
Saunders, D. 11
Scharf, R. 137
Scheuer, W. 87
Schill, W.-B. 363
Schmitt, H. 215
Schmoeckel, C. 363
Schuck, R. 233
Seipke, G. 3
Severin, E.S. 413
Shuvaeva, T.M. 195
Slavnova, T. 319
Smirnov, Y.V. 395
Soldatov, N.M. 301
Solovyev, V.D. 421
Steiner, M. 227
Sverdlov, E.D. 195, 387, 403, 421
Sychev, S.V. 291

Tashmukhamedov, M. 319
Thamm, P. 11
Thoenen, H. 171
Töpfer-Petersen, E. 363
Trakhanov, S. 209
Tsarev, S.A. 421
Tschesche, H. 343, 349
Tsetlin, V.I. 327
Turck, C. 97

Ulyashin, V.V. 47

Vaskovsky, B.V. 79
Vinokurov, L.M. 187
Voelter, W. 103
Voges, K.-P. 215
Vosekalna, I. 261
Vossler, G. 155

Wenzel, H.R. 343
Wieland, T. 63, 71
Wieneke, H.-J. 17
Wiesmüller, K. 87
Winkelmann, G. 333
Winter, W. 215
Wünsch, E. 29, 137, 255

Zahn, H. 17, 35
Zolotarev, A.S. 203

Ylyashin, V.V. 145

AUTHOR INDEX

containing names to be found in "References" at the end of each contribution

Abdukarimov, A. (6) 420
Abdukarimov, A. (7) 420
Abdulaev, N.G. (2) 85
Abdulaev, N.G. (1) 207
Abdulaev, N.G. (4) 247
Abdulaev, N.G. (13) 248
Abdulaev, N.G. (1) 254
Abita, J.P. (10) 260
Abita, J.-P. (6) 308
Abrahamson, E.W. (2) 247
Acher, R. (4) 348
Adelman, W.J. (5) 308
Adylbekov, M. (3) 326
Afanasyeva, G. (3) 266
Agarval, R.C. (2) 213
Agnew, W.S. (19) 308
Agnew, W.S. (4) 318
Akhtar, M. (11) 247
Akhtar, M. (6) 254
Alakhov, Yu.B. (2) 194
Alakhov, Yu.B. (3) 194
Alakhov, Yu.B. (4) 194
Alakhov, Yu.B. (5) 194
Alakhov, Yu.B. (6) 194
Alakhov, Yu.B. (7) 194
Alakhov, Yu.B. (8) 194
Aldanova, N.A. (3) 207
Almassy, R.J. (10) 318
Alpes, H. (9) 299
Alpes, H. (12) 299
Alpes, H. (13) 299
Altona, C. (1) 237
Alvarez, O. (3) 223
Amano, T. (17) 272
Amber, R.P. (6) 207
Ammerer, G. (8) 426
An, G. (10) 201
Anaokar, S. (8) 280
Ancans, J. (3) 266
Ancans, J. (4) 266
Anderegg, R.J. (2) 254
Anderegg, R.J. (5) 247
Anderegg, R.J. (12) 248
Anderer, F.A. (2) 362
Anderer, F.A.(3) 362
Anderson, J.C. (4) 34
Andreani, D. (9) 16
Andreani, D. (11) 16
Anfinsen, C.B. (2) 207
Anzinger, H. (9) 135
Apell, H.J. (9) 254
Apell, H.J. (9) 299
Apell, H.J. (12) 299
Apell, H.J. (13) 299
Argos, P. (7) 135
Aripdzhanov, A.A. (6) 420
Aripdzhanov, A.A. (7) 420
Armstrong, C.M (7) 308
Arnhold, M. (10) 386
Arseniev, A.S. (2) 53
Arseniev, A.S. (15) 54
Arseniev, A.S. (3) 332

Arseniev, A.S. (4) 332
Arseniev, A.S. (5) 332
Arseniev, A. (10) 332
Arzamazova, N.M. (3) 207
Asahi, K. (7) 61
Asanta-Poka, S. (14) 161
Aschauer, H. (11) 186
Aschauer, H. (12) 186
Aspetov, R.D. (13) 426
Atassi, M.Z. (8) 201
Atkinson, M.J. (5) 40
Audran, R. (6) 280
Avakov, A.E. (3) 393
Aydin, M. (8) 223
Aydin, M. (11) 223
Aydin, M. (13) 223

Babinet, C. (2) 410
Bailey, J.L. (2) 16
Bairamashvili, D.I. (15) 248
Baker, J.R. (7) 340
Balaram, P. (4) 223
Balaram, P. (30) 225
Balaram, P. (32) 225
Balasubramanian, T.M. (5) 223
Balasubramanian, T.M. (11) 288
Bali, J.-P. (3) 260
Bali, J.-P. (9) 260
Balmés, J.L. (9) 260
Balodis, J. (3) 266
Bamberg, E. (9) 254
Bamberg, E. (10) 254
Bamberg, E. (11) 254
Bamberg, E. (9) 299
Bamberg, E. (12) 299
Bamberg, E. (13) 299
Bamberg, E. (14) 299
Bamberg, E. (16) 299
Banquer, C. (6) 101
Baralle, F.E. (18) 186
Baranova, L.A. (4) 419
Barbashov, S.F. (6) 420
Barber, M. (8) 161
Barber, M. (10) 161
Barber, M. (12) 161
Barchi, R.L. (2) 308
Barde, Y.A. (3) 177
Baron, M.D. (10) 16
Barrett, A.J. (1) 355
Barry, G. (5) 393
Barton, M.A. (4) 34
Bartsch, H.-H. (5) 61
Bass, I.A. (6) 393
Bass, I.A. (8) 410
Bataille, D. (8) 144
Baumann, G. (5) 288
Bayer, E. (4) 135
Bayer, E. (9) 135
Bayer, E. (5) 144
Bayer, E. (4) 160
Bayer, E. (7) 260
Bayev, A.A. (5) 201
Bayev, A.A. (9) 410
Bayley, H. (3) 78
Bayley, H. (5) 78
Bayley, H. (9) 247
Bayley, H. (4) 254
Bayley, H. (11) 332
Beaven, G.H. (2) 185
Beaven, G.H. (5) 186
Beckey, H.D. (7) 161
Bedford, J.H. (4) 368

Begar, V.A. (17) 411
Bektas, M.S. (14) 186
Benedetti, E. (31) 225
Beneski, D.A. (11) 318
Beress, L. (9) 308
Béress, L. (6) 355
Béress, L. (7) 355
Bergmeyer, J. (16) 254
Berhanu, P. (13) 16
Berliner, L.J. (2) 348
Bermel, W. (1) 237
Berndt, H. (1) 101
Berndt, H. (3) 101
Berndt, H. (4) 101
Bérres, R. (7) 355
Bessler, W. (2) 95
Bessler, W. (3) 95
Bessler, W. (5) 95
Bessler, W.G. (6) 95
Bessler, W.G. (7) 95
Bessler, W.G. (8) 95
Bessler, W.G. (9) 96
Bethune, J.L. (2) 207
Beyermann, H.C. (7) 144
Bieber, F.A. (14) 186
Biemann, K. (5) 247
Biemann, K. (12) 248
Biemann, K. (2) 254
Birr, Chr. (11) 78
Birr, C. (3) 113
Birr, C. (10) 170
Bizzozero, S.A. (3) 120
Bláha, K. (43) 226
Bliznakow, E. (13) 114
Bloss, D.H. (11) 223
Blout, E.R. (7) 299
Blout, E.R. (8) 299
Blundell, T. (14) 16
Blundell, T. (2) 22
Blundell, T. (4) 28
Bodenmüller, H. (2) 70
Bodenmüller, H. (1) 170
Bodenmüller, H. (5) 170
Bodenmüller, H. (10) 170
Bodley, J.W. (1) 194
Bogomolni, R.A. (7) 232
Boheim, G. (1) 222
Boheim, G. (6) 223
Boheim, G. (7) 223
Boheim, G. (17) 224
Boheim, G. (21) 224
Boheim, G. (26) 224
Boheim, G. (27) 224
Boheim, G. (3) 288
Boheim, G. (4) 288
Boheim, G. (6) 288
Boheim, G. (8) 288
Boheim, G. (10) 288
Boheim, G. (16) 289
Boissonnas, R.A. (1) 126
Boll, W. (5) 426
Bolshakov, V.I. (3) 247
Bonora, G.M. (1) 135
Bonora, G.M. (2) 135
Bonora, G.M. (3) 135
Bordoli, R.S. (8) 161
Bordoli, R.S. (10) 161
Borin, G. (12) 386
Bosch, R. (34) 225
Bosch, R. (35) 225
Bosch, R. (36) 225
Bosch, R. (38) 225

Bosch, R. (40) 225
Bosch, R. (43) 226
Bosch, R. (44) 226
Bosch, R. (45) 226
Boucard, M. (9) 260
Bovermann, G. (27) 224
Bovey, F.A. (9) 238
Bradley, R. (12) 299
Brady, S.F. (7) 126
Bränden, C.I. (8) 340
Brandenburg, D. (3) 16
Brandenburg, D. (4) 16
Brandenburg, D. (6) 16
Brandenburg, D. (7) 16
Brandenburg, D. (7) 16
Brandenburg, D. (8) 16
Brandenburg, D. (9) 16
Brandenburg, D. (10) 16
Brandenburg, D. (11) 16
Brandenburg, D. (12) 16
Brandenburg, D. (13) 16
Brandenburg, D. (2) 22
Brandt, M. (6) 271
Brandt, M. (8) 272
Brauer, D. (5) 207
Braun, J.R. (15) 386
Braun, V. (1) 95
Braun, V. (2) 95
Braun, V. (4) 95
Braun, V. (11) 341
Braun, V. (1) 376
Braun, V. (3) 376
Braun, V. (4) 376
Braun, V. (5) 376
Braun, V. (6) 376
Braun, V. (7) 376
Braun, V. (8) 376
Braun, V. (9) 376
Braunitzer, G. (11) 186
Braunitzer, G. (13) 186
Braunitzer, G. (14) 186
Braunitzer, G. (15) 186
Braunitzer, G. (17) 186
Breslav, M. (4) 266
Brewster, A.I. (9) 238
Bricas, E. (10) 78
Brown, C.R. (8) 385
Brückner, H. (6) 95
Brückner, H. (8) 95
Brückner, H. (2) 222
Brückner, H. (9) 223
Brückner, H. (10) 223
Brückner, H. (11) 223
Brückner, H. (12) 223
Brückner, H. (34) 225
Brückner, H. (37) 225
Brückner, H. (41) 226
Brückner, H. (43) 226
Brückner, H. (9) 288
Brückner, H. (12) 289
Brufani, M. (9) 394
Brunfeldt, K. (4) 101
Brunfeldt, K. (9) 135
Brunfeldt, K. (8) 223
Brunfeldt, K. (31) 225
Brunner, J. (7) 78
Buchen, C. (6) 271
Buckley, J. (3) 385
Büllesbach, E. (3) 22
Büllesbach, E.E. (7) 28
Büllesbach, E.E. (8) 28
Büllesbach, E.E. (9) 28

Büllesbach, E.E. (10) 28
Bugg, C.E. (10) 318
Buku, A. (2) 70
Bulargina, T.V. (5) 420
Bundman, D. (2) 368
Bundule, M.A. (8) 194
Bundulis, Yu. P. (3) 194
Bundulis, Yu. P. (4) 194
Bundulis, Yu. P. (8) 194
Burck, P. (3) 28
Burgess, R.R. (4) 201
Burgess, R.R. (4) 401
Burgess, R.R. (15) 411
Burton, Z. (4) 201
Burton, Z. (4) 401
Burton, Z. (15) 411
Buse, G. (10) 232
Buse, G. (16) 386
Butler, E.A. (2) 185
Butters, T. (39) 225
Butz, R. (6) 40
Bystrov, V.F. (2) 53
Bystrov, V.F. (15) 54
Bystrov, V.F. (3) 332
Bystrov, V.F. (4) 332
Bystrov, V.F. (5) 332
Bystrov, V. (10) 332

Calker, D. van (4) 271
Calker, D. van (9) 272
Calker, D. van (10) 272
Calker, D. van (11) 272
Calker, D. van (12) 272
Calker, D. van (15) 272
Cam, A. le (12) 16
Capp, G.L. (3) 185
Carbone, E. (13) 318
Carpentier, J.-L. (12) 16
Carpousis, A.L. (8) 394
Carrano, C.J. (17) 341
Carrera, C. (6) 95
Casaretto, M. (5) 40
Casaretto, M. (6) 40
Casaretto, M. (7) 40
Catterall, W.A. (3) 308
Catterall, W.A. (9) 308
Catterall, W.A. (11) 318
Catterall, W.A. (13) 308
Catterall, W.A. (16) 308
Cavaggion, G. (12) 386
Cayrol, J.-P. (9) 260
Cech, C.L. (7) 394
Cerrini, S. (5) 120
Chachikyan, N.A. (7) 410
Chachmachtcheva, O.C. (9) 426
Chamberlin, M. (5) 401
Chamberlin, M.J. (5) 401
Chang, D. (14) 114
Chang, J.Y. (5) 207
Chang, K.J. (22) 272
Changeux, J.-P. (13) 332
Chauvet, J. (4) 348
Chen, R. (4) 207
Cherches, B.Z. (1) 213
Cherches, B.Z. (5) 213
Chertov, O.Yu. (1) 201
Chertov, O.Yu. (2) 201
Chertov, O.Yu. (1) 393
Chertov, O.Yu. (2) 401
Chertov, O.Yu. (3) 401
Chertov, O.Yu. (5) 410
Chessa, G. (12) 386

Chicheportiche, R. (6) 308
Chipens, G.I. (2) 46
Chipens, G. (3) 46
Chipens, G.I. (1) 266
Chipens, G.I. (2) 266
Chipens, G. (3) 266
Chipens, G. (4) 266
Chipens, G.I. (5) 266
Chipens, G.I. (6) 266
Chou, P.Y. (5) 135
Chowdry, V. (4) 78
Chupova, L.A. (5) 213
Claeys, H. (14) 386
Clark, J.L. (1) 28
Clegg, J.B. (4) 186
Cohen, S.A. (2) 308
Coopersmith, J. (16) 308
Cormick, J. (16) 186
Coulton, J.W. (5) 376
Court, D. (9) 201
Crea, R. (6) 426
Crea, R. (11) 426
Cuatrecasas, P. (22) 272
Cuervo, L.A. (5) 308
Culp, W.J. (14) 318
Cybulla, J. (8) 95
Czech, M.P. (1) 16

Dähn, U. (1) 61
Dahmen, J. (14) 114
Danho, W. (7) 40
Danho, W. (3) 85
Danilevskaya, O.N. (6) 393
Danilevskaya, O.N. (8) 410
Daves, D.G. (9) 114
Davis, D.G. (42) 226
Dawson, G. (21) 272
De Leeuw, F.A.A.M. (1) 237
De Riemer, L.H. (18) 411
Decker, G. (11) 386
Deffner, M. (1) 144
Dehler, W. (4) 61
Dehler, W. (6) 61
Deigin, V.I. (1) 53
Deigin, V.I. (4) 54
Deigin, V.I. (10) 54
Deigin, V.I. (11) 54
Deigin, V.I. (1) 153
Deimer, K.-H. (4) 260
Delcuve, G. (4) 393
Dencher, N.A. (9) 254
Denkewalter, R.G. (13) 54
Dennis, P.P. (4) 393
Deshko, T.N. (5) 54
Desiderio, D.M. (4) 160
Desiderio, D.M. (15) 161
Deslauriers, R. (6) 238
Deslongchamps, P. (2) 120
Di Bello, C. (1) 10
Diaconescu, C. (9) 16
Diddens, H. (10) 61
Diddens, H. (11) 61
Difton, M.J. (2) 332
Diordan, J.F. (13) 28
Dodson, G. (14) 16
Dodson, G. (4) 28
Dörsam, H. (9) 170
Dolphin, D. (3) 340
Dorgerloh, M. (11) 61
Dovgas, N.V. (7) 194
Dovgas, N.V. (8) 194
Downing, W. (4) 393

Drachev, L.A. (13) 248
Drees, F. (5) 34
Drees, F. (2) 144
Drefahl, G. (8) 61
Dreher, R. (7) 376
Drescher, H. (1) 185
Dress, F. (6) 34
Dress, F. (5) 260
Dubischar, N. (23) 224
Dugas, H. (6) 348
Dull, T.J. (4) 426
Dull, T. (6) 426
Dunford, H.B. (3) 340
Dupont, C. (8) 144
Dzieduszycka, M. (3) 85

Eaker, D. (4) 332
Ecsodi, J. (5) 426
Edgar, D. (3) 177
Efimov, V.A. (9) 426
Efremov, E.S. (1) 46
Ellena, J.F. (12) 332
Emery, L. (4) 340
Emery, T. (4) 340
Eng-Wilmot, D.L. (7) 340
Eng-Wilmot, D.L. (18) 341
Engelman, D.M. (3) 254
Epp, O. (7) 348
Erb, P. (3) 16
Ernst, J. (5) 340
Evans, S. (9) 161
Evanson, J.M. (2) 355

Fähnle, M. (4) 120
Fähnle, M. (6) 120
Fähnle, M. (5) 126
Faesel, J. (1) 78
Fahr, A. (10) 254
Fahr, A. (11) 254
Familletti, P.C. (6) 426
Farooqui, A.M. (9) 186
Fasman, G.D. (5) 135
Faulstich, H. (1) 70
Faulstich, H. (2) 70
Faulstich, H. (4) 70
Faulstich, H. (5) 70
Faulstich, H. (8) 70
Faulstich, H. (9) 70
Faulstich, H. (2) 78
Fawcett, D.W. (4) 368
Fedeli, W. (5) 120
Fedeli, W. (9) 394
Feher, G. (18) 289
Fehlmann, M. (12) 16
Feigina, M.Yu. (2) 85
Feigina, M.Yu. (1) 207
Feigina, M.Yu. (3) 207
Feigina, M.Yu. (4) 247
Feigina, M.Yu. (13) 248
Feigina, M.Yu. (15) 248
Feigina, M.Yu. (1) 254
Feinstein, G. (3) 355
Fessas, P. (6) 186
Fiedler, F. (11) 280
Fiedler, H.-P. (4) 61
Fiedler, H.-P. (6) 61
Fiers, W. (12) 426
Filatova, M.P. (1) 46
Fink, E. (4) 280
Fink, E. (9) 355
Finkelstein, A. (4) 326
Fischer, K. (6) 271

Fischer, U. (14) 254
Fischer, U. (15) 254
Flohé, L. (16) 386
Flynn, F.V. (2) 185
Fock-Nüzel, R. (13) 386
Föhles, J. (8) 28
Föhles, J. (9) 28
Folkers, K. (13) 114
Folkers, K. (14) 114
Fomina, J.P. (1) 213
Fontana, A. (6) 34
Fontecilla-Camps, J.C. (10) 318
Formanek, H. (7) 348
Forrester, J.D. (16) 341
Fossel, E.T. (7) 299
Frank, B.H. (3) 28
Frank, H. (5) 144
Frank, H. (1) 160
Frank, H. (2) 160
Frank, H. (3) 160
Frank, H. (4) 160
Frank, H. (5) 160
Frank, H. (7) 260
Franke, M. (10) 280
Freeman, R. (7) 238
Freidinger, R.M. (7) 126
Frenz, J. (6) 376
Freychet, P. (12) 16
Frien, M. (2) 368
Friedrich, A. (1) 237
Frisen, J.D. (10) 201
Fritz, H. (3) 280
Fritz, H. (5) 280
Fritz, H. (11) 280
Fritz, H. (9) 348
Fritz, H. (6) 355
Fritz, H. (9) 355
Fritz, H. (5) 385
Fritz, H. (6) 385
Fritz, H. (7) 385
Fritz, H. (9) 385
Fritz, H. (10) 386
Fritz, H. (11) 386
Fritz, H. (13) 386
Fritz, H. (19) 386
Fujiki, H. (10) 402

Gabibov, A.G. (3) 419
Galaktionov, S.G. (1) 46
Galanons, C. (1) 95
Gammeltoft, S. (2) 22
Ganten, D. (3) 177
Gario, A. (7) 326
Garry, P.J. (8) 280
Gattner, H.-G. (8) 10
Gattner, H.-G. (10) 10
Gattner, H.-G. (5) 22
Gattner, H.G. (6) 28
Gattner, H.G. (7) 28
Gattner, H.G. (8) 28
Gattner, H.G. (9) 28
Gavuzzo, E. (5) 120
Gehring-Müller, R. (15) 186
Geiger, R. (3) 10
Geiger, R. (5) 28
Geiger, R. (5) 101
Geiger, R. (18) 114
Geiger, R. (3) 144
Geiger, R. (2) 280
Geiger, R. (3) 280
Geiger, R. (4) 280
Geiger, R. (5) 280

Geiger, R. (10) 280
Gerber, G.E. (5) 247
Gerber, G.E. (12) 248
Gerber, G.E. (2) 254
Gertler, A. (3) 355
Gespach, C. (8) 144
Ghosh, B.K. (1) 376
Ghosh, M.N. (8) 260
Giacomello, G. (9) 394
Gierer, A. (4) 170
Gilbert, W. (6) 201
Gilbert, W. (10) 410
Gillespie, J.I. (12) 318
Gillessen, D. (3) 260
Gilon, C. (13) 254
Gisin, B.F. (42) 226
Glanville, R.W. (1) 177
Glaser, T. (8) 272
Glaser, T. (16) 272
Glickson, J.D. (6) 299
Gliemann, J. (2) 22
Gliemann, J. (6) 22
Glukhov, A.I. (8) 420
Goeddel, D.V. (4) 426
Goeddel, D.V. (6) 426
Goeddel, D.V. (7) 426
Goeddel, D.V. (8) 426
Goeddel, D.V. (11) 426
Göhring, W. (10) 61
Göring, W. (4) 280
Goldammer, E. von (3) 348
Goldammer, E. von (5) 348
Goldberg, R. (13) 254
Goodall, M.C. (6) 299
Goodman, M. (4) 160
Gordon, J. (12) 232
Gordon, L.G.M. (2) 288
Gorlenko, J.M. (7) 410
Grachev, M.A. (8) 402
Gralla, J.D. (8) 394
Grau, U. (6) 10
Graven, G.P. (2) 207
Gray, C.P. (5) 247
Gray, C.P. (12) 248
Gray, C.P. (2) 254
Gray, P.W. (4) 426
Gray, P.W. (7) 426
Gray, W.R. (11) 28
Gray, W.R. (7) 170
Gray, W.R. (8) 170
Greengard, P. (1) 362
Greenstein, J.P. (17) 114
Gregory, R.A. (1) 260
Gregory, R.A. (2) 260
Greiner, M. (9) 223
Greiner, M. (10) 223
Greiner, M. (9) 288
Grigorian G.C. (1) 46
Grigsby, R.D. (9) 161
Grillemeier, M. (17) 186
Grimmelikhuijzen, C.J.P. (6) 170
Grindstoff, Q.G. (9) 161
Grinkevich, Ch.A. (3) 207
Grinkevich, V.A. (2) 201
Grinkevich, V.A. (3) 207
Grinkevich, V.A. (1) 393
Grinkevich, V.A. (2) 401
Grinkevich, V.A. (5) 410
Grishin, E.V. (8) 308
Grishin, E.V. (15) 308
Grishin, E.V. (20) 308
Grishin, E.V. (21) 308

Grishin, E.V. (3) 318
Grishin, E.V. (5) 318
Grishin, E.V. (6) 318
Grishin, E.V. (7) 318
Grishin, E.V. (8) 318
Grishin, E.V. (9) 318
Gross, C.A. (4) 201
Gross, C.A. (4) 401
Gross, C.A. (15) 411
Gross, E. (3) 28
Gross, E. (8) 28
Gross, E. (10) 114
Gross, E. (5) 126
Gross, E. (4) 135
Gross, M. (4) 426
Gross, M. (6) 426
Grossman, M.I. (7) 144
Gubanov, V.V. (2) 201
Gubanov, V.V. (1) 393
Gubanov, V.V. (2) 393
Gubanov, V.V. (6) 393
Gubanov, V.V. (1) 401
Gubanov, V.V. (2) 401
Gubanov, V.V. (5) 410
Gubanov, V.V. (8) 410
Günzler, W.A. (16) 386
Gullis, R.J. (6) 271
Gullis, R. (18) 272
Gulyaev, N.N. (4) 419
Gundersen, C.B. (1) 326
Gurevich, A.I. (3) 393
Guryev, S.O. (2) 201
Guryev, S.O. (3) 201
Guryev, S.O. (5) 201
Guryev, S.O. (1) 393
Guryev, S.O. (1) 401
Guryev, S.O. (2) 401
Guryev, S.O. (5) 410
Guryev, S.O. (9) 410
Guryev, S.O. (2) 393

Haberland, G.L. (11) 280
Habermann, E. (17) 289
Habub, R.F.S.A. (8) 201
Hagenmaier, H. (1) 61
Hagenmaier, H. (3) 61
Hagenmaier, H. (5) 61
Hagenmaier, H. (6) 61
Hagenmaier, H. (1) 160
Hagenmaier, H. (2) 160
Hagenmaier, H. (3) 160
Hagenmaier, H. (4) 160
Hagenmaier, H. (5) 160
Hagie, F.E. (8) 426
Hagmann, W. (2) 362
Hagmann, W. (3) 362
Hall, A. (5) 426
Hall, B.D. (8) 426
Hall, J.E. (5) 223
Hall, J.E. (11) 288
Hall, Z. (9) 332
Hallermayer, K. (3) 271
Hamprecht, B. (2) 271
Hamprecht, B. (3) 271
Hamprecht, B. (4) 271
Hamprecht, B. (5) 271
Hamprecht, B. (6) 271
Hamprecht, B. (8) 272
Hamprecht, B. (9) 272
Hamprecht, B. (10) 272
Hamprecht, B. (11) 272
Hamprecht, B. (12) 272

Hamprecht, B. (13) 272
Hamprecht, B. (14) 272
Hamprecht, B. (15) 272
Hamprecht, B. (16) 272
Hamprecht, B. (17) 272
Hamprecht, B. (18) 272
Hancock, E. (5) 95
Hanke, W. (6) 223
Hanke, W. (7) 223
Hanke, W. (17) 224
Hanke, W. (10) 288
Hanke, W. (16) 289
Hantke, K. (5) 95
Hantke, K. (1) 376
Hantke, K. (6) 376
Hardy, P.M. (4) 34
Harmening, C. (3) 271
Harper, G.P. (1) 177
Harper, G.P. (2) 177
Harper, G.P. (3) 177
Harper, G.P. (5) 177
Harter, B. (12) 299
Hartley, B.S. (15) 386
Hartmann, A. (2) 33
Hartree, E.F. (8) 385
Hartshorne, R.P. (16) 308
Hartshorne, R.P. (11) 318
Hass, W. (4) 61
Hass, W. (9) 61
Haupt, H. (5) 355
Havinga, E. (9) 78
Haydon, D.A. (2) 288
Haydon, D.A. (4) 298
Heding, L.G. (2) 28
Hefti, F. (3) 177
Hehlein, W. (5) 238
Hehrmann, R. (5) 40
Hehrmann, R. (6) 40
Heil, A. (21) 411
Heimburger, N. (5) 355
Heimburger, N. (7) 385
Heinzel, W. (15) 114
Heinzel, W. (16) 114
Heitz, F. (22) 224
Heller, K. (5) 376
Helm, D. van der (7) 340
Helm, D. van der (18) 341
Helm, D. van der (19) 341
Henderson, R. (6) 247
Henderson, R. (3) 254
Henderson, R. (7) 254
Hendy, G.N. (1) 40
Henschen, A. (13) 386
Herlihy, W.C. (5) 247
Herlihy, W.C. (12) 248
Herlihy, W.C. (2) 254
Hesch, R.D. (6) 40
Hesch, R.D. (7) 40
Heumann, R. (3) 177
Heumann, R. (4) 177
Heumann, R. (4) 271
Heyneker, H.L. (6) 426
Heyward, R.S. (11) 411
Hider, R.C. (2) 332
Hille, B. (17) 308
Hilschmann, N. (15) 186
Hilse, K. (15) 186
Hirschmann, R. (13) 54
Hitzeman, R.A. (8) 426
Hladky, S.B. (4) 298
Hobom, G. (15) 186
Hodgkin, A.L. (1) 308

Hodgkin, D. (14) 16
Hodgkin, D. (4) 28
Höhne, H. (1) 61
Hökfelt, T. (1) 271
Höltje, J.-V. (9) 376
Hörhold, H.H. (8) 61
Holder, S. (4) 201
Holder, S. (4) 401
Holder, S. (15) 411
Holeysovsky, V. (7) 207
Holeysovsky, V. (18) 386
Holines, W. (6) 426
Holly, F.W. (7) 126
Holzer, H. (1) 348
Hossain, M.B. (7) 340
Hossain, M.B. (18) 341
Howard, B.D. (1) 326
Huang, K.S. (9) 247
Huang, K.S. (4) 254
Huber, R. (7) 348
Hucho, F. (8) 332
Hübner, K. (16) 272
Huehns, E.R. (2) 185
Huehns, E.R. (5) 186
Huehns, E.R. (9) 186
Hütter, P. (39) 225
Hull, W.E. (1) 237
Hurlbuf, W.P. (6) 326
Huxley, A.F. (1) 308
Huxley, A.F. (1) 318

Ilyina, T.S. (3) 410
Ilyina, T.S. (7) 410
Indulen, J. (3) 266
Inouye, K. (4) 10
Irmscher, G. (27) 224
Irmscher, G. (7) 288
Isaacs, A. (1) 426
Iselin, B. (2) 113
Iselin, B. (4) 114
Iserentant, D. (12) 426
Ishihama, A. (4) 410
Ishihawa, A. (14) 411
Isono, K. (7) 61
Isono, M. (9) 10
Ivakura, I. (4) 410
Ivanov, V.T. (9) 28
Ivanov, V.T. (1) 46
Ivanov, V. (3) 46
Ivanov, V.T. (1) 53
Ivanov, V.T. (2) 53
Ivanov, V.T. (3) 54
Ivanov, V.T. (4) 54
Ivanov, V.T. (5) 54
Ivanov, V.T. (6) 54
Ivanov, V.T. (7) 54
Ivanov, V.T. (8) 54
Ivanov, V.T. (9) 54
Ivanov, V.T. (10) 54
Ivanov, V.T. (11) 54
Ivanov, V.T. (15) 54
Ivanov, V.T. (1) 153
Ivanov, V.T. (3) 332
Ivanov, V. (10) 332
Ivanov, V.T. (10) 299
Ivanov, V.T. (11) 299
Ivanov, V.T. (4) 332
Ivanov, V.T. (5) 332
Ivanov, V.T. (6) 332
Ivanov, V. (10) 332
Iwasaki, M. (11) 10

Jaeger, E. (1) 144
Jaeger, E. (6) 144
Jaeger, E. (8) 144
Jaeger, E. (5) 260
Jaeger, E. (6) 260
Jäger, G. (18) 114
Janko, K. (26) 224
Janko, K. (8) 288
Jansen, A.C.A. (9) 78
Jochum, C. (8) 70
Jöhl, A. (2) 33
Johansson, O. (1) 271
Johnson, A.H. (11) 247
Johnson, A.H. (6) 254
Johnson, R.B. (7) 95
Johnson, M.H. (1) 368
Johnsrud, L. (5) 426
Johnsrud, L. (10) 426
Jonczyk, A. (8) 10
Jones, R.T. (3) 185
Jones, R.T. (16) 186
Jordanov, St. (10) 299
Jovin, T.M. (19) 411
Jovin, T.M. (20) 411
Jüppner, H. (5) 40
Jüppner, H. (6) 40
Jung, G. (10) 61
Jung, G. (6) 95
Jung, G. (7) 95
Jung, G. (8) 95
Jung, G. (2) 222
Jung, G. (6) 223
Jung, G. (7) 223
Jung, G. (9) 223
Jung, G. (10) 223
Jung, G. (11) 223
Jung, G. (12) 223
Jung, G. (13) 223
Jung, G. (14) 224
Jung, G. (15) 224
Jung, G. (16) 224
Jung, G. (17) 224
Jung, G. (20) 224
Jung, G. (21) 224
Jung, G. (23) 224
Jung, G. (24) 224
Jung, G. (25) 224
Jung, G. (26) 224
Jung, G. (27) 224
Jung, G. (28) 224
Jung, G. (29) 225
Jung, G. (33) 225
Jung, G. (34) 225
Jung, G. (35) 225
Jung, G. (36) 225
Jung, G. (37) 225
Jung, G. (38) 225
Jung, G. (39) 225
Jung, G. (40) 225
Jung, G. (41) 226
Jung, G. (43) 226
Jung, G. (44) 226
Jung, G. (45) 226
Jung, G. (7) 288
Jung, G. (8) 288
Jung, G. (9) 288
Jung, G. (10) 288
Jung, G. (12) 289
Jura, T. (4) 410

Kadner, R.J. (5) 376
Kahl, J.U. (6) 70

Kaiser, E.T. (12) 54
Kaji, Y. (2) 231
Kalamkarov, G.R. (3) 247
Kalbacher, H. (6) 114
Kalbacher, H. (7) 114
Kalbacher, H. (10) 114
Kalbacher, H. (11) 114
Kalbacher, H. (13) 114
Kalbacher, H. (14) 114
Kalinina, N.F. (3) 201
Kaltsova, A. (6) 186
Kamber, B. (2) 33
Kamber, B. (3) 34
Kamuzora, H. (8) 186
Kappel, M.J. (15) 341
Karelin, A.A. (6) 332
Karl, H.J. (4) 280
Karlsson, E. (2) 53
Karlsson, E. (15) 54
Karlsson, E. (1) 332
Karlsson, E. (3) 332
Karlsson, E. (4) 332
Karlsson, E. (5) 332
Karlsson, E. (10) 332
Kasai, N. (10) 96
Katayev, B.S. (2) 46
Katayev, B.S. (6) 266
Katz, E. (13) 223
Katz, E. (17) 224
Katz, E. (24) 224
Katz, E. (43) 226
Katz, E. (44) 226
Katz, E. (45) 226
Katz, E. (10) 288
Kaulen, A.D. (13) 248
Kavsan, V.M. (9) 426
Kawaoi, A. (7) 280
Keck, K. (3) 16
Keckeisen, A. (3) 61
Keckeisen, A. (6) 61
Keipert, B. (4) 286
Keller, O. (8) 34
Keller-Schierlein, W. (9) 340
Kemmler, W. (12) 28
Kemp, D.S. (6) 101
Kemper, W. (17) 272
Kendrick, N.C.E. (5) 223
Kenner, G.W. (4) 34
Kerling, K.E.T. (9) 78
Kerp, L. (6) 10
Kessler, H. (6) 126
Kessler, H. (1) 237
Kessler, H. (3) 238
Kessler, H. (4) 238
Kessler, H. (5) 238
Keutmann, H.T. (1) 40
Khesin, R.B. (6) 393
Khesin, R.B. (8) 410
Khitrina, L.V. (13) 248
Khodkova, E.M. (9) 426
Khodorov, B.I. (14) 308
Khokhlov, A.S. (1) 213
Khoklov, A.S. (5) 213
Khorana, H.G. (8) 232
Khorana, H.G. (5) 247
Khorana, H.G. (9) 247
Khorana, H.G. (12) 248
Khorana, H.G. (2) 254
Khorana, H.G. (4) 254
Kimura, T. (2) 40
Kirchner, H. (9) 96
Kirschbaum, J.B. (6) 410

Kiselev, A.V. (2) 85
Kiselev, A.V. (1) 207
Kiselev, A.V. (4) 247
Kiselev, A.V. (13) 248
Kiselev, A.V. (1) 254
Kisfaludy, L. (6) 260
Kiver, I.F. (6) 393
Kiver, I.F. (8) 410
Klauser, R. (9) 355
Klebensberger, W. (8) 272
Klee, W.A. (7) 271
Klee, W.A. (19) 272
Kleihauer, E. (10) 186
Klein, M.P. (6) 340
Klis, W.A. (10) 238
Klotz, G. (9) 16
Klusha, V.E. (2) 46
Kluša, V. (3) 266
Kluša, V.E. (6) 266
Knof, S. (6) 144
Knorre, D.G. (7) 402
Knowles, J.R. (2) 16
Knowles, J.R. (3) 78
Knowles, J.R. (5) 78
Knowles, J.R. (11) 332
Kobayashi, M. (11) 10
Kochetkov, S.N. (3) 419
Kochetkov, S.N. (5) 420
Koehler, J.K. (3) 368
Koehler, J.K. (6) 369
König, W.A. (1) 61
König, W.A. (2) 61
König, W.A. (3) 61
König, W.A. (4) 61
König, W.A. (5) 61
König, W.A. (6) 61
König, W.A. (9) 61
König, W. (5) 101
König, W. (3) 144
König, W.A. (8) 223
König, W.A. (9) 223
König, W.A. (10) 223
König, W.A. (11) 223
König, W.A. (12) 223
König, W.A. (13) 223
König, W.A. (26) 224
König, W.A. (8) 288
König, W.A. (9) 288
Kolaskar, A.S. (8) 135
Kolb, H.A. (1) 222
Kolb, H.A. (6) 288
Kolb, H.A. (16) 299
Kolb, H.J. (11) 280
Koloditskaya, T.A. (1) 213
Kolosov, M.N. (3) 393
Kondakov, V.I. (15) 54
Kondakov, V.I. (5) 332
Konisky, J. (2) 376
Konrad, E.B. (6) 410
Konz, W. (1) 78
Korneyev, A. (3) 326
Kosman, D.J. (10) 348
Kostetsky, P.V. (3) 213
Kostrzewa, M. (1) 120
Kostrzewa, M. (4) 126
Kostyuk, D.A. (1) 419
Kovalenko, V.A. (3) 318
Kovalenko, V.A. (5) 318
Kovalev, Yu.N. (7) 410
Kozlov, V.P. (8) 194
Kraas, E. (10) 61
Krack, G. (1) 237

Krail, G. (2) 22
Krainer, A. (5) 393
Krakow, J.S. (19) 411
Krasilnikov, O.V. (5) 326
Krauel, A. (8) 376
Kravchenko, Z.B. (3) 207
Krayev, A.S. (5) 201
Krayev, A.S. (9) 410
Krikis, A. (1) 266
Krikis, A. (2) 266
Krippahl, G. (3) 231
Krohn, H. (10) 223
Kronbach, T. (15) 114
Kruse, K. (12) 223
Kudelin, A.B. (15) 248
Kudryashova, V. (3) 266
Künzer, W. (1) 185
Kukla, D. (7) 348
Kumar, A. (2) 238
Kupfer, S. (9) 355
Kupfer, S. (5) 385
Kupryszewski, G. (6) 16
Kupryszweski, G. (2) 53
Kurbatov, V.A. (7) 402
Kusnetsov, V.P. (9) 426
Kustermann-Kuhn, B. (9) 96
Kuzin, A.P. (3) 213
Kyi, A.T. (5) 114

Läuger, H. (9) 254
Läuger, P. (10) 254
Lahm, H.W. (4) 101
Lai, M.C.S. (5) 186
Lang, P. (11) 78
Lankford, C.E. (1) 340
Lanyi, J.K. (3) 231
Lanyi, J.K. (5) 231
Lanyi, J.K. (6) 231
Lanyi, J.K. (9) 232
Laskowski, M. (4) 355
Latorre, R. (3) 223
Lauger, P. (14) 299
Lauro, R. (9) 16
Lauro, R. (11) 16
Lawn, R.M. (4) 426
Lazdunski, M. (10) 260
Lazdunski, M. (6) 308
Lazdunski, M. (8) 348
Leban, J. (14) 114
Lee, C.Y. (1) 332
Lee, H.M. (13) 28
Leibfritz, D. (16) 224
Leibfritz, D. (23) 224
Leibfritz, D. (26) 224
Leibfritz, D. (28) 224
Leibfritz, D. (8) 288
Lehmann, W.D. (13) 161
Lehmann, H. (8) 186
Lehnert, P. (1) 144
Lehnert, P. (6) 144
Lemke, H.-D. (13) 232
Lemke, H.D. (10) 247
Lemke, H.D. (5) 254
Lemke, H.D. (8) 254
Lemke, H.D. (11) 254
Lemke, H.D. (16) 254
Leong, J. (12) 341
Leong, J. (13) 341
Leong, J. (14) 341
Leong, J. (20) 341
Leung, D.W. (4) 426
Leung, D. (7) 426

Leung, D. (11) 426
Levine, H.L. (8) 426
Lewis, H. (4) 393
Liepins, E. (3) 266
Liepins, E. (4) 266
Liepkaula, I.K. (5) 266
Lim, V.I. (7) 135
Lin, J. (4) 201
Lin, J. (4) 401
Lin, J. (15) 411
Lindenmann, J. (1) 426
Lindley, E.V. (4) 231
Lindsay, D.C. (2) 22
Lipkin, V.M. (1) 201
Lipkin, V.M. (2) 201
Lipkin, V.M. (3) 201
Lipkin, V.M. (7) 201
Lipkin, V.M. (1) 393
Lipkin, V.M. (2) 393
Lipkin, V.M. (6) 393
Lipkin, V.M. (1) 401
Lipkin, V.M. (2) 401
Lipkin, V.M. (3) 401
Lipkin, V.M. (8) 402
Lipkin, V.M. (5) 410
Lipkin, V.M. (8) 410
Lipp, W. (4) 280
Livengood, D.R. (20) 272
Ljungdahl, A. (1) 271
Llinás, M. (6) 340
Lobanov, N.A. (2) 85
Lobanov, N.A. (1) 207
Lobanov, N.A. (4) 247
Lobanov, N.A. (1) 254
Lochinger, W. (11) 78
Löffler, F. (11) 272
Löffler, F. (12) 272
Löw, M. (6) 260
Loffet, A. (1) 53
Loghry, R.A. (7) 340
Loghry, R.A. (18) 341
London, E. (8) 232
Losick, R. (5) 401
Lottspeich, F. (13) 386
Lucente, G. (5) 120
Lucente, G. (7) 120
Ludescher, U. (2) 126
Lüben, G. (1) 78
Lundanes, E. (13) 114
Lundanes, E. (14) 114
Lundberg, J.M. (1) 271

MacDonald, R.E. (4) 231
MacLeod, J.K. (4) 34
Machleidt, W. (6) 355
Maeda, S. (6) 426
Maelicke, A. (13) 318
Mästle, W. (4) 120
Mästle, W. (6) 120
Magnusson, S. (14) 386
Mair, G. (11) 280
Makarova, I.A. (2) 201
Makarova, I.A. (1) 393
Makarova, I.A. (2) 401
Makarova, I.A. (5) 410
Makarova, N. (3) 266
Makofske, R.C. (3) 85
Malon, P. (43) 226
Malygin, A.G. (6) 401
Mann, K. (4) 780
Mantei, N. (3) 426
Marchenko, T.V. (2) 201

Marchenko, T.V. (1) 393
Marchenko, T.V. (2) 401
Marchenko, T.V. (5) 410
Marchiori, F. (12) 386
Marignan, R. (9) 260
Markussen, J. (7) 10
Markussen, J. (2) 28
Marschner, I. (6) 40
Marshall, G.R. (5) 223
Marshall, G.R. (11) 288
Martonosi, A.N. (11) 299
Masaki, T. (9) 10
Mashimo, J. (10) 96
Massaglia, A. (1) 22
Masui, Y. (2) 40
Matheson Jr., R.R. (6) 348
Matoni, G. (3) 101
Matoni, G. (4) 101
Matzanke, B. (15) 341
Mauro, A. (7) 326
Maxam, A. (6) 201
Maxam, A. (10) 410
May, L. (6) 426
Mayers, D.F. (6) 299
Mayr, W. (25) 224
Mayr, W. (28) 224
Mayr, W. (29) 225
Mayr, W. (33) 225
McCandliss, R. (4) 426
McCandliss, R. (6) 426
McClure, W.R. (7) 394
McClure, W.R. (20) 411
McKenzie, D.T. (14) 318
McLachlan, A.D. (3) 254
McNamee, M.G. (12) 332
McRorie, R.A. (5) 369
McRorie, R.A. (2) 385
Meazes, C. (18) 411
Meienhofer, J. (8) 28
Meienhofer, J. (10) 114
Meienhofer, J. (5) 126
Meienhofer, J. (4) 135
Meienhofer, J. (2) 160
Meienhofer, M. (4) 160
Meinhofer, J. (3) 85
Melchers, F. (1) 95
Melderis, H. (7) 186
Mendlowski, M. (7) 126
Mercola, D. (14) 16
Mercola, D. (4) 28
Mesrob, B. (7) 207
Methfessel, C. (17) 224
Methfessel, C. (10) 288
Meves, H. (12) 318
Meyer, H. (2) 160
Mikes, O. (18) 386
Mikhaleva, I.I. (1) 53
Mikhaleva, I.I. (2) 53
Mikhaleva, I.I. (3) 54
Mikhaleva, I.I. (4) 54
Mikhaleva, I.I. (5) 54
Mikhaleva, I.I. (6) 54
Mikhaleva, I.I. (10) 54
Milkowski, J.D. (13) 54
Miller, R.J. (22) 272
Mindlin, S.Z. (6) 393
Mindlin, S.Z. (3) 410
Mindlin, S.Z. (7) 410
Mindlin, S.Z. (8) 410
Miozzari, G. (6) 426
Miroshnikov, A.I. (10) 299
Mishlyakova, N. (3) 46

Mishlyakova, N. (3) 266
Mishlyakova, N. (4) 266
Mishlyakova, N.V. (6) 266
Misina, I.P. (2) 46
Misins, I.P. (6) 266
Misler, S. (6) 326
Mock, W.L. (3) 120
Modyanov, N.N. (1) 201
Modyanov, N.N. (2) 201
Modyanov, N.N. (7) 201
Modyanov, N.N. (1) 393
Modyanov, N.N. (2) 401
Modyanov, N.N. (3) 401
Modyanov, N.N. (8) 402
Modyanov, N.N. (5) 410
Mollan, R.C. (6) 101
Monastyrskaya, G.S. (2) 201
Monastyrskaya, G.S. (3) 201
Monastyrskaya, G.S. (5) 201
Monastyrskaya, G.S. (1) 393
Monastyrskaya, G.S. (2) 393
Monastyrskaya, G.S. (6) 393
Monastyrskaya, G.S. (1) 401
Monastyrskaya, G.S. (2) 401
Monastyrskaya, G.S. (5) 410
Monastyrskaya, G.S. (8) 410
Monastyrskaya, G.S. (9) 410
Monastryrskaya, G.S. (9) 426
Moody, A.J. (6) 22
Moore, D. (4) 201
Moore, D. (4) 401
Moore, D. (15) 411
Morihara, K. (5) 10
Morihara, K. (11) 10
Morikawa, T. (2) 40
Moroder, L. (3) 260
Moroder, L. (5) 260
Moroder, L. (5) 271
Moroder, L. (6) 271
Morris, G.A. (7) 238
Motuz, L.P. (2) 194
Motuz, L.P. (3) 194
Motuz, L.P. (4) 194
Motuz, L.P. (5) 194
Motuz, L.P. (6) 194
Motuz, L.P. (7) 194
Motuz, L.P. (8) 194
Moule, M.L. (15) 16
Moulin, A. (10) 260
Mozhayeva, G.N. (6) 318
Mozhayeva, G.N. (7) 318
Mozhayeva, G.N. (8) 308
Mozhayeva, G.N. (11) 308
Mozhayeva, G.N. (10) 308
Mozhayeva, G.N. (20) 308
Müller, J. (8) 114
Müller, J. (9) 114
Müller, M. (10) 272
Müller, M. (15) 272
Mueller, P. (18) 224
Mueller, P. (1) 288
Mueller, P. (5) 288
Mueller, P. (13) 289
Müller-Esterl, W. (5) 385
Müller-Esterl, W. (6) 385
Müller-Esterl, W. (11) 386
Müller-Esterl, W. (12) 386
Müller-Esterl, W. (13) 386
Müller-Esterl, W. (19) 386
Mukohata, Y. (2) 231
Muller, E. (11) 247
Muller, E. (6) 254

Muller, P. (3) 298
Munekata, E. (5) 70
Murtuzaev, I.M. (4) 419
Mutter, M. (1) 135
Mutter, M. (2) 135
Mutter, M. (3) 135
Mutter, M. (4) 135
Mutter, M. (9) 135
Mutulis, F.K. (2) 46
Mutulis, F. (3) 46
Mutulis, F.K. (6) 266
Myagkova, M.A. (1) 53
Myagkova, M.A. (3) 54
Myshlyakova, N.V. (2) 46
Myasoedov, N.F. (3) 318
Myers, P.R. (20) 272

Naegeli, H.-U. (9) 340
Nagaraj, R. (4) 223
Nagaraj, R. (30) 225
Nagaraj, R. (32) 225
Nagata, S. (3) 426
Nagata, S. (5) 426
Nagayama, K. (2) 238
Naithani, V.K. (6) 28
Naithani, V.K. (8) 28
Naithani, V.K. (9) 28
Nakamura, K. (9) 10
Nakane, P.K. (7) 280
Narahashi, T. (2) 318
Nassal, M. (6) 78
Nassal, M. (8) 78
Nathanson, N. (9) 332
Naujoks, K.W. (3) 177
Naumov, A.P. (8) 308
Naumov, A.P. (10) 308
Naumov, A.P. (11) 308
Naumov, A.P. (12) 308
Naumov, A.P. (20) 308
Naumov, A.P. (6) 318
Naumov, A.P. (7) 318
Navashin, S.M. (1) 213
Nazimov, I.V. (7) 201
Negulyaev, Yu.A. (11) 308
Negulyaev, Yu.A. (12) 308
Negulyaev, Yu.A. (10) 318
Neher, E. (19) 289
Neher, E. (15) 299
Neil, A. (1) 410
Neilands, J.B. (2) 340
Neilands, J.B. (6) 340
Neilands, J.B. (14) 341
Nesterova, M.V. (1) 419
Nesterova, M.V. (2) 419
Nesterova, M.V. (4) 419
Nesterova, M.V. (6) 420

Nesterova, M.V. (7) 420
Nesterova, M.V. (8) 420
Nesvadba, H. (1) 120
Nesvadba, H. (4) 126
Netzelmann, U. (3) 348
Netzelmann, U. (5) 348
Nevskaya, N.A. (10) 299
Newman, A. (11) 411
Nicholson, G.J. (5) 144
Nicholson, G.J. (12) 223
Nicholson, G.J. (7) 260
Niepel, B. (5) 40
Nihei, K. (5) 247
Nikiforovich, G. (4) 266
Nirenberg, M. (7) 271

Nirenberg, M. (19) 272
Nisman, B.Ch. (20) 308
Norrestam, R. (8) 340
Nosick, D.N. (13) 426
Nosyreva, E.D. (11) 308
Nosyreva, E.D. (8) 308
Nosyreva, E.D. (12) 308
Nosyreva, E.D. (7) 318
Novochatsky, A.S. (13)
Nuriddinov, A.R. (2) 53
Nuriddinov, A.R. (7) 54
Nuriddinov, A.R. (8) 54
Nuriddinov, A.R. (9) 54

O'Rand, M.G. (4) 368
O'Riordan, J.L.H. (1) 40
Obermeier, R. (3) 10
Obermeier, R. (6) 10
Obermeier, R. (10) 238
Oekonomopulos, R. (14) 224
Oekonomopulos, R. (15) 224
Oekonomopulos, R. (16) 224
Oekonomopulos, R. (28) 224
Oekonomopulos, R. (29) 225
Oesterhelt, D. (1) 85
Oesterhelt, D. (1) 231
Oesterhelt, D. (3) 231
Oesterhelt, D. (9) 232
Oesterhelt, D. (13) 232
Oesterhelt, D. (10) 247
Oesterhelt, D. (5) 254
Oesterhelt, D. (8) 254
Oesterhelt, D. (11) 254
Oesterhelt, D. (14) 254
Oesterhelt, D. (15) 254
Oesterhelt, D. (16) 254
Ötting, F. (16) 386
Offord, R.E. (1) 10
Ohgaku, S. (11) 10
Ohlsson, K. (10) 386
Ohta, M. (14) 114
Oka, T. (5) 10
Oka, T. (11) 10
Okhanov, V.V. (8) 54
Olefsky, J. (9) 16
Olefsky, J. (11) 16
Olefsky, J.M. (13) 16
Ooka, T. (26) 224
Ooka, T. (8) 288
Orci, L. (12) 16
Ornston, L.N. (2) 376
Osawa, T. (10) 96
Ostertag, W. (7) 186
Ostrowsky, M.A. (3) 247
Oswald, R.E. (13) 332
Ottenbreit, B.P. (3) 95
Ottenheym, H. (1) 78
Ottolenghi, M. (13) 254
Ovchinnikov, Yu.A. (9) 28
Ovchinnikov, Yu.A. (1) 46
Ovchinnikov, Y. (3) 46
Ovchinnikov, Y. (10) 54
Ovchinnikov, Yu. (11) 54
Ovchinnikov, Yu. A. (15) 54
Ovchinnikov, Yu.A. (2) 85
Ovchinnikov, Yu.A. (1) 153
Ovchinnikov, Yu. A. (2) 194
Ovchinnikov, Yu. A. (5) 194
Ovchinnikov, Yu. A. (7) 194
Ovchinnikov, Yu. A. (8) 194
Ovchinnikov, Yu. A. (2) 201
Ovchinnikov, Yu. A. (1) 201

Ovchinnikov, Yu. A. (3) 201
Ovchinnikov, Yu.A. (5) 201
Ovchinnikov, Yu.A. (7) 201
Ovchinnikov, Yu.A. (1) 207
Ovchinnikov, Yu.A. (4) 247
Ovchinnikov, Yu.A. (13) 248
Ovchinnikov, Yu.A. (14) 248
Ovchinnikov, Yu.A. (15) 248
Ovchinnikov, Yu.A. (1) 254
Ovchinnikov, Yu.A. (20) 308
Ovchinnikov, Yu.A. (21) 308
Ovchinnikov, Yu.A. (3) 318
Ovchinnikov, Yu.A. (9) 318
Ovchinnikov, Yu.A. (3) 332
Ovchinnikov, Yu.A. (5) 332
Ovchinnikov, Yu.A. (10) 332
Ovchinnikov, Yu.A. (1) 393
Ovchinnikov, Yu.A. (2) 393
Ovchinnikov, Yu.A. (6) 393
Ovchinnikov, Yu.A. (1) 401
Ovchinnikov, Yu.A. (2) 401
Ovchinnikov, Yu.A. (3) 401
Ovchinnikov, Yu.A. (5) 410
Ovchinnikov, Yu.A. (8) 410
Ovchinnikov, Yu.A. (9) 410
Ovchinnikov, Yu.A. (9) 426

Packer, L. (12) 254
Palan, J. (7) 135
Palm, P. (21) 411
Panem, S. (3) 426
Parrish, R.F. (21) 386
Pashkov, V.S. (1) 46
Pashkov, V.S. (2) 53
Pashkov, V.N. (3) 318
Pashkov, V.N. (5) 318
Pashkov, V.S. (3) 332
Pashkov, V.S. (4) 332
Pashkov, V. (10) 332
Pauls, P. (39) 225
Pauly, H. (4) 22
Pedone, C. (31) 225
Pestka, S. (6) 426
Pestka, S. (11) 426
Petersen, K.-G. (6) 10
Petersen, T.E. (14) 386
Peterson, J.D. (12) 28
Peterson, R.N. (2) 368
Pettee, J.M. (3) 28
Pfaff, K.-P. (5) 61
Pfeifle, J. (2) 362
Pham-Tham-Chi, T.N. (9) 260
Pheiffer, B.H. (10) 394
Phodes, G. (19) 411
Pillai, V.N.R. (3) 135
Pillai, V.N.R. (9) 133
Pinnen, F. (5) 120
Pirro, de R. (9) 16
Pirro, de R. (11) 16
Pletnev, V.Z. (3) 213
Pletnev, A.G. (8) 402
Pluzhnikov, K.A. (15) 54
Pluzhnikov, K.A. (3) 332
Pluzhnikov, K.A. (4) 332
Pluzhnikov, K. (5) 332
Pluzhnikov, K.A. (6) 332
Pluzhnikov, K. (10) 332
Pober, J.S. (8) 247
Polakoski, K.L. (20) 386
Polakoski, K.L. (21) 386
Polevaya, L.K. (1) 266
Polevaya, L.K. (2) 266

Poling, M. (19) 341
Polovnikova, I.N. (2) 201
Polovnikova, I.N. (1) 393
Polovnikova, I.N. (2) 401
Polovnikova, I.N. (5) 410
Porunkevich, E.A. (5) 266
Possani, L.D. (13) 318
Potapenko,N.A. (3) 207
Potts, J.T. (1) 40
Potts, J.T. (3) 40
Potts, J.T. (4) 40
Pressman, B.C. (2) 298
Prestipino, G. (13) 318
Preston, J. (4) 34
Probst, F. (5) 271
Prosolova, T.K. (15) 308
Proudfoot, N.J. (18) 186
Prox, A. (1) 78
Pudill, R. (1) 120
Pudill, R. (4) 126
Pudill, R. (5) 126
Pullen, R.A. (2) 22

Quelle, M.-J. (12) 299

Rabussay, D. (21) 411
Radhakrishnan, R. (9) 247
Radhakrishnan, R. (4) 254
Raftery, M.A. (19) 308
Raftery, M.A. (4) 318
Raftery, M.A. (7) 332
Ramabrahman, V. (8) 135
Rampold, G. (13) 114
Rampold, G. (14) 114
Ratkevich, M.P. (5) 266
Raymond, K.N. (3) 340
Raymond, K.N. (12) 341
Raymond, K.N. (13) 341
Raymond, K.N. (14) 341
Raymond, K.N. (15) 341
Raymond, K.N. (17) 341
Raymond, K.N. (20) 341
Rehfeld, J. (1) 271
Reich, E. (14) 386
Reiser, G. (4) 271
Reiser, G. (13) 272
Reiser, G. (14) 272
Reissmann, S. (1) 46
Resch, K. (5) 95
Resch, K. (9) 96
Reshetov, P.D. (1) 213
Reshetov, P.D. (5) 213
Reusser, F. (5) 223
Reutova, T.O. (1) 46
Rialdi, G. (1) 22
Rich, D.E. (3) 28
Rich, D.H. (4) 120
Richards, F.M. (7) 78
Richardson, J.S. (4) 213
Rifkin, D. (14) 386
Rigas, D.A. (3) 185
Rinehart, K.L. (17) 161
Riniker, B. (2) 33
Rittel, W. (2) 34
Robertson, R.T. (4) 385
Rodionov, A.V. (14) 248
Rodionov, A.V. (15) 248
Rödl, J. (9) 170
Rohen, J.W. (11) 280
Rohrbach, M.S. (1) 194
Rohrer, H. (3) 177
Rohrschneider, S. (10) 280

Romani, S. (7) 34
Romanovskis, P.J. (5) 266
Romeo, A. (7) 120
Rosa, U. (1) 22
Rosenbach, V. (13) 254
Rosenberg, M. (9) 201
Rosenblatt, M. (3) 40
Rosenblatt, M. (4) 40
Ross, A.H. (9) 247
Ross, A.H. (4) 254
Rosselin, G. (8) 144
Rossi, C.A. (1) 22
Roth, J. (9) 16
Roth, J. (11) 16
Rothe, I. (3) 126
Rothe, M. (1) 120
Rothe, M. (4) 120
Rothe, M. (6) 120
Rothe, M. (3) 126
Rothe, M. (4) 126
Rothe, M. (5) 126
Rubin, L.L. (4) 326
Rudin, D.O. (1) 288
Rudin, D.O. (3) 298
Rüegg, U. (5) 22
Rühlmann, A. (7) 348
Rudinger, J. (8) 34
Rudloff, V. (15) 186
Russell, L. (2) 368
Ruttenberg, M.A. (2) 10

Saari, J.C. (7) 247
Sabbatini, J.Z. (15) 161
Sadykov, A. (3) 326
Sakakibara, S. (2) 40
Sakmann, B. (4) 288
Sakmann, B. (19) 289
Sakur, N. (14) 114
Salardi, S. (1) 135
Salikhov, Sh. (3) 326
Saling, P.M. (1) 385
Salomatina, I.S. (3) 201
Salomatina, I.S. (2) 393
Salomatina, I.S. (1) 401
Samukov, V.V. (7) 402
Sandorfy, C. (18) 308
Sanguansermsri, R. (11) 186
Sanguansermsri, T. (12) 186
Sarges, R. (1) 298
Sasaki, A.N. (10) 78
Sashchenko, L.P. (2) 419
Sashchenko, L.P. (3) 419
Sashchenko, L.P. (5) 420
Sauer, M.M. (1) 40
Saunders, D. (4) 16
Saunders, D. (7) 16
Saunders, D. (8) 16
Saunders, D. (9) 16
Saunders, D. (10) 16
Saunders, D. (12) 16
Saunders, D. (13) 16
Savige, W.E. (7) 70
Savko, P. (19) 224
Savko, P. (20) 224
Savko, P. (14) 289
Schaal, H. (37) 225
Schaal, H. (43) 226
Schachner, M. (21) 411
Schachter, M. (1) 280
Schäfer, W. (12) 186
Schaller, H.C. (1) 170
Schaller, H.C. (2) 170

Schaller, H.C. (3) 170
Schaller, H.C. (4) 170
Schaller, H.C. (5) 170
Schaller, H.C. (6) 170
Schaller, H.C. (10) 170
Schaller, K. (3) 376
Schaller, K. (4) 376
Schaller, K. (6) 376
Schaller, K. (7) 376
Schaller, K. (8) 376
Schaller, K. (9) 376
Scharf, R. (1) 144
Scharf, R. (6) 144
Scheit, K.H. (12) 411
Scheppele, S.E. (9) 161
Scheraga, H.A. (6) 135
Scheraga, H.A. (6) 348
Scherrer, P. (12) 254
Schiessler, H. (9) 385
Schiessler, H. (10) 386
Schild, H.O. (8) 260
Schill, W.-B. (7) 369
Schill, W.-B. (8) 369
Schill, W.-B. (9) 369
Schill, W.-B. (10) 369
Schill, W.-B. (7) 385
Schindler, W. (1) 120
Schindler, W. (4) 126
Schindler, W. (5) 126
Schindler, H. (18) 289
Schleuning, W.-D. (7) 385
Schleuning, W.-D. (9) 385
Schlüter, K.J. (6) 10
Schmalle, H. (5) 61
Schmid , J. (1) 78
Schmidt, D.E. (14) 161
Schmidt, G. (3) 70
Schmidt, T. (6) 170
Schmitt, E. (10) 10
Schmitt, E.W. (7) 28
Schmitt, E.W. (9) 28
Schmitt, H. (2) 222
Schmitt, H. (39) 225
Schmitt, H. (40) 225
Schmitt, H. (43) 226
Schmitt, H. (44) 226
Schmitt, H. (45) 226
Schmitt, H. (12) 289
Schmoeckel, C. (7) 369
Schmoeckel, C. (8) 369
Schneuer, V.W. (9) 96
Schobert, B. (5) 231
Schrank, B. (17) 186
Schroeder, W.A. (16) 186
Schuck, R. (4) 238
Schuck, R. (5) 238
Schuck, R. (8) 238
Schüttler, A. (3) 16
Schulten, H.R. (13) 161
Schultzberg, M. (1) 271
Schulz, I. (6) 144
Schumann, I. (1) 126
Schurrs, A.W.W.M. (9) 280
Schuster, P. (18) 308
Schwab, M. (4) 177
Schwarz, G. (19) 224
Schwarz, G. (20) 224
Schwarz, G. (14) 289
Schwarz, M.A. (3) 368
Schwarzstein, M. (3) 426
Schweitz, H. (6) 308
Schweitz, H. (8) 348

Schwenk, D.A. (7) 126
Schwyzer, R. (5) 114
Schwyzer, R. (2) 126
Sedgwick, R.D. (8) 161
Sedgwick, R.D. (10) 161
Seeburg, P.H. (4) 426
Seeburg, P.H. (6) 426
Segre, G.V. (3) 40
Segre, G.V. (4) 40
Seifert, W. (21) 411
Seipke, G. (6) 10
Sekacis, I. (4) 266
Seltzer, S. (12) 254
Senn, H. (7) 78
Server, A.C. (6) 177
Seth, V.S. (21) 411
Severin, E.S. (1) 419
Severin, E.S. (2) 419
Severin, E.S. (3) 419
Severin, E.S. (4) 419
Severin, E.S. (5) 420
Severin, E.S. (6) 420
Severin, E.S. (7) 420
Severin, E.S. (8) 420
Shain, W. (20) 272
Shamala, N. (30) 225
Shamotienko, O.G. (15) 308
Sharma, S.K. (19) 272
Shaw, E. (14) 386
Sheldrick, G.M. (39) 225
Shelton, J.B. (16) 186
Shelton, J.R. (16) 186
Shemyakin, M.F. (6) 401
Shen, W.-F. (5) 393
Shepard, G.L. (4) 40
Shepard, H.M. (11) 426
Shepel, E.N. (10) 299
Sheppard, R.C. (4) 16
Sheppard, R.C. (4) 34
Shevchenko, V.P. (3) 318
Shigeta, G. (11) 10
Shkrob, A.M. (14) 248
Shkrob, A.M. (15) 248
Shooter, E.M. (6) 177
Shoulders, C.C. (18) 186
Shuvaeva, T.M. (3) 201
Shuvaeva, T.M. (7) 201
Shuvaeva, T.M. (2) 393
Shuvaeva, T.M. (1) 401
Sieber, P. (2) 33
Sieber, P. (2) 113
Sieber, P. (4) 114
Sieber, P. (2) 126
Siegmeier, R. (4) 238
Sieker, L. (3) 340
Siekevitz, P. (2) 326
Siemion, Z. (6) 16
Siemion, I.Z. (2) 53
Siemion, I.Z. (10) 238
Sigworth, F.J. (4) 308
Siskov, I.V. (5) 266
Skar, D.C. (1) 194
Skryabin, K.G. (5) 201
Skryabin, K.S. (9) 410
Skujins, A.A. (5) 266
Skulachev, V.P. (13) 248
Sloma, A. (6) 426
Sloma, A. (11) 426
Smirnov, I.V. (3) 419
Smirnov, V.V. (1) 213
Smirnov, Yu.V. (1) 201
Smirnov, Yu.V. (3) 401

Smirnova, G.M. (1) 213
Smith, I.C.P. (6) 238
Smith, J.F. (8) 170
Smulkowski, M. (3) 85
Soejima, M. (9) 10
Sönksen, P.H. (10) 16
Soerensen, U.B. (3) 260
Sokatch, J.R. (2) 376
Soldatov, N.M. (20) 308
Soldatov, N.M. (21) 308
Soldatov, N.M. (6) 318
Soldatov, N.M. (9) 318
Solomon, T.E. (7) 144
Solomoniya, R.O. (1) 419
Solovyev, V.D. (9) 426
Soman, K.V. (8) 135
Sorm, F. (18) 386
Sorokin, A.V. (3) 201
Sorokina, I.B. (1) 213
Sosnov, A. (4) 266
Sottrup-Jensen, L. (14) 386
Spach, G. (22) 224
Spalding, B.C. (4) 308
Sperling, W. (9) 254
Squires, C.L. (5) 393
Staehlin, T. (12) 232
Standefer, J.C. (8) 286
Standring, D. (11) 332
Stangl, A. (17) 186
Stahnke, G. (11) 78
Stambaugh, R. (3) 385
Stan, M. (6) 22
Stan, M.A. (6) 22
Starkey, P.M. (1) 355
Staros, V.T. (11) 332
Stavropolous, A. (6) 186
Stebbing, N. (6) 426
Steer, W. (13) 186
Steers, E. (2) 207
Steffen, K. (3) 126
Steffens, G. (10) 232
Steffens, G.J. (16) 386
Stein, J.L. (15) 161
Steinbach, J.H. (19) 289
Steinberger, R. (1) 120
Steinberger, R. (4) 126
Steinbuch, M. (6) 280
Steiner, D.F. (1) 28
Steiner, D.F. (12) 28
Steinheider, G. (7) 186
Stender, W. (12) 411
Stengrevics, O.A. (2) 194
Stengrevics, O.A. (3) 194
Stengrevics, O.A. (5) 194
Stengrevics, O.A. (6) 194
Stensland, B. (8) 340
Stewart, W.E. (2) 426
Stieve, H. (9) 254
Stoeckenius, W. (1) 231
Stoeckenius, W. (1) 247
Stoyachenko, I.A. (5) 213
Strähle, J. (33) 225
Strähle, J. (37) 225
Strähle, J. (43) 226
Straub, J. (16) 254
Streuli, M. (3) 426
Streuli, M. (5) 426
Stryer, L. (8) 247
Stuckstedte, U. (5) 280
Sturm, G.P. (9) 161

Stutz, A.A. (12) 411
Subramanyam,
Sucharda-Sobczyk, A. (10) 238
Suddath, F.L. (10) 318
Surin, A.M. (2) 53
Surin, A.M. (15) 54
Surin, A.M. (3) 332
Surin, A.M. (4) 332
Surin, A.M. (5) 332
Surin, A. (10) 332
Suzuki, S. (7) 61
Suzuki, T. (11) 280
Sverdlov, E.D. (2) 201
Sverdlov, E.D. (3) 201
Sverdlov, E.D. (5) 201
Sverdlov, E.D. (1) 393
Sverdlov, E.D. (2) 393
Sverdlov, E.D. (6) 393
Sverdlov, E.D. (1) 401
Sverdlov, E.D. (2) 401
Sverdlov, E.D. (8) 402
Sverdlov, E.D. (9) 402
Sverdlov, E.D. (5) 410
Sverdlov, E.D. (8) 410
Sverdlov, E.D. (9) 410
Sverdlov, E.D. (16) 411
Sverdlov, E.D. (17) 411
Sverdlov, E.D. (9) 426
Sychev, S.V. (10) 299
Sychev, S.V. (11) 299

Tabor, J.M. (6) 426
Taira, H. (5) 426
Takagaki, Y. (9) 247
Takagaki, Y. (4) 254
Takai, T. (2) 40
Takashima, S. (18) 224
Takashima, S. (13) 289
Tan, N.H. (12) 54
Tanaka, S. (6) 135
Tashmukhamedov, M. (3) 326
Tashmukhamedov, B.A. (5) 326
Tautz, C.H. (10) 186
Taylor, L.C.E. (9) 161
Taylor, M. (5) 223
Templeton, D.H. (16) 341
Terao, T. (10) 96
Ternovsky, V.I. (5) 326
Tetler, L.W. (10) 161
Tetlow, L.C. (2) 355
Thamm, P. (5) 16
Thamm, P. (6) 16
Thamm, P. (7) 16
Thamm, P. (9) 16
Thamm, P. (10) 16
Thamm, P. (12) 16
Thamm, P. (13) 16

Theyson, R. (1) 120
Theysohn,R. (4) 126
Thoenen, H. (1) 177
Thoenen, H. (2) 177
Thoenen, H. (3) 177
Thoenen, H. (4) 177
Thoenen, H. (5) 177
Tickle, I.J. (2) 22
Töpfer-Petersen, E. (7) 369
Töpfer-Petersen, E. (8) 369
Töpfer-Petersen, E. (9) 369
Töpfer-Petersen, E. (10) 369
Todd, D. (5) 186
Tomasek, V. (18) 386

Tonelli, A.E. (11) 238
Toniolo, C. (1) 135
Toniolo, C. (2) 135
Toniolo, C. (3) 135
Toniolo, C. (9) 135
Toniolo, C. (31) 225
Torff, H.J. (9) 170
Tosteson, D.C. (15) 289
Tosteson, M.T. (15) 289
Towbin, H. (12) 232
Traber, J. (8) 272
Traber, J. (18) 272
Tracy, H.J. (1) 260
Trakhanov, S.D. (3) 213
Trautschold, I. (9) 348
Travers, A.A. (13) 411
Tregear, G.W. (3) 40
Tregear, G.W. (4) 40
Trindler, P. (3) 16
Trischmann, H. (4) 70
Trudelle, Y. (22) 224
Truscheit, E. (11) 280
Tsai, P. (13) 16
Tsarev, S.A. (8) 402
Tsarev, S.A. (9) 402
Tsarev, C.A. (16) 411
Tsarev, S.A. (17) 411
Tsarev, S.A. (9) 426
Tschesche, H. (11) 280
Tschesche, H. (1) 348
Tschesche, H. (3) 348
Tschesche, H. (5) 348
Tschesche, H. (8) 355
Tschesche, H. (9) 355
Tschesche, H. (11) 386
Tsetlin, V.I. (2) 53
Tsetlin, V.I. (7) 54
Tsetlin, V.I. (8) 54
Tsetlin, V.I. (9) 54
Tsetlin, V.I. (15) 54
Tsetlin, V.I. (3) 332
Tsetlin, V.I. (4) 332
Tsetlin, V.I. (5) 332
Tsetlin, V.I. (6) 332
Tsetlin, V. (10) 332
Tsuzuki, H. (5) 10
Tubiana, M. (9) 260
Tudge, H. (9) 161
Tunitskaya, V.L. (4) 419
Turck, Ch. (1) 101
Turck, Ch. (2) 101
Turk, V. (2) 280
Turner, J.E. (3) 177
Tyler, A.N. (8) 161
Tyler, G. (4) 40
Tzeng, M.C. (2) 326
Tzeng, M.C. (4) 326

Uhl, R. (2) 247
Ullrich, A. (4) 426
Ullrich, A. (6) 426
Ulmasov, Kh.A. (7) 420
Ulyashin, V.V. (10) 54
Ulyashin, V.V. (11) 54
Ulyashin, V.V. (1) 153
Unwin, P.N.T. (6) 247
Unwin, P.N.T. (7) 254
Urry, D.W. (5) 299
Urry, D.W. (6) 299
Urry, D.W. (12) 299
Uschkoreit, J. (9) 16
Uschkoreit, J. (11) 16

Utkin, Yu.N. (2) 53
Utkin, Yu.N. (6) 54
Utkin, Yu.N. (15) 54
Utkin, Yu.N. (3) 332
Utkin, Yu.N. (4) 332
Utkin, Yu.N. (5) 332
Utkin, Yu.N. (6) 332
Utkin, Yu. (10) 332

Vaciago, A. (9) 394
Varga, S. (13) 54
Varga, S.L. (7) 126
Vasiliev, V.Yu. (2) 419
Veatch, W.R. (7) 299
Veatch, W.R. (8) 299
Veber, D.F. (13) 54
Veber, D.F. (7) 126
Velkov, V.V. (3) 410
Veretennikova, N.I. (2) 266
Veretennikova, N. (4) 266
Vincent, J.-P. (8) 348
Vinokurov, L.M. (2) 194
Vinokurov, L.M. (3) 194
Vinokurov, L.M. (6) 194
Vinokurov, L.M. (7) 194
Vinokurov, L.M. (8) 194
Vitale, Lj. (2) 280
Vodyanoy, I. (5) 223
Vodyanoy, I. (11) 288
Voelter, W. (9) 28
Voelter, W. (3) 46
Voelter, W. (10) 54
Voelter, W. (6) 114
Voelter, W. (8) 114
Voelter, W. (9) 114
Voelter, W. (10) 114
Voelter, W. (11) 114
Voelter, W. (13) 114
Voelter, W. (14) 114
Voelter, W. (15) 114
Voelter, W. (16) 114
Voelter, W. (2) 222
Voelter, W. (12) 289
Voelter, W. (10) 332
Voeykova, T.A. (3) 410
Voges, K.-P. (21) 224
Voges, K.-P. (35) 225
Voges, K.-P. (43) 226
Voges, K.-P. (44) 226
Voges, K.-P. (45) 226
Volpina, O.M. (1) 53
Volpina, O.M. (2) 53
Volpina, O.M. (4) 54
Volpina, O.M. (5) 54
Volpina, O.M. (6) 54
Vosekalna, I. (4) 266
Voskamp, D. (7) 144
Vossler, G. (5) 160
Vossler, G. (6) 160
de Vries, J.X. (1) 78

Wabl, M.R. (3) 376
Wachter, C. (6) 40
Wadström, T. (4) 332
Waehneldt, T.V. (11) 232
Wagner, G. (3) 231
Wagner, G. (2) 238
Wakefield, C. (9) 161
Wallace, B.A. (3) 254
Walter, R. (2) 160
Walter, U. (14) 272
Wan, Y.-P. (14) 114

Wang, S.-W. (6) 101
Wanke, E. (13) 318
Warwas, M. (11) 280
Watt, D.D. (10) 318
Weatherall, D.J. (4) 186
Weber, H.J. (6) 231
Weber, H.J. (7) 232
Weck, P. (7) 426
Weemann, van B. K. (9) 280
Weimann, H.-J. (3) 16
Weissman, C. (3) 426
Weissmann, C. (5) 426
Weitzel, G. (11) 114
Weitzel, G. (14) 114
Weitzel, G. (2) 222
Weitzel, G. (12) 289
Wendt, V. (7) 385
Wenzel, H.R. (3) 348
Wenzel, H.R. (5) 348
Werle, E. (9) 348
Westheimer, F.H. (4) 78
Wieland, Th. (1) 70
Wieland, Th. (2) 70
Wieland, Th. (3) 70
Wieland, Th. (5) 70
Wieland, Th. (6) 70
Wieland, Th. (8) 70
Wieland, Th. (9) 70
Wieland, Th. (1) 78
Wieland, Th. (2) 78
Wieland, Th. (8) 78
Wiesmüller, K. (7) 95
Williams, D.H. (11) 161
Williams, D.H. (16) 161
Williams, W.L. (5) 369
Williams, W.L. (2) 385
Williams, W.L. (4) 385
Wilmsen, H.-U. (17) 224
Wilmsen, H.-U. (10) 288
Wilson, K.A. (4) 355
Wiman, B. (17) 386
Winitz, M. (17) 114
Winkelmann, G. (3) 340
Winkelmann, G. (5) 340
Winkelmann, G. (10) 341
Winkelmann, G. (11) 341
Winkelmann, G. (15) 341
Winter, A. (23) 272
Winter, W. (34) 225
Winter, W. (35) 225
Winter, W. (36) 225
Winter, W. (37) 225
Winter, W. (38) 225
Winter, W. (39) 225
Winter, W. (40) 225
Winter, W. (43) 226
Winter, W. (44) 226
Winter, W. (45) 226
Wipperfürth, Th. (8) 95
Wisher, M.H. (10) 16
Wittig, B. (11) 386
Witkop, B. (1) 298
Wittmann-Liebold, B. (15) 186
Wittmann-Liebold, B. (5) 207
Witzeman, V. (7) 332
Wolf, G. (1) 61
Wolff, H. (4) 376
Wollmann, H. (12) 114
Wollmer, A. (2) 22
Wong, G.B. (15) 341
Wong, H.B. (4) 186
Wood, G.W. (14) 161

Wood, S.P. (2) 22
Wood, W.G. (6) 40
Woolley, D.E. (2) 355
Wünsch, E. (9) 28
Wünsch, E. (1) 33
Wünsch, E. (5) 34
Wünsch, E. (6) 34
Wünsch, E. (7) 34
Wünsch, E. (3) 46
Wünsch, E. (10) 54
Wünsch, E. (1) 113
Wünsch, E. (19) 114
Wünsch, E. (1) 144
Wünsch, E. (2) 144
Wünsch, E. (4) 144
Wünsch, E. (6) 144
Wünsch, E. (8) 144
Wünsch, E. (3) 260
Wünsch, E. (4) 260
Wünsch, E. (5) 260
Wünsch, E. (6) 260
Wünsch, E. (5) 271
Wünsch, E. (6) 271
Wünsch, E. (10) 332
Wüthrich, K. (2) 238
Wunderer, G. (6) 355
Wunderer, G. (7) 355

Yang, C.C. (14) 54
Yantorno, R.E. (18) 224
Yantorno, R.E. (13) 289
Yelverton, E. (4) 426
Yelverton, E. (6) 426
Yelverton, E. (7) 426
Yelverton, E. (11) 426
Yip, C.C. (15) 16
Yeung, C.W.T. (15) 16
Yokoyama, K. (10) 96
Yura, T. (14) 411

Zachmann, B. (10) 170
Zähner, H. (1) 61
Zähner, H. (3) 61
Zähner, H. (4) 61
Zähner, H. (6) 61
Zähner, H. (10) 61
Zähner, H. (11) 61
Zähner, H. (12) 61
Zahn, H. (2) 22
Zahn, H. (8) 28
Zahn, H. (5) 40
Zahn, H. (6) 40
Zahn, H. (7) 40
Zaichikov, E.F. (8) 402
Zakharyev, V.M. (5) 201
Zakharyev, V.M. (9) 410
Zalkin, A. (16) 341
Zaneveld, L.J.D. (4) 385
Zaneveld, L.J.D. (20) 386
Zanotti, G. (5) 120
Zechel, K. (21) 411
Zervas, L. (17) 114
Zhigis, L.S. (5) 213
Zhukova, G.F. (3) 54
Zhukova, G.F. (7) 54
Zillig, W. (1) 410
Zillig, W. (21) 411
Zimmerman, S.B. (10) 394
Zimmerman, R.E. (3) 28
Zingsheim, H.P. (15) 299
Zubov, A.N. (20) 308
Zundel, G. (18) 308

Zurek, G. (10) 402
Zweifel, B.O. (3) 120
Zwilling, R. (9) 170

Subject Index

Acetylcholine	267
Acetylcholine receptor	327
N-Acetyltryptamine	47
Aconitine	302,304,305,309,310,311
[^{3}H] Aconitine	311
Acrosin	277,278,363,377,378,379,380, 382,385
Acrosin, biosynthesis of	385
Acrosin-sepharose	378
Acrosomal cap	363
Acrosomal enzymes	363
Acrosomal membrane	363,364,365,366,368
Acrosome	363,365,379
F-Actin	63
Actinomyces globisporus	209
Actinoxanthin	209,210
Actinoxanthin, backbone folding of	209
Actinoxanthin structure	211
Acylamidines	123
3-N$^{\alpha}$-Acylpeptidyloxy-2-hydroxy-N-methylbenzamides	101
1-(1-Adamantyl)-1-methylethoxycarbonyl	104
Adenosine	269
Adenosine-5'-chloromethylpyrophosphonate	415
Adenosine-5'-trimethaphosphate	395,396,397,401
Adenylate cyclase activity	143
Adenylate cyclases	267

Adipocytes	15
N-Adipylgramicidin	298
Adpoc-F	111
Adpoc-Ala-Lys(Boc)-Leu-OMe	110
Adpoc amino acids	109,110
Adpoc-Arg(Adpoc)$_2$-OH	111
Adpoc dipeptides	110
Adpoc group	104,105,109,110
Adpoc-Leu-OH	109
Adpoc-Leu-Trp-OH	109,110
Adpoc-Leu-Trp-OMe	110
Adpoc-Lys(Boc)-OH	109
Adpoc-Lys(Z)-OH	109
Adpoc-Pro-OH	109
Adpoc protection	103
Adpoc-Trp-Lys(Boc)-OH	110
Adpoc-Trp-OH	110
Adrenal medullary cells	172
Adrenal medullary tumor	172
Adsorption chromatography	148
Adsorption mode	82
Adult haemoglobin	179,180
Ala-desoxo-viroidin	64
Ala-virodin	64
Alamethicin	215,217,218,219,222,281,283,284,286,287
Alamethicin,analogues of	282
Alamethicin F	216
Alkaloid neurotoxin receptor	303
Alkaloid toxins	302
N-Alkyl-maleimides	31

All-L-cyclotripeptides 116
All-trans-C_{15}-^{3}H-retinal 228
Allo-hydroxyproline 65
Amanita mushrooms 63
Amanita phalloides 71
Amanita virosa 63
Amatoxin 63,71
Amino acid analysis 14,21,22
L-2-Amino-6-t-butyloxy-carbonylamino-4-hexynoic acid 76
α-Amino-γ-hydroxy acids 57,58,60
α-Amino-γ-hydroxy butyric acid 55
Amino protecting groups, recently developed 103
Aminoacyl diketopiperazines 115,116,118,123
5-Aminohexuronic acid 55
α-Aminoisobutyric acid 287
Aminopeptidase M 169,256
Aminopolyoxyethylene glycol 31
cAMP 269,270,271,358
cAMP-binding protein of bacteria 417
cAMP-binding proteins 418
cAMP-dependent protein kinase 413,414,417,418
cAMP-independent protein kinase 413
Anemonia sulcata 349,352,354
Angiotensin 261,262,263,264
Angiotensin, antagonistic effect of 264
cis-Anhydromevalonic acid 334

trans-Anhydromevalonic acid	334
Antamanide	71
Antamanide analogue, photolabile	71,72
Anthopleura elegantissima (sea anemone)	166
Anti-boar acrosin-IgG	378
Antibody production	274
Anti-kallikrein-IgG peroxidase	276
Anti-kallikrein-IgG peroxidase conjugate	274,275
Antiviral activity	424
Apamin	47
Asolectin bilayers	286
Astacus fluviatilis	169
Astroblasts	268
Astrocytes	268
ATP, synthesis of	358,241
Autophosphorylation	361
Avogadro's number	296
Axoplasm	301
4-Azido-phenylacetyl groups	12
p-Azidobenzoyl groups	329
Azidophenyl insulins	15
2,2'-Azino-di(3-ethyl-benzthiazoline-6-sulfonate)	279
Azoc group	103
Azurin	210
Bacillus subtilis	60
Bacterial cell wall	87
Bacterial lipoproteins	87,95

Bacterial mitogens 92
Bacterial plasmid 423
Bacteriocins 371
Bacteriorhodopsin 79,203,227,228,241,245,246,247,249,251
Bacteriorhodopsin fragments 79
Bacteriorhodopsin segments, CD spectra of 85
Bacteriorhodopsin segment Z-34-50-OMe 83
Bacteriorhodopsin segment Z-34-65-OH 84
Bacteriorhodopsin transmembrane strands, α-helical structure of 85
Batrachotoxin 309
Bee venom constituent melittin 282
Bence-Jones protein 210
N^{α}-Benzoyl-arginine-ß-naphthylamide 378
N^{α}-Benzoyl-D,L-arginine-ß-naphthylamide 381
N^{α}-Benzoyl-L-arginine-4-nitroanilide 380,381
1-Benzyl-1-methyl-ethoxycarbonyl (Bzpoc) group 104
2-Benzyl-2-propanol 112
Benzylisopropyl phenylcarbonate 112
3-[N^{α}-Benzyloxycarbonyl peptidyloxy]-2-hydroxy-N-methyl-benzamides 97
2-Benzylpropyl-(2)-oxycarbonyl 104

Bilayer phospholipid membranes	325
Biogel HT	229
Biological membranes	291
Biorex 70	52,312
2-(4-Biphenylyl)-isopropyloxycarbonyl	20
2-p-Biphenylylisopropyloxycarbonyl group	103,104
Blood clotting	343
Blood pressure	262
Blood pressure regulation	273
Boar acrosin	378,380,381,385
Boar acrosin heavy chain	383,384
Boar acrosin,primary structure of	382
Boar spermatozoa	364,366
μ-Bondapak C18	31,256
Bovine chymotrypsin	383,384
Bovine chymotrypsinogen	382
Bovine hypothalamus	166
Bovine nerve growth factor	171
Bovine prothrombin A-chain	382
Bovine serum albumin	274
Bovine trypsin	277,278,383,384
Bovine trypsin inhibitor	247,344
Bpoc group	21,103,104,105
Bradykinin	170,261,263,264,265,270
Bradykinin,antagonistic effect of	264
Bradykinin conformers	41
Bradykinin cycloanalogues	263
Bradykinin,myotropic effect of	265

Bradikinin, quasi-cyclic structure of the natural hormone	41
Bradikinin, synthesis of cycclic analogues of	41
Brain cell cultures	269
Brain synaptosomal membranes	304
4-Bromo-phenylethanol, silylated	73
t-Bumeoc amino acids	107
t-Bumeoc group	104,105
t-Bumeoc-Lys(Boc)-OH	106
t-Bumeoc-Phe-OH	106,107
t-Bumeoc protection	108
t-Bumeoc-Ser(Bu^t)-OH	106
α -Bungarotoxin	47
Buthus eupeus venom	306
n-Butyllithium	73
t-Butyl-dimethyl-silyl group	74
Bzpoc-Ala.DCHA	113
Bzpoc derivates	112,113
Bzpoc-Gly-OH	113
Bzpoc group	104,105,111,112,113
Bzpoc-Phe-OH	113
Calcitonin	269
Calcium ions	372
Calf lymphocytes	87
Capillary endothelial cells	268
N-Carboxygramicidin channels	295
N-Carboxygramicidins	292
Carboxypeptidase A	4,25,204,242,244
Carboxypeptidase B	27,204,242,244
Carboxypeptidase Y	27
Casein	413,414
Casein phosphorylation	358
Catechol ester method	98
Cell death	349

Cell lysis	372
Cell membrane barrier	59
Cell membrane receptor	173
Cellular hypertrophy	172
Cellulose acetate electrophoresis	14
Cellulose D-52	320
CM-Cellulose	19,25,37,182
DEAE-Cellulose	19
Centruroides sculpturatus	316,317
Chick embryo	172
Chirasil-Val column	138
Chloramine T method	14
4-Chloro-7-nitrobenzofurazan	31
4-Chloro-7-sulfobenzo-furazan	31
Choline acetyltransferase	172
Cholinergic neurons	172
Chromatin	417,418
Chromic complexes of ferrichrome	336
Chromic complexes of ferrichrysin	336
Chromoproteins	231
Chymotrypsin	182,189,203,206,207,243,246, 274,277,278,344,382
Chymotroypsin A	375
Chymotrypsin, titration curves for	236
Chymotrypsin cleavage	246
A-cis-chromic desferri ferrichrome	336
Citraconic anhydride modification of lysine residues	200

Citric acid	333
Cleavage constants of different protecting groups	105
Clones	424,426
Cloning of the human leucocyte interferon gene	421
cCMP	358
Cobra venom neurotoxins	327
Codon	200
Coelenterates	165,166
Colicin M	371,372,374,375
Colicin M action	373
Colicin-specific antibodies	372
Colicins	371
Si60 columns	231
Complement immune reactions	343
Concanavalin A	364
Concanavalin-A Sepharose	242
Concavaline	313
Conformational preferences of model peptides	127
Coomassie Blue	378,396
Coprogen	334,338,339
Corticotropin	269,270
Counter current distribution	36,45
Crab axons	312
Cu proteins	210
Cyanogen bromide cleavage	182,189,190,191,200,203, 204,243,353
Cyclic peptides	63,233
Cyclic tripeptides	121
ε-Cyclokallidin	45,262
Cyclo-ε-kallidin, tryptic digestion of	46
Cyclo-[$(N^{\varepsilon 1}$-Lys$^{1})$-angiotensin]	262

Cyclo-[ε-(Lys^1,Gly^6)-bradykinin] 41
Cyclo[Pro-Phe-Gly-Phe-Gly] 233,236
Cyclo[Pro-Pro-Phe-Phe-Leu-Ile-Ile-Leu-Val] 234,237
Cycloaddition 58
Cyclohexapeptides 121
Cyclols 124,125
Cyclopeptides 41,43
Cyclotripeptides 115,117,123,124,125,126
Cystine peptides,unsymmetrical 29
Cytoplasmic membrane 371,374
Cytosol 359,360
Dansyl chloride 14
Dbz group 104,105
Dbz-Phe-OH 107
Ddz 103
Dehydroacylamidines 117,120
Dehydrocyclols 120
Demethylphalloin 68
H-Demethylphalloin-actin complex 68
Deoxyribonuclease 371
Des-Ala-B30-insulin 4
Desamidoinsulin 5,6
Desoxo-viroisin 64
3,5-Di-tert-butylbenzyloxycarbonyl group 104
2-(3,5-Di-tert-butylphenyl)-propyloxycarbonyl group 104
Diacylimide group 123
D-Diaminobutyric acid 14
Diaminocarboxylic acids 333
Diaminopimelate 373

Diazirine absorption 75
Difference spectroscopy 233
Digestion, proteolytic process 343
Dihydroxyproline 65
3,5-Diiodo-tyrosine 17
3,5-Dimethoxy(α,α-dimethyl)- 103
benzyloxycarbonyl group 103
Dimethyl disulfide 67
α,α-Dimethyl groups in pro- 104
tecting groups of the ure-
thane type
α,α-Dimethyl-4-phenylazobenzyl- 103
oxycarbonyl group
4-N,N-Dimethylaminoazoben- 204
zene-4-thiohydantoins
1-Dimethylaminonaphthalene- 204
5-sulfonyl derivatives
Dimethylaminopyridine 88
α,α-Dimethylbenzyloxycar- 103
bonyl group
Dimyristoylphosphatidylcho- 291
line
Dioleylphosphatidyl choline 294
Dipalmitoylphosphatidylcho- 291
line
Diphenylphosphoryl azide 125
Diploid zygote 377
1,3-Dipolar cycloaddition 57,58
Dipole flip-flop gating model 286,287
Dipole moment 282
Disulphide links 23
DNA-dependent RNA-Polymerase
Dopamine 267
Double helices 291

Edman degradation	13,168,182,190,204,207
Egg white	349
Ejaculation	363
Elastase	349
Elastase inhibitors	349,350,351,352,354
Elastin	134
Electrically excitable membranes	310,316
Electrophysiological experiments	309
Electrostatic repulsion	282
ß-Elimination	73
Elisa technique	378
Elman's reagent	191
Ellman's test	30,33
Elongation factor G	187
Elongation factor G, amino acid sequence of	188
Embryonic haemoglobin Portland	181
Embryonic haemoglobins	180,182
Embryonic haemoglibin Gower	181
Embryonic haemoglibins in mammals	179
Enantio-ferrichrome	335
Endogenous phophorylation	358,359
Endogenous phosphorylation on the cell surface	357
Endonucleases	196
Endoproteases	374
Endorphins	269
Enkephalins	269
Enzakryl K-2	149,150
Enzymatic digestion	46
Enzymatic transamidation	9

Enzyme immunoassay 275,276,277,278
Enzyme immunoassay for human tissue kallikrein 273
Ependymal cells 268
Ergot alkaloids 115
Escherichia coli 87,371
E.coli RNA polymerase 195,196
Eukaryotic microorganisms 334
Eucaryotic proteins 422
Eukaryots 413
Exocytosis 349
FAB-MS of H_2N-Ser(Bzl)-Cys(MOB)-ODCBxHCl 156
FAB-MS of H_2N-Thr(Bzl)-Phe-Thr(Bzl)-Ser(Bzl)-Cys(MOB)-ODCB x TFA 159
Fast atom bombardment mass spectrometry(FAB-MS) 155
FD-MS of H_2N-Ser(Bzl)-Cys(MOB)-ODCBxHCl 156
FD-MS of H_2N-Thr(Bzl)-Phe-Thr-(Bzl)-Ser(Bzl)-Cys(MOB)-ODCBxTFA 159
Ferric rhodotorulic acid 338
Ferric siderophores 338
Ferrichrome 335,338
Ferrichrome A 334,338
Ferrichrome C 334
Ferrichrome ligand 335
Ferrichrome transport 334
Ferrichrome-type siderophores 334
Ferrichrysin 334,338
Ferricrocin 334,338
Ferrioxamine B 338,339
Ferrioxamine E 338,339

Ferrirhodin	334
Ferrirubin	334
Ferritin	366
GAR-Ferritin	366
Fertilization	343,363,364,368
Fertilization processes	273
Fetal haemoglobin	180
Fibroblasts	268,418
Field desorption mass spectrometry (FD-MS)	155
Flip-flop gating model	215
Fluorescence labels	327
Fluorescence microscopical methods	364
Fluorescence spectroscopy	291
5'- N-Fluorosulphonylphenyl-phosphate	415
Fluram	190
4-Formyl-4-imidazolin-2-one	55
Fucose	380
^{3}H-Fucose	358
Fusarinine	338
Galactose	380
Galactose oxidase	367
ß-Galactosidase structural gene	424
Gamete	377
Garnet Blue	378
Gastrin	255
Gastrin analogues	256,257
Gel electrophoresis	230
Gene expression	425
Gene technology	23
Genome expression	413

Glia cells 268,271
Glucagon 269
Glucosamine 380
Glucose metabolism 273
Glucose transport 15
N-Glutarylgramicidin 298
Glycerylmonooleate membrane 297
cGMP 270,358
Goldfish 172
Gramicidin A 291,298
Gramicidin A channel 297
Grayanotoxins 309
Grignard reagent 73
GTP 358
Guanidination 344
Guanylate cyclases 267
Guest amino acid residues 128,131
Guest-host model 127
Guest-host peptides 128,133
Haemoglobin Barts 182
Haemoglobin components in human and pig embryos 179
Haemoglobin Gower 182
Haemoglobin Heide 182
Haemoglobin Portland 179,189,182
Halo-opsin 228
Halobacterium halobium 79,227
Halophilic microorganisms 241
Halorhodopsin 227,228,229
Hamster fibroblasts 357
Head activator peptide 165,166,168,170
HeLa cells 357
α-Helical nucleation 130
Helical structure 128,130

α-Helix 286,287
α-Helix dipoles 282
Helix-inducing power 129
Hemorrhagia of the liver 63
Hepatocytes 15,72
Heteronuclear two-dimensional NMR spectroscopy 233
N-Hexadecanoyl-L-cysteine-tert-butyl ester 90
High-efficientcy columns 149
High performance liquid chromatography of protected peptides 145
Histamine 269
Histone 414
Histone H1 413
Homo-L-Ala_{10}-peptide 129
Homo-oligopeptides 127
Horseradish peroxidase 274
Host peptide 127,133
HPLC 5,7,8,31,32,33,230,367
Human acrosin 380
Human embryonic haemoglobin 179
Human embryos 180
Human hypothalamus 166
Human insulin 3
Human leucocyte interferon 425,426
Nle^{15}-human-little gastrin 259
Nle^{15}-human-minigastrin 259
Human plasmin 383,384
Human urinary kallikrein 275,276,277,278,279
Human urokinase 382
Hybrid cells 271
Hydra 165
Hydra attenuata 166

Hydrophobic binding sites 385
Hydrops foetalis blood 181
Hydroxamate groups 334
7-Hydroxy-2-methylbenzisoxazo-liumtetrafluoroborate 98
5-Hydroxy-2-propenylpyridine 59
Hydroxyapatite column 229,350,351,360
Hydroxyproline 65
N-Hydroxysuccinimido ester 31
N-Hydroxysuccinimidyl p-azido-benzoate 328
MOS Hypersil 182
Hypotensive action 41
Immune cells 88
Immunofluorescence 362
Immunofluorescent staining of acrosin 379
Immunoglobulin G 279
Immunosympathectomy 173
Indole functions 29
INEPT technique 294
Inflammatory processes 349
Insulin 5,14,375
A19-(3,5-diiodo-tyrosine) porcine Insulin 17,21
Insulin A-chain 24
Insulin analogs 5,7
Insulin B-chain 3,8,24
Insulin B-chain,modification of 26
Insulin B-chain,porcine 4
Insulin derivatives,photo-sensitive 11
Insulin,di-S-sulphonated-B-chain 19

Insulin, direct iodination 17
Insulin,human 4,6,7,8,10
Insulin,immunoassay 17
Insulin,iodination of the tyrosines 17
Insulin molecule,binding sites 11
Insulins,photo-reactive 11
Insulin,porcine 4,5,6,8,9
P-Insulin,radio-iodinated 11
^{125}I-P-Insulin-receptor complexes 11
Insulin,spatial structure of 15
Insulin,tertiary structure of 8
Insulin,X-ray studies of 23
Interferon production 422
Interferons 421
Intestinal absorption 273
Iodinated insulin 21
Ion channels 267
Ion-conducting channel 310
Ion exchange chromatography 6
Ion selectivity 309
Ionophores 340
Iron-chelating cyclic peptides 334
Iron ionophores 334
Isotope cluster 158
Isotope pattern of the molecular ion 157
Isoxazoline 59
Isoxazoline carboxylic acid azide 60
Isoxazoline derivatives 57

Kallidin	261
Kallikrein	273,275,277,278
Kallikrein assays	278
Kallikrein, dog blood pressure assay	279
Kallikrein enzyme immunoassay	277
Kallikrein of human tissue	273
Kallikrein substrates	279
Kazal inhibitors	354
KCK C-8 column	146
Kidney function	273
Kinin-liberating assays	273
Kinins	261
Kunitz inhibitor	343,354
ß-Lactam antibiotics	373
Latrodectus Dahli Levi	320
Latrodectus pallidus pavlovski Charit	320
Latrodectus spider toxins	320
α-Latrotoxin	322,325
Lead citrate staining	365,366
Lectin	313,364
Lectin binding ability of the acrosomal membranes	365
Lectin binding studies	366
Lectin-membrane interaction	366
Lectin staining procedures	367
Leiurus quinquestriatus	317
Leucine, monohydroxylated	64
Leucine, trihydroxylated	64
L-Leucyl-oxymethylpolystyrene	31
Leukocyte elastase	348,352,355
Leukocyte tissue damage	349

LH-RH 103
Lichroprep RP 18 257
RP2 LiChrosorb 182
LiChrosorb RP-8 167,168,169,182
LiChrosorb RP-18 83,84,106,107,167
Lipid bilayers 215,244,281,283,291
Lipid vesicles 282
Lipogenesis 15
Lipopeptide fragments 87
Lipophilic α-helix 287
Lipopolysaccharide 92,94
Lipoprotein 92,93,94
Lipoprotein-binding protein 95
Lipoproteins from the outer membrane 87
Liposomes 228,291,293,294
Liver receptor protein 63
Lung emphysema 349
Lymphocyte plasma membranes 87,93
B-Lymphocyte activation 92
ß-Lymphocyte mitogens 87
B-Lymphocytes 92,94,95
T-Lymphocyte mitogen 92
T-Lymphocytes 95
Lysylendopeptidase 3,5
Lytic activity 282
Macrophages 268
Maleimide spin label reagent 344
3-Maleimidopropionic acid 31
3-Maleimidopropionic acid amido-PEG 32
Malonichrome 334
Mammalian embryos 180
Mammals 166
Manganese dioxide 228

Mannose	380
Maxam-Gilbert technique	196
Melanotropins	269,270
Melittin	216,282,283,284,285,286, 287,288
Melittin-1-21	216
Membrane conductance to sodium and potassium ions	301
Membrane lipids	306
Membrane receptor	88,301
Membrane vesicles	228
2-Mercaptoethanol	47
Meta-chloro-perbenzoic acid	59
Metastasizing cells	357
Metastasizing tumor cells	362
4-Methoxy-benzyl(MOB) group	158
Methyl glutaconic acid	334
2'-Methylthio-tryptophan	68
5-Methylcytosine	199
α-Methylmannoside	243,244
N-Methylmorpholine	68
Methylsulfenyl chloride	67
Methylsulfonyl group	68
2'-Methylsulfonyl-tryptophan	65
Microbial chelating agents	333
Microelisa reader	275
Mini-gastrin analogues	256
Mitogenic activity	362
Mitogenicity	87
Mitogens	87,94
Molluscan neuropeptide	103,106,108
Monoclonal antibodies	174

Morphogen	165
Mouse fibroblasts	357,360,418
Murein	373
Murein biosynthesis	373
Myotropic effect on rat uterus	264
Naja naja oxiana	327
Naja naja oxiana cobra venom	145
Naja naja siamensis	329
Neocarzinostatin	212
Neonatal rats	172,173
Neoplastic transformation of a cell	362
Nerve fibre outgrowth	172
Nerve growth factor (NGF)	171
Neural cells	171
Neuroeffectors	267
Neurohumoral compounds	267
Neuromodulators	267
Neuron cells	268
Neuroregulators	267
Neurohormones	267
Neurospora crassa	334,338
Neurotoxic alkaloids	302
Neurotoxic phospholipases	319
Neurotoxin "active center"	327
Neurotoxin II from naja naja oxiana cobra venom	47
Neurotoxin receptors	310
Neurotoxin-Sepharose	329
Neurotoxin surface	328
Neurotoxins	47,301,306,309,310, 316,319
Neurotoxins from micro-organisms	319

Neurotoxins from the venom of Elapidae snakes 319
Neurotoxins from the venom of latrodectus spiders 319
Neurotoxins of Latrodectus spiders 322
Neurotransmitter-synthesizing enzymes 172
Neurotransmitters 267
Nikkomycin B 56
Nikkomycin E 56
Nikkomycins 60
Nikkomycin Z 56
Nikkomycin X 57
2-Nitro-4-azido-phenyl-acetyl group 12
Nitrostyrene 31
^{13}C NMR T_1-measurements of Boc-(L-Ala-Aib-Ala-Aib-Ala)$_3$-OMe 221
^{13}C NMR spectra of tripalmitoyl-Cys-t -butyl ester 91
^{13}C NMR spectrum of Boc-(L-Ala-Aib-Ala-Aib-Ala)$_3$-OMe 221
NOE effects 236
Noradrenaline 267,269
15-Norleucine-human-little gastrin 257,258
Nucleosil 7C-8 83,84
Nucleotide sequence of the rpoC gene 201
Oligodendroblasts 268
Oligodendrocytes 268
Operon 199

Opioid receptors 271
Opioids 269,271
Opsin kinase 245
Ovum 377
Oxacyclols 115
Pal-Cys(CH_2-CHOH-CH_2)-OMe 89
N-Palmitoyl-cysteine methyl ester 89,92
1-Palmitoyl-2-oleoyl-glycero-3-phosphocholine 283
Pancreas 137
Pancreatic elastase 352
Pancreatic juice 277
Pancreozymin 181
Papain 242
Paper electrophoresis 21,24
Parasympathetic neurons 172
Parathyrin 269,270,271
Parathyrin, human 35
Parathyrin metabolism 35
Parathyrin segments 35
Parathyrin sequence (32-43), synthesis of 36
Parathyrin sequence Tyr-(43-55), synthesis of 38
Parathyrin sequence Tyr-(62-68), synthesis of 38,39
PEG-reagent 33
Pentafluorophenol 82
Peptide antibiotics 281
Peptides as iron ionophores 333
Peptide backbone 130,334
Peptide bioregulators 261
Peptide cyclols 115

Peptide dehydrocyclols 117
Peptide α-helices 282
Peptide hormones,influence on neural cells 267
Peptide-induced membrane conductivity 283
Peptidic iron ionophores 334
Peptide-nucleoside antibiotics 55
Peptide synthesis at elevated temperatures 98
Peptidoglycan 373
Peroxidase 14
Peroxidase activity 274
Peroxidase conjugate dilution 276
Peroxidase IgG conjugate 274
Periodate oxidation 311
Phagocytosis 343,349
Phalloidin 69,71
Phallotoxins 63,70,71
α-Phenylethylamides 57
Phenylsepharose 360
Phenylthiohydantoins 204
Pheochromocytoma PC12 cells 172
Phosphatidylethanolamine 285
Phosphodiesterase 245
Phospholipid carbonyl band 291
Phospholipid membrane permeability 323
Phospholipid vesicles 85
Phosphoproteins 358,359
Phosphorylated membrane proteins 357
Phosphorylating agents 396
Phosphorylation mechanism 358

Phosphotransferase reaction 415
Phosvitin 413,414
Photo-affinity labelling 11
Photo-insulins 11,12,15
Photo-labels 12
Photo-reactive insulins, potency of 14
Photo-sensitive labels 12
Photoactivable derivatives 329
Photoactivable derivatives of postsynaptically acting neurotoxins 328
Photoactivable label 328
Photoaffinity labeling 71,74
Photoinduced crosslinking 331
Photolabeling experiments 78,327
Photolabile residue 72
Photoreceptor membranes 242,245
Physarum polycephalum 414
Pig brain 414
Pig brain protein kinase 417
Pig embryos 180,182
Pig pancreatic kallikrein 278
N-Pimelylgramicidin 298
Plant lectins 364
Plasmatic proteinases 384
Plasma membrane 366
Plasmid 196,405,422,424
Plasmin 382
ß-Pleated sheet 291
Polyacrylamide gels 181
Polyarthritis 349
Polyclonal antibodies 230
Polyoxin B 56
Polyoxins 55

Polypeptide antibiotics	215,217
Porcine insulin	3
Porcine insulin, semisynthesis of	3
$[Tyr(I_2)^{A19}]$porcine insulin	18,19
Postsynaptic membrane	327
Ppoc group	103
Preinterferons	422
Pressure regulation	343
Presynaptic membrane	319,323
Proacrosin	385
Proinsulin	23
Proinsulin analogues	24
Proinsulin analogues, semisynthesis of	23
Proinsulin model	26
Prostaglandin	269
Prostaglandin E_2	269
Protease from St.aureus	242
Protein kinases	413,414
Protein phosphorylation systems	357
Protein-protein association	343
Proteinase-inhibitor complexes	343
Proteinase inhibitors	343,349,354
Proteinases	278,343,374,382,384
Pyroglutamase	230
Racemization	27
Racemization test with L-amino acid oxidase	46
Rad Pak C18 column	139,140,141,256
Radio-ligand analysis	310
Radioimmunoassay of human urinary kallikrein	278,279

Radioimmunoassay of kallikrein 273
Radioimmunochemical measurement of parathyrin 35
Radioiodination of parathyrin 40
Rast's method 45
Rat brain membranes 314
Rat brain synaptosomes 312
Rat fundic glands 143
Rat intestine 166
Receptor-labelling 15
Receptor protein 371,372
Recirculation chromatography 44
Recombinant plasmid 196
Resonance splitting of the amide I band 291
Restriction endonucleases 196,197,404,405
Restriction endonuclease sites 423
Restriction map 422
Retinal 228
Retinal binding component 231
Retinal binding pigments 227
Retinal binding protein 227
Retinal-containing proteins 247
Retinal explants 172
Retinyl peptides 230
Reverse phase chromatography 44,52,82,145,148,151,152, 165
Rheumatoid arthritis 349
Rhodopsin 203,241,242,243,245
Rhodopsin sequencing 203
RIA 7
Ribonuclease 371
Ricinus communis agglutinin-120 364
Rifamicin 403,409

Rifamicin-restistancy of E. coli RNA polymerase	403
Rifamicin-resistant mutations	408
Rifampicin	405
Rifamycin	382,388,390,391
Rifamycin-resistant RNA polymerase	389
Rigid peptide bonds	121
RNA polymerase	195,201,387,389,390,392, 393,396,397,398,403,410, 417
RNA-polymerase complex	401
RNA polymerase-DNA complex	395
Saliva	277
Saxitoxin	309
Scatchard plots	313
Scavengers	47
Scorpion neurotoxin	306
Scorpion venom	316
SDS gel electrophoresis	15,175,229,367
Sea anemone	349
Sea anemone toxins	310
Secondary structure	127
Secretin	137,139,143,269,270,271
D-Phe^{6}-Serectin	140
HOBt-Secretin	137,139,143
HONSu-Secretins	138,139
Secretin, HPLC of	140,141
Secretin, HPTLC	141
Secretin,synthesis of	137
Seminal plasma	277
Sensory neurons	172
Serine proteinase superfamily	384
Serine proteases	175,278,382

Serotonin	267
DEAE-Sephacel	180,181,182
DEAE Sephadex A-25	165,350
QAE-Sephadex A-25	190
SP-Sephades C-25	350,351
Sephadex G-10	165
Sephadex G-25	19,190
Sephadex G-25f	25
Sephadex G-25 sf	329
Sephadex G-50f	25,26,350
Sephadex G-100	274
Sephadex LH-20	37,147,149
Sephadex LH-60	246
SP-Sephadex	25
Sepharose CL-6B	315
DEAE-Sepharose	313,315
DEAE Sepharose CL-6B	314
WGA-Sepharose	313,314,315
Sepharon P-1000	150
Si-40 , silica cel	147
Sialic acid	363
Siderophores, absolute configuration of	339
Siderophore transport	335
Siderophoreses	333,335,336,338
Silasorb 600	147,150
Silica gel "L"	44,146,147,150,153
Silica gel Si-40	150
Size exclusion conditions	148
^{3}H-Sodium borhydride reduction	367
Sodium channels	301,302,309,310,316
Sodium cyanoborohydride	228
Sodium dodecyl sulphate	94,372
Sodium ions, transport of	310

Solid-phase fixed antibodies	275,276
Solid-phase immunoglobulin-fixed human urinary kallikrein	276
Solid phase synthesis	35
Somatostatin	103,155,269,270
Somatostatin segment 10-14 (H_2N-Thr(Bzl)-Phe-Thr(Bzl)-Ser(Bzl)-Cys(MOB)-ODCB x TFA)	158
Somatostatin segment 13-14	155
FD Spectra	157,160
Sperm head	379
Sperm penetration	377
Sperm proteinase	377
Sperm surface	364
Spermatozoa	363,364,365,368,377,378
Spermatozoan membranes	363,364,366
Sphaerogena	334
Spherical silica gel	147
Spheron P-40	150,151
Spheron P-300	150
Spheroplasts	372
Spin-label technique	343
Spin-labeled N-hydroxysuccinimides	328
Spleen cells	94,95
Staphylococcus aureus-V8 protease	230
ß-Structure, antiparallel	132
ß-Structure, critical chain lengths for	132

ß-Structures, disruption of 134
ß-Structure formation 127,131,134
ß-Structure-forming residues 133
ß-Structures 128,129,133
N-Suberylgramicidin 294,296,298
Subtilysin 230
Succinic acid 333
N-Succinylgramicidin 294,298
Sulfenohydrazide method 29,33
Sulfenohydrazide procedure 30
Sulfenyl halides 29
Sulfuryl chloride 67
Superoxide dismutase 210
Suzukacillin 216,217,218,219
Suzukacillin A 281
Sympathetic neurons 172,173
Synaptosomal membranes 303,313
Synaptosomal receptor 302
Target cells 11
Testicular compartment 385
Tetrodotoxin 303,309,311,312,313
Tetrodotoxin analog 312
Tetrodotoxin binding analysis 303
Tetrodotoxin receptor 306,312,313
α-Thalassaemia 179
Thermolysine 189,242
Thioanisole 47
Thiol-activated sepharose 4B column 191
Threonine esters 5
^{3}H-Thymidine, incorporation of 93,360

Thymopoietin II 111
Thymopoietin pentapeptide 103
Thymosin ß9 103
Toxin aggregation 311
Toxin binding 311
Toxin radioactive derivatives 310
Toxins 261,301,316
Torpedo marmorata 329
Transamidation 5,7,8
Transcription 417,424
Transcription mechanism 387
Transcription terminator 199,200
Transducin 245
Translation 200,425
Translocation 417
Transmembrane channel 291
Transpeptidation 150
ß-Transpeptidation 143
Triacetylfusarinin 339
Trichoderma viride 217
Trichotoxin 217,218
Trichotoxin A 216,219,281,282
Trifluoroacetylpiperidine 73
4-(2,2,2-Trifluoro-1,1-azoethyl)-benoic acid 72,74,77
4-2(2,2,2-Trifluoro-1,1-azoethyl)-styrene 73
3-Tri-fluoromethyl-3-phenyl-diazirine 72
Tripalmitoyl-cysteine 88,93
Tripalmitoyl-L-cysteine-tert-butyl-ester 88
Tripalmitoylpentapeptide 87,88,89,90,92,94

Tripeptide cyclols 115,116,117,118
Tris(hydroxamato)iron complexes 337
Tris(hydroxamato)iron complex, optical isomers of 336
Tritiated borohydride 228
TRH 103
Trypsin 3,5,8,9,27,189,200,203, 246,274,344,372,381,382
Trypsinogen 258
Trypsinolysis 188,189,190,191,329
Tryptic cleavage 182
Tryptic digests 139
Tryptic fragments of secretin 142
Tryptic transamidation 9
Tryptophan-containing peptides 29
Tryptophan, oxidation of 65
TSK 3000 column 367,368
Tumor cells 359,418
ß-Turn 134,135
Two-dimensional NMR methods (2D NMR) 233
UPD-N-Acetyl glucosamine 373
UDP-N-Acetyl-muramyl pentapeptide 373
Ulcer 137
Uracil 55
Uranyl citrate staining 365,366
^{3}H-Uridine, incorporation into RNA 92
Urine 277
Ustilago sphaerogena 334,336

Vasoactive intestinal peptide 269,270
Venom of latrodectus mactans tredecim-guttatus 320
Venom of latrodectus spider 323
Veratridine 302,303,305,306,309
Veratridine receptors 301,306
Veratridine-sepharose 306
Veratridine-sepharose affinity 306
Veratridine-sepharose affinity chromatography 303,304
Vesiculation 363
Viral resistance 421
Viroidin 64
Viroisin 63,64
Virotoxin analogues, synthesis of 66,67
Virotoxins 63,64
Wittig reaction 57
Woodward's reagent 41,43
X-ray analysis of oligopeptides containing Aib residues 219
X-ray structure of Boc-Aib-L-Ala-Aib-Ala-Aib-OMe 220
X-ray structure of Boc-L-Ala-Aib-Ala-Aib-Ala-Glu(OBzl)-Ala-Aib-Ala-Aib-Ala-OMe 220
Zona pellucida 363,364,377
Zorbax ODS column 146,147
Zorbax PSM-60 column 82
Zorbax Sil column 82,147